Understanding The Outboard Motor

Second Edition

Eugene W. Stagner

Prentice Hall
Upper Saddle River, New Jersey   Columbus, Ohio

Library of Congress Cataloging-in-Publication Data

Stagner, Eugene.
    Understanding the outboard motor/Eugene W. Stagner.—2nd ed.
       p.   cm.
    Includes index.
    ISBN 0-13-861980-8
    1. Outboard motors.   I. Title
  VM348.S73   1999
  623.8'7234—DC21

98-15673
CIP

Cover photo: © Mercury Marine Brunswick Corporation
Editor: Ed Francis
Production Editor: Stephen C. Robb
Design Coordinator: Karrie M. Converse
Cover Designer: Brian Deep
Production Manager: Patricia A. Tonneman
Production Supervision: Custom Editorial Productions, Inc.
Marketing Manager: Danny Hoyt

This book was set in Times Roman and Gill Sans by Custom Editorial Productions, Inc., and was printed and bound by R.R. Donnelley & Sons Company. The cover was printed by Phoenix Color Corp.

 © 1999, 1985 by Prentice-Hall, Inc.
Simon & Schuster/A Viacom Company
Upper Saddle River, New Jersey 07458

Printed in the United States of America

10 9 8 7 6 5 4 3 2 1

ISBN   0-13-861980-8

Prentice-Hall International (UK) Limited, *London*
Prentice-Hall of Australia Pty. Limited, *Sydney*
Prentice-Hall Canada Inc., *Toronto*
Prentice-Hall Hispanoamericana, S. A., *Mexico*
Prentice-Hall of India Private Limited, *New Delhi*
Prentice-Hall of Japan, Inc., *Tokyo*
Simon & Schuster Asia Pte. Ltd., *Singapore*
Editora Prentice-Hall do Brasil, Ltda., *Rio de Janeiro*

Contents

Preface

"It's not the fixing that is so hard, it's finding out just what needs the fixing." The person who can do this is called a technician. The technician has put in hours upon hours of study to gain a thorough knowledge of the product and has developed an understanding of principles of operation. The technician approaches each mechanical or electrical problem as a challenge, systematically testing to determine the extent of repairs needed. With a good attitude toward the customer, honest repair practices, and technical studies, you can acquire a knack for making repairs.

For the student outboard technician, this manual explains the technical theory of operation necessary to prepare for training and certification at the factory training center. Reading this manual will be profitable because the basic understanding gained will help you troubleshoot outboards. With greater understanding of electronic systems, you will be able to make accurate repairs more quickly.

The novice mechanic should use this text in conjunction with a factory service manual. First, read the theory in this text that relates to the repairs to be made. Then, with the aid of the factory service manual for the outboard, make the repairs, applying the theory learned. This text makes following the technical repair procedures given in the service manual easier to understand because you will know what must take place in the outboard at the completion of repair. Second, this text will help clarify the reasons for technical instructions given by the factory.

For the vocational instructor, this manual presents outboard theory in a logical sequence for better student comprehension. Each chapter contains theory troubleshooting to apply the theory learned. Also included in each chapter is a set of review questions that can be discussed in class and also given for homework assignments. Pictures and drawings clearly illustrate the topics discussed and bring into the classroom actual problems that develop and are seen as service work is performed.

Acknowledgments

The author thanks the following companies and corporations, which have provided training through the years and allowed photographs to be taken in their training centers and/or have permitted the use of copyrighted material:

American Boat and Yacht Council
Battery Council International
Boating Industry Association
Brunswick Corporation
Champion Spark Plug Company
Exide Corporation
General Tool Manufacturing Co., Inc.
International Dyno Corporation
Lisle Corporation
Merc-O-Tronic Instruments

Mercury Marine Corporation
National Marine Manufacturers Association
Neway Manufacturing Corporation
Outboard Marine Corporation
Specialty Manufacturing Company
Stevens Instruments Company
Tecumseh Products Company
Trail Rite, Inc.
Unique Functional Products
United States Coast Guard

Safety:
You Are in Charge

Ultimately you are responsible for your own safety when working. Regulations enforced by teachers, supervisors, and the Occupational Safety and Health Administration (OSHA) make us aware of hazards in the school shop and workplace. Pay close attention to instructions in the service manuals and from teachers or supervisors. Learning the safe, correct way to use tools and equipment may prevent injury or death.

Safety in the Shop

Use eye protection when operating grinders, wire wheels, and other power tools or equipment that throw debris.

Hands and fingers hold and control the material with which you are working. Keep your hands and fingers as far as possible from grinding wheels and running outboard flywheels.

Loose clothing and long hair can be windblown or air-blown into a rotating flywheel. Dress appropriately for the job.

If running the outboard on a flushette, remove the propeller to prevent injury.

Learn about tools and equipment before you use them. If you need to use a certain tool, ask an experienced person to show you how to operate the tool safely.

Accidental powerhead starts may injure you. Always disable the ignition system by removing spark plug wires, placing them in a spark tester or disconnecting the primary ignition circuit (turn key switch to Off).

When disconnecting the manual recoil starter spring, follow the manufacturer's instructions and wear safety glasses.

Batteries contain sulfuric acid, and explosive hydrogen gas may be present. Keep sparks and flame away from the battery.

Do not use electrical tools or equipment that have damaged electrical wiring cords. Do not disconnect (cut) the ground (third terminal) on a drop cord or power tool. If necessary, use an appropriate adapter.

When working on the outboard, do not touch electrical wires while the engine is running. A high-voltage shock to your hand may occur, causing you to jerk your hand or arm into the moving flywheel.

Gasoline explosions can cause fires and burns. Ensure that fuel tanks and connectors and approved fuel hoses are in good condition and pass the pressure test. Fuel the portable fuel tank away from the boat. Keep the gas pump nozzle in contact with the fuel tank at all times!

Before fueling a boat with a fixed fuel tank, evacuate the boat and close the lower compartments. Give a "sniff test" for fumes in lower area compartments before starting the engine.

Be prepared to use a Coast Guard approved fire extinguisher (Figure A) when needed.

There are currently three types of fire extinguishers approved and available on the market for marine use: carbon dioxide (CO_2), dry chemical, and halon. Check local and state laws for the size and number of extinguishers required for your boat.

A Coast Guard approved
ABC Fire Extinguisher
with multipurpose dry
chemical (Ammonium
Phosphate Base)

All fire extinguishers operate on internal pressure and must be checked at regular intervals.

To use a fire extinguisher, hold it upright and pull the pin. Squeeze the lever and direct the discharge toward the base of the flames, sweeping back and forth while shouting "Fire! Fire!"

Familiarize yourself with the safety policy and locations of the first aid kit and fire extinguishers at school and at work. Are there warning signs or equipment operating instructions posted? If so, familiarize yourself with them before using the equipment.

Avoid horseplay or foolish activities while you are in the shop. Clean up oil or fuel spills immediately. Do not lay out an air hose or an electrical drop cord where someone may trip over it. Route them appropriately to the work area. Remove them when they are no longer needed. Do not leave equipment running unattended. Do not play with fire extinguishers. Do not work with greasy, oily tools. Wipe them off before use. Do not use a hardened hammer or punch on a hard surface. Chips may fly into surrounding area, injuring someone. Use the right tools for the job. Never use a tool that is damaged or in poor condition. Inform your instructor about the tool condition.

When using wrenches, pull the wrench toward your body. If you push on a wrench and it slips, you may lose your balance and be injured.

Place all oil-soaked dirty rags in a special metal safety container.

When washing parts with solvents, work in a well-ventilated area and use safety gloves. Keep flames and sparks away from the work area. Never wash parts in gasoline.

Keep your work area clean.

Carbon monoxide in exhaust gases from running engines is colorless and odorless and can cause death if continuously inhaled for even a short period of time. Always run outboards in water (flushette) and in a well-ventilated area.

Secure transom brackets before running an outboard in a test tank. Remove the propeller if running an outboard on a transom using a flushette.

Compressed air can forcefully move debris through the air, so always use safety glasses. Always direct the air flow away from yourself and others.

When lifting heavy outboards, keep your back straight. It may be necessary to squat or kneel and use your legs to lift the load. When working on an outboard installed on a boat, block the trailer wheels so the trailer won't move.

Hazards When Working on the Outboard

- Carbon monoxide
- High voltage electricity
- Loose transom brackets
- Unexpected boat/trailer movement
- Test instrument leads get caught in rotating flywheel
- Excessive engine RPMs, no load
- Rotating flywheel
- Spinning propeller (remove it before you start)
- Flushette comes off lower unit
- Lifting smaller outboards
- Lifting larger outboards without proper lifting eye and hoist

Remember, SAFETY is your job!

Carefully—Fill'er up

Filling up a portable fuel tank sitting in the bed of your pick-up truck or in the boat can be dangerous. Several explosions or fires have been linked to filling gas cans/portable tanks sitting on a plastic bed liner or carpet. The problem is, static electricity builds up as gasoline flows through the pump hose/nozzle and into the container. Fires are more likely on days when the air is dry. High humidity acts as a conductor of sorts and can disperse static.

Whether the container is metal or plastic, problems can occur, but they are more likely with a metal portable fuel tank. If the container is not properly grounded while it is being filled—and it's not grounded while sitting on plastic or fiberglass—a little arc can ignite the gasoline fumes and cause a fire. If this should occur, put the tank cap back on immediately, smothering the fire. Unfortunately, the arc can also cause an explosion, and nothing can be done to stop an explosion.

To avoid this danger, place the portable fuel tank directly on the ground (cement slab) before filling it with gasoline. Always keep the pump nozzle in contact by touching the tank. This keeps the tank grounded in two ways. First, static electricity can discharge into the ground (cement slab). Second, with the pump nozzle in contact, touching the tank opening, static electricity can be discharged back through the pump hose (spring) and to the ground.

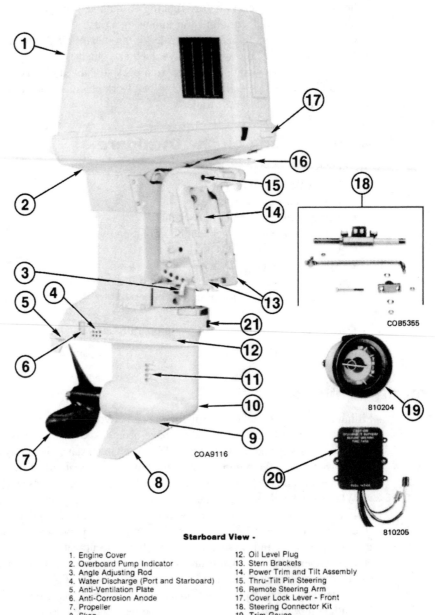

Starboard View -

1. Engine Cover
2. Overboard Pump Indicator
3. Angle Adjusting Rod
4. Water Discharge (Port and Starboard)
5. Anti-Ventilation Plate
6. Anti-Corrosion Anode
7. Propeller
8. Skeg
9. Oil Drain/Fill Plug
10. Gearcase
11. Water Intake (Port and Starboard)

12. Oil Level Plug
13. Stern Brackets
14. Power Trim and Tilt Assembly
15. Thru-Tilt Pin Steering
16. Remote Steering Arm
17. Cover Lock Lever - Front
18. Steering Connector Kit
19. Trim Gauge
20. Junction Box and Cable Assembly
21. Pitot Tube Hose Nipple

COA9116

COB5355

810204

810205

Outboard Marine Corporation Model #J235STLCT 2.6 GT. *(Courtesy of Outboard Marine Corporation)*

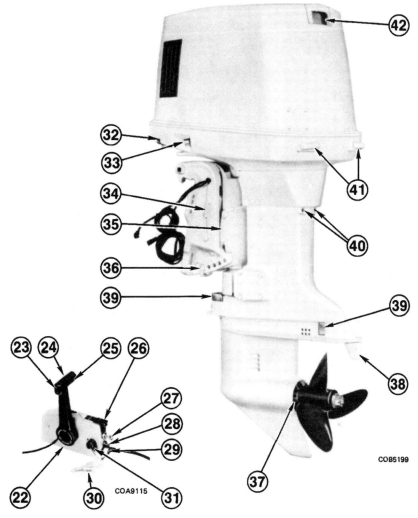

COB5199

COA9115

Port View -

22. Remote Control and
    Electric Cable Assembly
23. Trim/Tilt Switch
24. Remote Control Handle
25. Neutral Lock Slide
26. Fast Idle Lever
27. Starter/Primer Switch and Key
28. Throttle Friction Adjustment
29. Accessory Plug Connector
30. Lanyard
31. Emergency Ignition Cut-Off Switch

32. Fuel Line Retainer
33. Fuel Connector
34. Model and Serial Number Plate
35. Trail Lock
36. Angle Adjusting Rod Retainer
37. Over the Hub Exhaust Outlet
38. Steering Trim Tab
39. Anti-Corrosion Anode
40. Exhaust Relief
41. Cover Lock Levers, Aft (2)
42. Tilt Grip and Air Inlet

Outboard Marine Corporation Model #J235STLCT 2.6 GT. *(Courtesy of Outboard Marine Corporation)*

CHAPTER **1**

General Use and Maintenance of Outboard Motors

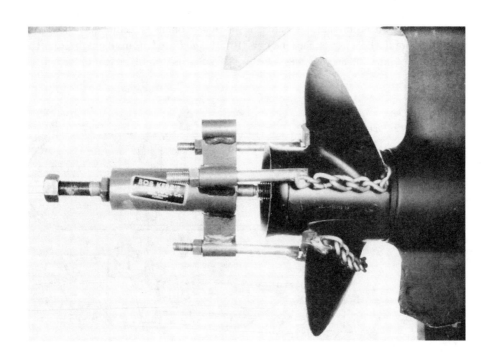

1

After studying this chapter, you will know

- What the sides, front, and rear of the boat are called.
- Proper outboard installation and persons safety.
- Outboard steering system basics and persons safety.
- Proper handling and transporting of the outboard.
- The importance of maintenance for safety and longevity.
- Lubrication points on the outboard.
- Operational checks.
- Proper trailering of the boat.
- How to measure the tongue weight of the trailer.
- Techniques for motor storage.
- How to perform preseason checks.
- Corrosion control techniques.
- Maintenance on recovered submerged outboards.

1.1   Which Side Is Which?

Using the terms *left* and *right* side of an outboard can be confusing; therefore, the sides are designated as port and starboard (see the figures on the preceding pages). **Port** means the left side of the boat as you face the bow (front). **Starboard** means the right side. The **stern** is the rear of the boat, and the **bow** is the front. The **aft** is toward the rear of the boat (Figure 1–1).

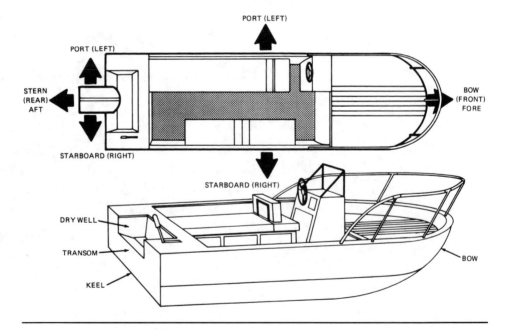

Figure 1–1   Which side is which?

1.2   **Outboard Installation**

Before installing the outboard, check the boat plate for the U.S. Coast Guard capacity information, which lists the maximum horsepower rating for the boat. Use only that horsepower or a smaller outboard. If there is no plate, find the number stamped in the hull (since 1973), call the boat manufacturer, and ask for the horsepower rating. Don't say, "Oh, it will be okay" and put the outboard on! It might just be a life or death decision! Other information given on the plate may be the maximum number of persons allowed, maximum allowable weight, lights, compartment ventilation, flotation, and a serial number.

When placing the outboard on the ***transom*** (part of the hull across the stern) in a fixed bolt-on location or with transom brackets (such as those used on the lower horsepower fishing motors), follow some basic considerations. Mount the outboard in the center of the transom. Insure this centering by measuring an equal distance in from the chimes or use intersecting arcs (Figure 1-2). Mark this center line so small outboards can be put on-center repeatedly.

The transom height of the boat also needs to be determined. Transom height determines how high or low the outboard sits on the transom. To find out the boat's transom height, measure as shown in Figure 1–3. Do not include a keel in your measurement. Now apply this dimension of the transom to the outboard (Figure 1–4).

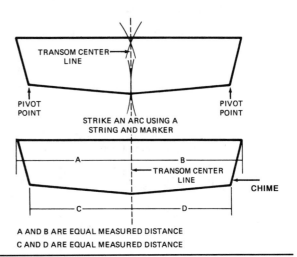

Figure 1-2 Methods used to determine center line.

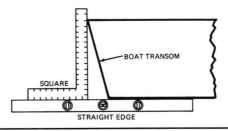

Figure 1–3 Measuring transom height.

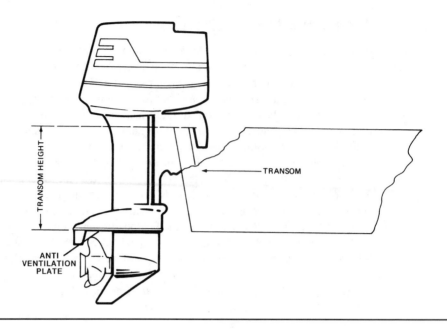

Figure 1–4   Transom height measurement applied to the outboard.

The anti-ventilation plate is needed to keep the air out of the prop while the boat is getting on plane. As a general rule, the anti-ventilation plate should be at and parallel to the bottom of the boat. (There are some exceptions to this rule, of course.) In performance boating you may want the anti-ventilation plate a bit higher to reduce drag, which is caused by the extension of the lower unit below the boat. Raising the anti-ventilation plate above the bottom of the boat is largely a trial-and-error procedure. About six inches or more above the bottom of the boat on some high-performance hulls may gain one MPH per inch. Each boat and outboard engine combination will have to be set up individually. This is done in the water using a tachometer and boat speedometer when selecting the correct prop and tilt (trim) and transom-height position. Be careful about setting it too high, because of ventilating the propeller when bringing the boat into a turn or drawing air into the water pick-up tube. Even small amounts of air entering the cooling system can cause overheating of the powerhead. Extreme caution should be exercised if you decide to raise the outboard up so the anti-ventilation plate is above the bottom of the boat. If the outboard is raised, it is possible that it will require a change of propeller, which can run nearer the water surface. It is also suggested that a water pressure gauge be installed to monitor the cooling system.

If the anti-ventilation plate is lower than the bottom of the boat, a piece of one-inch hardwood spacer can be attached to the top of the transom to bring a small horsepower motor to the correct transom height (Figure 1–5).

Bolting the adjustable stern bracket to the transom with stainless steel bolts and nuts should be done on larger outboards. On new outboards use only the attaching hardware sent with the motor. They're special and won't rust or weaken. The bolts should be put

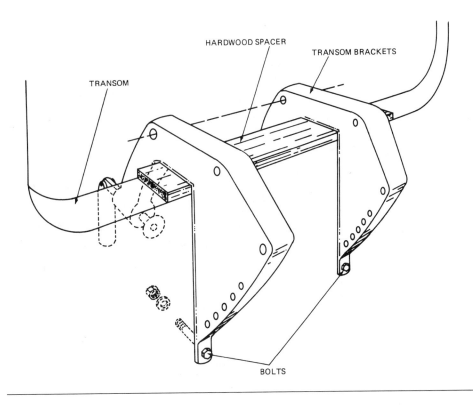

TRANSOM

HARDWOOD SPACER

TRANSOM BRACKETS

BOLTS

Figure 1–5   Correcting transom height.

through the transom from the outside (Figure 1–6). Coat bolts and washers with RTV silicon sealer to prevent water from entering the transom. *Safety* must be the main consideration. You may want to consider a different length outboard; available transom heights include 15, 20, and 25 inches. If the transom is curved, a spacer which is flat on the outboard side will have to be installed between the transom and outboard. Total length of the spacer must be longer than the transom brackets.

The positioning of the outboard motor is critical to maintain good steering and reduce drag of the lower unit as much as possible. Better boat performance is the object of correct positioning of the outboard motor. Hull design permitting, a one-inch rise will increase performance by approximately $1^1/_2$ miles per hour.

1.3   The Steering System

The steering system is a safety system. What happens at the steering wheel must direct propeller thrust direction. The modern boat has come a long way since the days when just being "self-propelled" was enough to satisfy the boat owner. Improvements to the steering system brought the operator away from the steering handle and up front to the driver's seat.

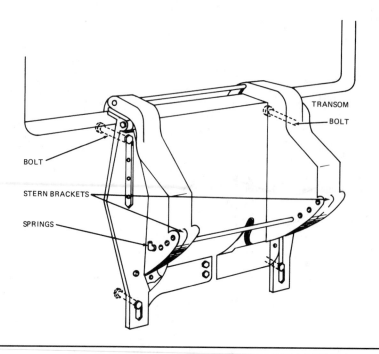

Figure 1–6   Transom mounting of adjustable stern bracket.

Tremendous improvements have been made in steering wheel design, including tilt or fixed position, with rack-and-pinion and rotary steering systems. A push/pull force is transmitted by plastic-covered armored steering cables rated for the horsepower of the outboard. All parts, like the drag link, are also rated according to horsepower. Do not substitute them with those of a lesser rating. Stay with Original Equipment Manufacturer (***OEM***).

The steering starts at the steering wheel; by movement of the steering wheel a positive force is transmitted throughout the steering cable to the steering link and on to the outboard steering arm, thus to the propeller and trim tab. This cable movement then turns the propeller thrust in the desired direction. What happens at the steering wheel must happen at the propeller without undue resistance, grabbing, seizing, or lost motion. It must happen in any tilt or trim position without binding.

There is special mounting hardware that won't rust or come loose. These parts and mounting hardware should not be substituted with those of lesser grades of material that do not meet the safety standards of ***NMMA/ABYC*** (National Marine Manufacturers Association, American Boat and Yacht Council). If primary boating is in salt water, stainless steel parts are used. The consequences of a failed steering system can be disastrous. If the cable comes loose, the boat can suddenly turn and circle, possibly throwing people into the water. If the steering system locks up or comes apart, the operator may not be able to control the boat and may panic (Figure 1–7).

Outboard Steering System and Persons Safety

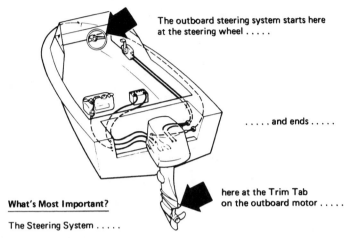

The outboard steering system starts here at the steering wheel . . . . .

. . . . . and ends . . . . .

here at the Trim Tab on the outboard motor . . . . .

What's Most Important?

The Steering System . . . . .

- must not come apart

- must not jam

- must not be sloppy or loose

What Could Happen?

- If steering system comes apart, boat might turn suddenly and circle . . . . . persons thrown into water could be hit.

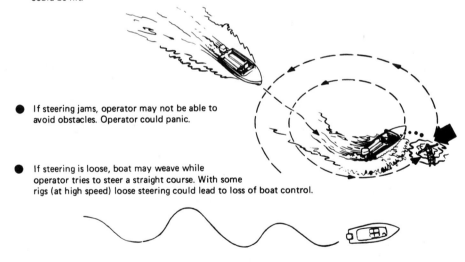

- If steering jams, operator may not be able to avoid obstacles. Operator could panic.

- If steering is loose, boat may weave while operator tries to steer a straight course. With some rigs (at high speed) loose steering could lead to loss of boat control.

Figure 1–7   Outboard steering system and persons safety (continued on next page).
(Courtesy of Outboard Marine Corporation)

How Can Loss of Steering Control be Minimized?

- Choose steering system which meets Marine industry Safety Standards (BIA/ABYC).
- <u>Read</u>, <u>Understand</u>, and <u>Follow</u> manufacturers <u>Instructions</u>.
- Follow warnings marked " ⚠ " closely.

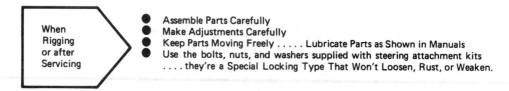

When Rigging or after Servicing	

- Assemble Parts Carefully
- Make Adjustments Carefully
- Keep Parts Moving Freely . . . . . Lubricate Parts as Shown in Manuals
- Use the bolts, nuts, and washers supplied with steering attachment kits
. . . . they're a Special Locking Type That Won't Loosen, Rust, or Weaken.

- When Transom Mounted steering systems (see picture) are used, check to uncover possible Trouble!

Tilt motor into boat . . . then turn it.

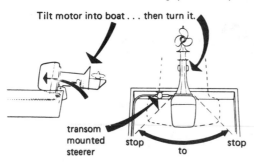

transom mounted steerer   stop   to   stop

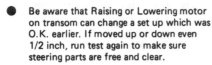

During this procedure, steering parts

- Must Not Bind

- Must Not <u>Touch</u> Other boat, motor or accessory Parts in Transom Area.

<u>Why?</u> A hard blow to the motor's lower can result in damage to steering parts here

- Be aware that Raising or Lowering motor on transom can change a set up which was O.K. earlier. If moved up or down even 1/2 inch, run test again to make sure steering parts are free and clear.

- Check for Damaged parts . . . . . Blows to the Motor like this

or this can put Heavy Loads on steering parts. Look for . . . . .

- Cracked parts
- Bent parts
- Loose nuts and bolts

- Replace damaged parts. If weakened, parts could fail later . . . . on the water . . . . when least expected.

Figure 1–7   Continued.

The steering system should be checked and lubricated every 60 days of freshwater operation or every 30 days of salt water operation and just before going into storage. Ninety percent of the steering problems relate directly to lack of use (movement) and improper lubrication. The cable may seize up because of the marine environment. Many cables are replaced each year because they are not properly cared for. So, if your boat is in storage or on a trailer in the yard, take time to work the steering several times from lock to lock during the storage season.

 Safety

When repairs are made, such as cable replacement, every effort should be made to stay with parts made by the same manufacturer as the original steering system. In this way, system function, reliability, and component compatibility will allow each component to work together as a system, thus maintaining the required safety standards.

Once underway, the boat may want to pull to the left or right. This can be caused by the condition of the hull, but more likely it is caused by the outboard. If pulling does occur, check to see that the outboard is mounted in the center of the transom. If off-center, correct it and test again.

Consider the raised outboard and particularly the propeller. The normal rotation for the propeller is in a clockwise direction. The mere fact that the propeller is turning and the lower part of the propeller is operating in less disturbed water than the upper half can cause the stern of the outboard to walk to the starboard and steer the boat to the port. A propeller that is turning counterclockwise will walk the stern of the outboard to the port and the boat will steer to the starboard (Figure 1–8).

Because of this effect and varied hull conditions, many outboards come equipped with an adjustable trim tab bolted to the anti-ventilation plate. If the boat pulls to the port, adjust the trim tab to the port for necessary steering correction. Other motors may offset the lower unit slightly off-center to compensate for this walking tendency. If you are going to make repairs on the water pump or lower unit, mark the position of the trim tab before unbolting it. This will save time-consuming adjustments once back in the water. The customer will appreciate not having the hassle of a steering problem again. The trim tab can be part of the corrosion control system. When made of zinc, it becomes an anode that is sacrificed to save the more expensive castings. To be effective, the trim tab and the area it contacts, should remain unpainted.

Steering torque may be caused by propeller torque and affects the attitude and handling of the boat. See the detailed information given under "Tilt Angle Adjustment" in Chapter 13.

1.4   Handling The Outboard

There is nothing simpler than taking the small outboard motor on or off the transom; yet, there are some procedures to follow. Safety is always first, so lift the motor correctly and be careful when stepping off the dock into the boat. The boat has a way of moving when

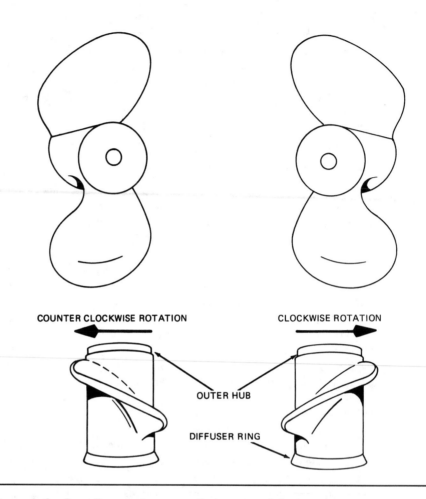

Figure 1–8   Propeller rotation can walk the stern of the boat.

least expected, balance is lost, and the outboard somehow misses the transom and is "deep sixed" (Figure 1–9).

Probably the best way to get the motor onto the transom is to lay the outboard on the dock (spark plugs up) next to the boat. Get yourself into the boat and then pick up the outboard and install it on the transom. Don't forget to hook up the safety rope or chain so you won't lose the motor if it comes loose.

After the fishing trip is over and the motor is taken from the transom, how do you position the outboard for transport? There have been many outboards ruined right at this point. When removing the outboard from the transom always keep the powerhead up so any water remaining in the cavities of the powerhead or exhaust housing will not pour into the cylinders via the exhaust system. If you consider it, the exhaust system has a direct

Figure 1–9   The one that got away!

opening right to the water. Laying an outboard down with the propeller higher than the powerhead, it is possible for any water remaining in the housings to pour right into the cylinders. This water then sits there causing corrosion between pistons and cylinder walls, and seizes the piston to the cylinder wall. Some water may even find its way through the intake port or piston ring gaps into the crankcase, where rusting of rods and bearings can take place. And it doesn't take much water— any water at all is too much in these locations. The first indication of this water may be the next time you try to use the outboard. You attempt to start the engine and it won't turn over—it is seized up! How can this be? It ran so good last time out! Then you take the motor to the dealer, who says that water got in it and it will take money to make the repair. Just being aware that this water can go down the exhaust is worth the cost of an overhaul. Handling and positioning the outboard for transport with the powerhead (spark plugs) up, therefore, is very important. While not always convenient, the best means of transport and storage is in the upright position. Always go with the powerhead (spark plugs) up!

Many of the larger outboards weigh more than a person. When lifting a large outboard on or off the boat for repairs, lift from the bracket provided or attach a lifting eye to the flywheel (see Figures 1–10 and 1–11).

Safety

Lifting eyes are available through your dealer or there is a special flywheel puller that can be adapted with a lifting eye. Always use the hardened bolts that come with the puller because others may be of poor quality and break. Make sure the hoist is capable of lifting the weight of the outboard and that cables, hooks, etc. are in good condition. *Again, always keep the outboard up!*

1.5   Maintenance for Safety

Boating is not only fun but it is expensive. To maintain your investment and for safety in boating, maintenance needs to be kept up-to-date. It is certainly discouraging when your dealer-mechanic tells you that the lower unit needs to be overhauled for $800, and that it could have been prevented if you had only brought the outboard in for the seasonal service. Or, maybe you are out for a weekend of boating when all of a sudden the powerhead starts making funny noises. Again, the dealer informs you that preventive maintenance could have avoided this costly repair. Much preventive maintenance can be done by the owner if he takes the time to learn about the outboard. Learning about the maintenance required, how the outboard operates, and what makes the steering work smoothly can pay big dividends to the owner. Let us consider some basic maintenance items, as well as some operational checks.

Figure 1–10   Powerhead lifting eye. *(Courtesy of Mercury Marine)*

Figure 1–11   Powerhead lifting eye using universal puller. *(Courtesy of Outboard Marine Corporation)*

1.5.1   Rated Parts

The marine environment is hostile and boat parts must be able to withstand abuse and not fail prematurely. You could substitute OEM parts. However, do not substitute with a lesser grade part than the original manufacturer installed on the unit. There are even some nuts and bolts that will get you in trouble if substituted with a lesser grade. Check the characteristics of the part: Do they look alike, have the same strength, are they made of the same material, and are they the same size? Threads on powerhead bolts may be of an "odd pitch" and therefore different than hardware bolts. They must be purchased through the OEM dealer. Bolts are marked on the head to identify the grade, strength, and type of material used (but not pitch). This is an indication of the tensile strength of the bolt. Look for marks or figure numbers (Metric) on the head and re-assemble using the same grade of bolt- and nut-style removed. See Figure 1–12 for various bolt markings. The bolts are either stainless steel or are plated to resist rust and corrosion, so stay with correct bolts. Do not substitute unless *you know* they are the same in *all* characteristics.

When installing bolts in the lower unit or in the exhaust cover, cylinder head, outer exhaust cover, cylinder assembly, or exhaust housing, coat the entire bolt (except head) with the recommended sealing compound per service manual instructions. These are non-hardening gasket compounds, and help control corrosion that can develop between the bolt and aluminum casting. (Do not use lubricants containing graphite compounds, as they create a galvanic cell that causes aluminum to corrode.) If you don't coat the bolt and do your

Figure 1–12   SAE bolt grade
indications.

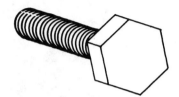

SAE GRADE 0-1-2
TENSILE STRENGTH 74,000 PSI
LOW CARBON STEEL

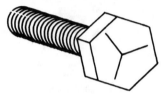

SAE GRADE 5
TENSILE STRENGTH 120,000 PSI
MEDIUM CARBON HEAT TREAT STEEL

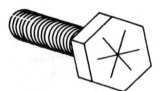

SAE GRADE 8
TENSILE STRENGTH 150,000 PSI
MEDIUM CARBON ALLOY STEEL

boating in brackish or saltwater, corrosion will develop, seizing the bolt to the casting. This
will make it nearly impossible to remove without using a torch to heat the casting. Even so,
the bolt may still break.

1.6   Lubrication Points on the Outboard

On each outboard there are several lubrication points requiring periodic attention. Each
manufacturer gives a detailed list on what and when to lubricate and with what type of lubrication. If you don't have a service manual for your outboard, visit your local dealer and
purchase one. The cost is minimal and savings will be gained through preventive maintenance. Lubrication of specific parts should be done every 60 days in freshwater operation

Lubrication Chart – 1 thru 10

FIGURE	LUBRICATION POINT	TYPE OF LUBRICANT	FREQUENCY † (PERIOD OF OPERATION)	
			FRESH WATER	SALT WATER †
1	Throttle Shaft and Linkage, Rear Motor Cover Latch	A	60 days	30 days
2	Shift Lever Shaft	A	60 days	30 days
3	Vertical Throttle Shaft	A	60 days	30 days
4	Carburetor Linkage, Cam and Shifter Starter Lockout	A	60 days	30 days
5	Clamp Screws and Tilt Tube	A	60 days	30 days
6	Swivel Bracket, and Tilt Lock (Port and Starboard)	A	60 days	30 days
7 8	Gearcase Capacity: 20, 25, S25RP, 30: 11 fl. oz. (325 ml) 25RS, 25RW, A25RP, 28: 8 fl. oz. (245 ml)	B	Change after first 20 operating hours. Check level every 50 operating hours. Change every 100 operating hours or once each season, whichever comes first.	
9	Electric Starter Pinion Helix	C	As required	As required
10	Tilt Lever Shaft and Reverse Lock	A	60 days	30 days

† Some areas may require more frequent lubrication.

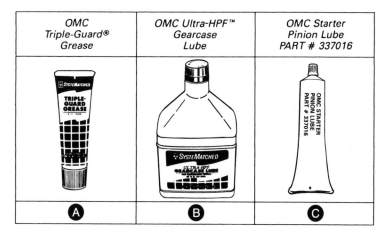

OMC Triple-Guard® Grease	OMC Ultra-HPF™ Gearcase Lube	OMC Starter Pinion Lube PART # 337016
A	B	C

Figure 1–13   Lubrication points (continued on next page). *(Courtesy of Outboard Marine Corporation)*

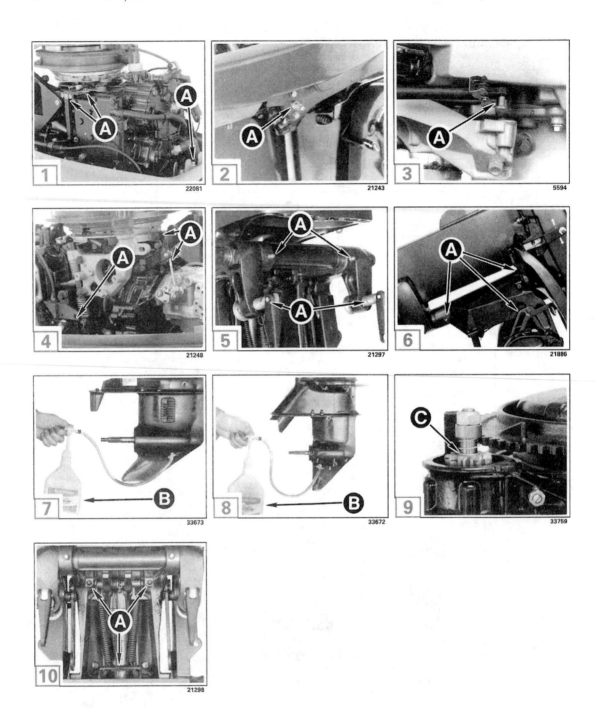

Figure 1–13   Continued.

Application Guide for Outboard Maintenance Products

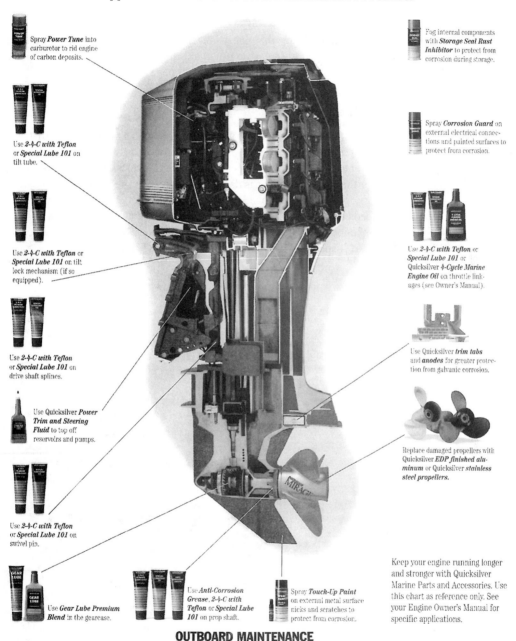

Spray **Power Tune** into carburetor to rid engine of carbon deposits.

Use **2-4-C with Teflon** or **Special Lube 101** on tilt tube.

Use **2-4-C with Teflon** or **Special Lube 101** on tilt lock mechanism (if so equipped).

Use **2-4-C with Teflon** or **Special Lube 101** on drive shaft splines.

Use Quicksilver **Power Trim and Steering Fluid** to top off reservoirs and pumps.

Use **2-4-C with Teflon** or **Special Lube 101** on swivel pin.

Use **Gear Lube Premium Blend** in the gearcase.

Use **Anti-Corrosion Grease, 2-4-C** with **Teflon** or **Special Lube 101** on prop shaft.

Spray **Touch-Up Paint** on external metal surface nicks and scratches to protect from corrosion.

Fog internal components with **Storage Seal Rust Inhibitor** to protect from corrosion during storage.

Spray **Corrosion Guard** on external electrical connections and painted surfaces to protect from corrosion.

Use **2-4-C with Teflon** or **Special Lube 101** or Quicksilver **4-Cycle Marine Engine Oil** on throttle linkages (see Owner's Manual).

Use Quicksilver **trim tabs** and **anodes** for greater protection from galvanic corrosion.

Replace damaged propellers with Quicksilver **EDP finished aluminum** or Quicksilver **stainless steel propellers**.

Keep your engine running longer and stronger with Quicksilver Marine Parts and Accessories. Use this chart as reference only. See your Engine Owner's Manual for specific applications.

OUTBOARD MAINTENANCE

Figure 1–13   Continued. *(Courtesy of Brunswick Corporation)*

and every 30 days in salt water operation. Lubrication of the drive shaft splines is a seasonal job. This includes removing the lower unit to apply OMC Moly Lube 1 to the drive shaft splines. Change the gear case oil seasonally or every 50 hours. Examine the gear oil for being milky, which indicates that water is leaking into the gear case. The water pump impeller should be changed at the same time, and the water pump housing inspected for damage by corrosion and or scoring (Figure 1–13, 3 pages)

1.6.1   Propeller Shaft

Another seasonal job is removing the propeller and applying Special Lube 101 (Mercury) or anti-corrosion grease to the splines of the propeller shaft and propeller hub (Figure 1–14). This prevents corrosion and salts from seizing the propeller hub to the propeller shaft. Remember that if the propeller is damaged severely or the shear pin is sheared when you are out on the water, you may need to put your spare propeller on, or shear pin in, and the propeller needs to come off easily (Figure 1–15). Make sure that you carry the necessary tools to do the job. This spare propeller or shear pin may be the only way back to port! When back in port inspect the propeller shaft for twisted splines, which can be caused by the impact.

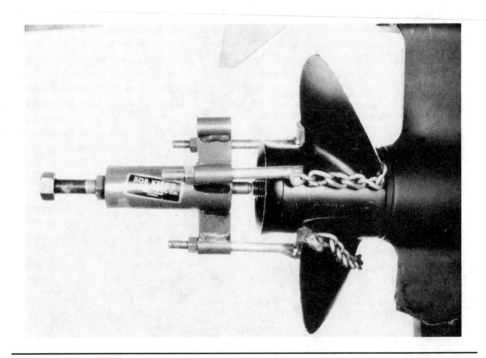

Figure 1–14   Pulling seized propeller from corroded propeller shaft splines.

Figure 1–15   Damaged propeller with hub removed.

1.6.2   Propeller

Check the skeg and propeller for visual damage. Dress them using a file, being careful to follow the existing contour. The propeller blades are balanced with equal pitch. Check each blade to see if it is bent. Chips on the leading edge of the propeller or skeg can cause *cavitation*, which is water vapor bubbles flowing along the propeller blade and condensing back, causing a burn on the propeller blade. If the nicks and chips are bad enough, they can cause severe cavitation and affect performance (Figure 1–16).

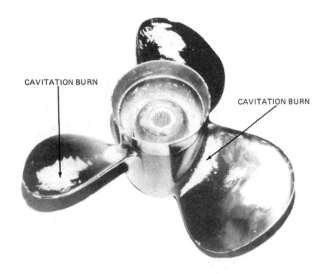

Figure 1–16   Propeller showing cavitation burns.

1.6.3   **Corrosion**

The trim tab located on the anti-ventilation plate may be somewhat deteriorated, because it is made of zinc and is acting as a sacrificial anode. Deterioration of this part is normal, and the fact that it is eroding means that it is doing its job of protecting the lower unit (Figure 1–17). This part is sacrificed to save the more expensive ones. Change the trim tab only when it is two thirds gone and begins to look extremely bad (Figure 1–17a, b). Instead of a zinc trim tab, an anode may be attached to the lower unit bearing housing (Figure 1–17c). (See Section 1.12 Corrosion—What Can Help in this chapter.)

(a)

(b)

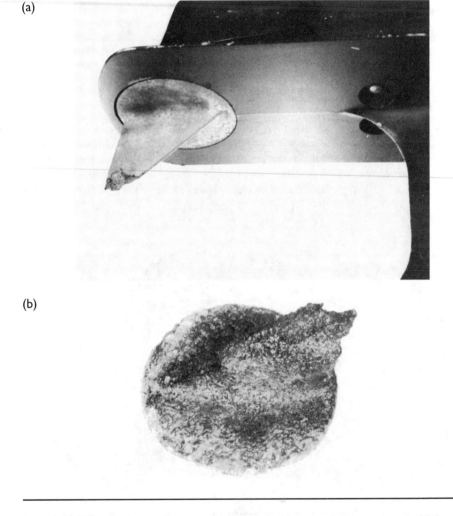

Figure 1–17   Trim tab may be a sacrificial anode (a); expended zinc trim tab (b) (continued on next page).

(c)

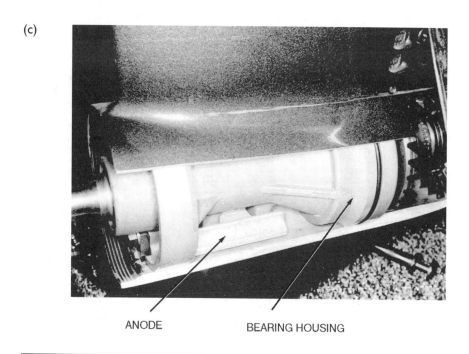

ANODE   BEARING HOUSING

Figures 1–17 (continued)   Cutaway showing lower unit bearing housing and anode (c).

1.7   **Operational Checks**

There are easy operational checks on the charging, ignition, and fuel systems. These systems insure that the battery stays charged, and that you get the rated RPM for the powerhead.

Don't ignore tune-up services because the motor was running well all season. The two-stroke motor is unforgiving when it comes to damage caused by tune-up adjustments that have changed.

If synchronization of linkage to the carburetor and timing are not correct, then detonation or preignition can cause major piston damage and cylinder wall scoring (Figure 1–18).

When you are boating keep your eye on the tachometer, which reads the powerhead RPMs. RPM is one indication of the load applied to the outboard. At wide-open throttle (**WOT**), the tachometer should read in the upper one-quarter of the operating range for the powerhead. If it is out of the operating range, then repair work is needed, perhaps a simple propeller change or a tune-up (Figure 1–19). The key is that troubles (adjustments) are preventing the powerhead from running at the rated RPM range, and that something should be done right away to prevent major damage from developing. For more details see Chapter 13.

The portable fuel tank should be examined at least once a season. Check the cap sealing gasket and the vent valve for proper sealing and opening. If these leak, fuel can run into your car or truck during transport, when internal pressure develops because of the sun's heat. Look for cracks at the squeeze bulb and pressurize the hose. Leakage at the motor

Figure 1–18   Damaged piston
caused by preignition.

Figure 1–19   Checking operating RPM at wide-open
throttle.

connector indicates that an O-ring is leaking and should be replaced. This O-ring leak will
allow air bubbles into the fuel line, reducing fuel flow. Look into the metal tank; if rust is
present at the bottom, the tank will need to be replaced with a plastic tank. There is no effective way to repair a rusted tank short of replating it. The rust can be cleaned off, but will
reappear and get into the carburetor. This will cause flooding and fuel stoppage, and soon
the carburetor will need to be overhauled. A tank adapter is available from Outboard Marine Corporation (OMC) which fits the tank opening. Short bursts of air pressure are ap-

plied from a hand pump, pressurizing the tank, bulb, hose, and connector to 10 PSI. By submerging one end of the tank at a time, bubbles will indicate any leaks.

 **Warning**

Any leaks in the tank itself should not be repaired but the tank should be replaced. Any other leaks in lines/connectors should be repaired (Figure 1–20). If there are problems with the old metal tank, lightweight plastic tanks with special inhibitors and plasticizers to protect against ultraviolet rays and aging are available. They meet ABYC standards (Figure 1–21).

1.7.1 Cooling System

Some outboards are equipped with a thermostat with a pressure relief spring and possibly a pressure-relief valve which helps to cool the motor at a high RPM. On a V-6 Mercury the thermostat controls temperature at idle for better combustion and controls pressure from 1–3 pounds. The water pressure relief valve (poppet valve) will control the pressure from 4–6 pounds at 2500 RPM.

Corrosion can cause the thermostat to stick, especially when operating in salt water, and can cause overheating. If the outboard is designed for a thermostat, it should not be removed and discarded. Doing so allows the powerhead to run cooler than normal and causes

Figure 1–20   Using air pressure to test portable fuel tank and hose for fuel leaks.

Figure 1–21   Lightweight plastic fuel tanks. *(Courtesy of Mercury Marine)*

problems with fuel delivery, carbon, plug fouling, and poor running when cold. Also, cold lake or ocean water coming into critical areas of the powerhead causes undue stress on some of the castings. Additionally a warmed-up outboard just runs better (Figure 1–22).

1.7.2 Exterior

Nicks and scratches received over the boating season should be touched up with a matching paint available from your favorite dealer. The paint is part of the corrosion protection given the outboard and should be maintained. Especially important is the lower unit that is always in the water. The lower units of some outboards have drain holes that drain water from cavities in the casting. Check to see that these are open. If they are not open, freezing or corrosion can develop in the casting cavity and can apply pressure to the casting and it may crack.

Some seasonal services are involved procedures, so mechanical abilities and a service manual are necessary. If you don't have these, the outboard should be taken to a dealer,

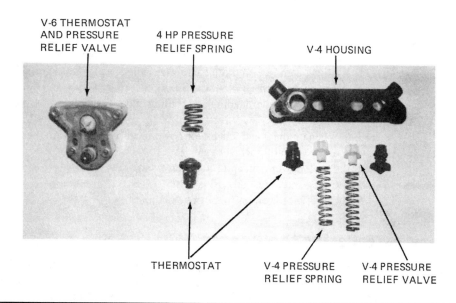

V-6 THERMOSTAT
AND PRESSURE
RELIEF VALVE

4 HP PRESSURE
RELIEF SPRING

V-4 HOUSING

THERMOSTAT

V-4 PRESSURE
RELIEF SPRING

V-4 PRESSURE
RELIEF VALVE

Figure 1–22   OMC thermostats and pressure relief valves.

who can perform the services. Also check with your local community college or technical institute for a class to improve your own skills.

1.7.3   Midseason 50-Hour Check-Up

How often should you change the oil or lubricate your automobile? Most of us can probably answer that question very quickly. Can you answer as quickly when it comes to the outboard? Certain maintenance items should be cared for on the outboard lower unit every 50 hours of operation. Other items need to be lubricated every 60 days in freshwater operation and every 30 days in saltwater operation. Many of these items can be done by the do-it-yourselfer. The following is a brief list of some of the more important maintenance items:

1. Lubricate the steering arm pivot and all cables.
2. Lubricate the carburetor linkages.
3. Check gear case oil level; if oil is milky-looking there is a water leak. Drain, flush, and refill oil. (Get it pressure/vacuum tested and the water leak repaired.)
4. Inspect portable fuel tank and clean or replace fuel filter.
5. Check synchronization of the linkages and timing to the carburetion. (This should be done by a qualified technician.)
6. Clean and adjust/replace spark plugs.
7. Clean battery, test state of charge, and load-test it for capacity (MCA).
8. Work remote control through full travel, checking for binding and deterioration.

9. Work steering from lock to lock, checking for binding and/or stiffness.
10. Check water pump operation and powerhead for overheating, using a Thermo-melt Stik and watching exhaust outlet or "tell-tale" outlet.
11. Adjust carburetor mixture screw and idle speed. (This should be done by a qualified technician.)
12. Inspect propeller leading and trailing edges for nicks and scores and cavitation burn.
13. When on the water, check for powerhead running within the upper one-quarter of the operating RPM range.
14. Check power trim and tilt oil reservoir.
15. Check and adjust any drive belts.
16. Check motor oil on four-stroke outboard.

1.8   Trailering the Boat

The purpose of the trailer is to transport the boat, but at what price? The trailer is used not only to transport, but it should give *good support* to the boat bottom while in transport or storage. Choose your trailer so there is good support all the way along the boat bottom (Figure 1–23).

The boat is made to transport people and gear on the water. While on the trailer, don't load the boat down with all kinds of gear, as if it were a utility trailer. The boat trailer is hard pressed to support the boat bottom properly with the weight of the boat on it, without the additional weight of camping gear.

1.8.1   Boat Bottom

Without proper support the boat bottom can be shaped into a rocker while in storage and/or in transit. A rocker is a bulge or convex condition in the bottom of the stern of the boat (Figure 1–24).

If you are in the market for a used boat, make sure you look at the bottom. Put a long straight edge on it to determine if the condition is satisfactory. The last several feet of the boat is the wetted area of the boat bottom used for planing. A rocker in this location will affect handling and may cause the boat to porpoise.

Another condition in the bottom of the boat is a ***hook.*** This concave condition near the stern of the boat may be built in the bottom by the manufacturer. A built-in hook will reduce the tendency to ***porpoise*** (move up and down). You will need to know whether the hook was put in the bottom by the manufacturer or did the prior owner cause it by improper trailering? A call to the boat manufacturer might just be a good investment before purchasing the boat.

The boat bottom with either a rocker or a hook can be corrected with a lot of hull work. It is expensive and probably should be done by a repair facility that specializes in hull repair. It is easier to avoid these problems than to cure them.

Several trailer designs support the boat on rollers or on carpet plank (pads) and have different methods for tie-downs. Select one recommended by your boat manufacturer that

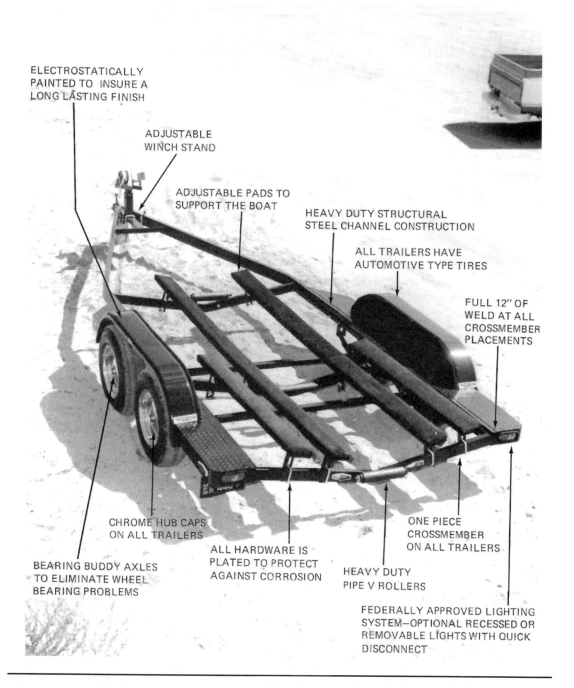

ELECTROSTATICALLY
PAINTED TO INSURE A
LONG LASTING FINISH

ADJUSTABLE
WINCH STAND

ADJUSTABLE PADS TO
SUPPORT THE BOAT

HEAVY DUTY STRUCTURAL
STEEL CHANNEL CONSTRUCTION

ALL TRAILERS HAVE
AUTOMOTIVE TYPE TIRES

FULL 12" OF
WELD AT ALL
CROSSMEMBER
PLACEMENTS

CHROME HUB CAPS
ON ALL TRAILERS

ONE PIECE
CROSSMEMBER
ON ALL TRAILERS

ALL HARDWARE IS
PLATED TO PROTECT
AGAINST CORROSION

BEARING BUDDY AXLES
TO ELIMINATE WHEEL
BEARING PROBLEMS

HEAVY DUTY
PIPE V ROLLERS

FEDERALLY APPROVED LIGHTING
SYSTEM—OPTIONAL RECESSED OR
REMOVABLE LIGHTS WITH QUICK
DISCONNECT

Figure 1–23   Good boat trailer design. *(Courtesy of Trail Rite, Inc.)*

THE "HOOK" FORCES THE BOW DOWN BY LIFTING THE TRANSOM

Figure 1–24 Any hook or rocker will have a negative effect on boat speed.

A "ROCKER" IS A BULGE IN THE FORE-AND-AFT DIRECTION

gives support in the right places, depending on the boat bottom configuration. If the trailer is put into brackish or saltwater, the trailer needs to be rinsed off after each launch and retrieval of the boat, if city water is available. Otherwise the finish of the trailer will quickly deteriorate.

1.8.2 Electric Circuits

Electric circuits on the trailer deteriorate from repeated launchings, and the lights always seem to be a source of trouble. Standard automotive lights just don't do the job, when put under water. If you use this type of light, a light board should be attached to the stern of the boat or rear of the trailer and removed prior to launching the boat. In this way, the lights will not be submerged and become inoperative. There are some sealed lights that may be used as fixed installations, but they are more expensive than standard lights. All connections which are made in the wiring should be soldered. After electrical connections (repairs) are made, coat all terminal connections and repairs with Liquid Neoprene to avoid corrosion. Crimping may be fast and expedient, but wiring corrodes and comes loose, causing problems. Electricity is being carried a long distance from the tow vehicle; therefore, use 12-gauge wire in all the circuits, including a ground circuit from the trailer frame to the tow vehicle frame. Stay away from cheap receptacles (plugs) as they probably will not stand the marine environment.

1.8.3 Wheel Bearings and Bearing Buddy II

Wheel bearings need to be serviced frequently because of repeated submersions in water. Use water-resistant grease for boat trailer wheel bearings. This is not a waterproof grease but grease that resists the water. Hand pack the wheel hub totally full to keep the water out. There are special seals to be used but they may leak, and water could seep into a pocket deteriorating the lubrication. The wheel bearings warm up as they are rolling. They are probably warm when you arrive at the boat ramp. The bearings are then submerged while

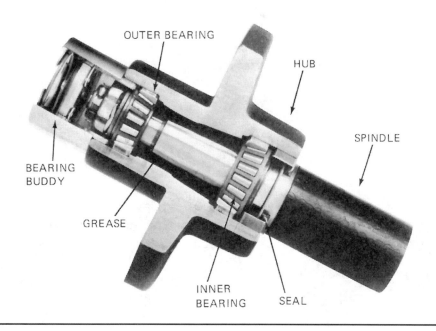

Figure 1–25   Bearing buddy axle. *(Courtesy of Unique Functional Products)*

launching the boat. The bearing hubs are cooled quickly, possibly drawing in water and/or condensation, contaminating the lubricant. This is a normal situation; if maintenance is neglected, the bearing may burn out while traveling and disrupt the trip. A product known as Bearing Buddy II by Unique Functional Products will prevent these problems.

1.8.4   It Can Happen To Any Trailer*

Without proper lubrication, hubs on all types of trailers can overheat, causing delay, inconvenience, and expensive repair. In short, it can ruin your trip (see Figures 1–25 and 1–26).

Trailers equipped with Bearing Buddy II by Unique Functional Products can be submerged in water or left in storage without wheel-bearing damage. You can float a boat on and off your trailer, without the threat of frozen bearings and unexpected cost.

Bearing Buddy II works by replacing the dust cap in the axle hub. The hub is filled with grease through a fitting in the piston of the Bearing Buddy II. When the wheels are submerged, the pressure from the spring-loaded piston, combined with the grease, keeps out water. Any grease coming out the front of a Bearing Buddy II is due to overfilling (Figure 1-26b).

Trailering even for short distances heats the hubs. When wheels on a boat trailer are submerged during launching (Figure1-26c), the hubs suddenly cool and air inside unprotected hubs contracts, forming a vacuum that draws in water through the rear seals. Water and grit drawn into the hubs will destroy the bearings.

*This section courtesy of Unique Functional Products.

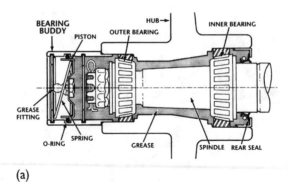

(a)

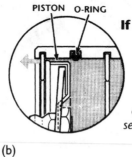

If You Overfill

The rear seals are protected from rupture by an overfill feature which allows excess grease out around the piston, limiting pressure in the hub to 3 psi. With Bearing Buddy, the bearings are assured of vitally needed lubrication and the rear seals last. They ride on a lubricated surface.

(b)

(c)

PROPER LUBRICANT LEVEL

Blue ring extended

LOW LUBRICANT LEVEL

Blue ring recessed

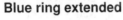

(d)

Figure 1-26   Bearing buddy continuous wheel-bearing lubrication. *(Courtesy of Unique Functional Products)*

Check for low grease level or overfilling when the hubs are warm. For boat trailers, make a Bearing Buddy II check right before launching. Check the grease level by gently pressing on the edge of the piston. If you can move it, there is plenty of grease inside. If the piston won't rock, use a hand grease gun to add enough lubricant to push the piston out about 1/8 inch.

Bearing Buddy II protects wheel bearings by maintaining a slight pressure inside a grease filled hub (Figure 1–26d). Without positive pressure, water will enter the hub and damage the wheel bearings.

When the hub is properly filled, a spring loaded steel piston pushes against the grease and maintains positive pressure. The blue indicator ring rests outward from the steel snap ring. This is the correct operating position.

If the hub lubricant is low, the blue ring recedes until it is flush with the steel snap ring. Add a small amount of grease to move the blue ring outward about 1/8 inch. It's that simple!

If the blue ring moves out beyond 1/4 inch, the hub is overfilled. The rear seals are protected from rupture by an overfill feature which allows excess grease out around the piston, limiting pressure in the hub to 3 PSI. With Bearing Buddy II, the bearings are assured of vitally needed lubrication and the rear seals last. They ride on a lubricated surface.

When selecting a trailer, consider the type of wheels and tires. Tires must be large enough to support the load they are expected to carry. They must absorb or cushion, by deflecting part of the shock from road irregularities. A tire of passenger-car size will rotate approximately 800 revolutions for each mile traveled. Such a tire should be balanced the same as a car tire. An improperly balanced tire can shake the boat enough to cause hull problems. If you plan to tow the boat at freeway speeds, nothing smaller than automotive-size wheels and tires should be considered.

When you initially purchase a trailer, it should be checked for how it matches with the boat. Are contact points, bolster, and rollers adjusted to the boat contour? As the towing vehicle makes sharp turns, will any of the trailer or boat parts come in contact with the tow vehicle? Is the electrical equipment working and is there a separate ground wire installed through the plug? This ground wire will prevent the blinking of the lights as you travel down the road. Do all the lights meet legal requirements? Many of the lighter trailers are not required to have brakes. If they are required on your trailer, do they work with application of the tow vehicle brake pedal?

1.8.5   Tongue Weight

The tongue weight (hitch load) of a loaded trailer can be measured with your bathroom scale, as shown in Figure 1–27. (Use the tow vehicle for support while you position the scale, etc.) Place a piece of wood of approximately the same thickness as the bathroom scale on the ground in line with the trailer hitch jack. It should be spaced so that a short piece of pipe or other round piece will lay exactly one foot from the center line of the jack extension. Place the scale so that another round piece of pipe is exactly two feet from the center line of the jack extension in the other direction. Place a 2 × 4 or a 4 × 4 on the two round pieces of pipe, and screw down the jack extension on top of the 2 × 4 until the tongue

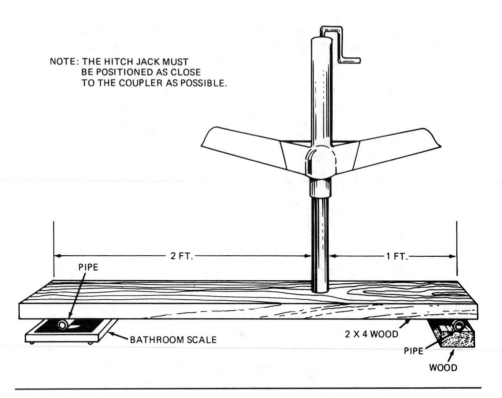

NOTE: THE HITCH JACK MUST
BE POSITIONED AS CLOSE
TO THE COUPLER AS POSSIBLE.

2 FT.   1 FT.

PIPE

BATHROOM SCALE   2 X 4 WOOD

PIPE

WOOD

Figure 1–27 Measuring tongue weight.

of the trailer is supported by it. (Trailer hitch is released and clear of the tow vehicle.) Multiply the scale reading by three. This is the tongue weight of the trailer. If you exceed the capacity of the bathroom scale, increase the two-foot dimension to three or more feet, but always multiply the scale reading by the total number of feet between the pipe and wood block on the ground and the pipe and scale. The trailer must be level when you read the scale.

The trailer hitch on the tow vehicle must be capable of supporting this weight. If you are not sure of the make and capacity of your tow-vehicle hitch, visit your local trailer hitch specialist for the answers. Bolts used to mount the hitch need to be grade 5 (3 marks on head) or higher. Always use a higher grade bolt rather than a lower grade.

1.8.6 Loading

Proper positioning of the boat is important so that you have the right amount of tongue weight. Too much weight forward is hard on the tow vehicle suspension system. Too little weight forward may give the trailer the tendency to fishtail, and this side-to-side movement is hard on the truck, trailer wheels, tires, and suspension. It can also be dangerous.

Proper loading depends on the weight of the loaded trailer. Boats and trailers should have about 5 to 10 percent of the total weight on the tongue. It may be possible to move the boat forward or backward to adjust the tongue load.

1.8.7 Pre-Highway Trailer Check

Repeated launchings of the boat take its toll on the trailer. Many times the trailer is just taken for granted before going on the road. The trailer should be checked for the following before pulling out:

1. Is the hitch properly hooked up, including safety chains and break-away cable?
2. Are all the lights functioning?
3. Are the tires properly inflated for the load they are to carry? Do you have a spare?
4. Do the brakes operate properly? Double-check the trailer brake hand control and the brake pedal.
5. Are the wheel bearings properly lubricated?

The boat should be inspected for these conditions:

1. Are all tie-downs properly secured?
2. Is the outboard secured to the transom, and did you push the power trim down button to put pressure on the trailer lock (trim cylinder all the way in), and relieve the weight from the hydraulics? (This takes the bounce off of the hydraulics.)
3. Are you using a transom-saver support between the outboard lower unit and the trailer frame? If so, is it tightly secured?
4. Is there road clearance between the skeg and propeller and the road for pulling in and out of service stations or driveways?
5. Are the locks on the winch, bunk, or tilt mechanisms in their proper position?
6. Are the items in the boat properly stowed? (Don't overload the boat with camping gear and damage the bottom of the boat; it is not a utility trailer.)
7. Are all fuel tanks tightly closed?

 Safety

Tanks should be filled at destination. Remove portable tanks from the boat and place them on the ground/slab for a static ground. Always keep pump nozzle in contact with fuel tank, so static electricity can also be grounded to the gasoline pump.

8. Are all Coast Guard safety requirements aboard? Preparation makes for a safe trip!

1.8.8 Electrical Color Code on the Trailer

White: Ground (trailer frame to tow-vehicle body or frame)
Brown: Running lights and/or tail lights
Yellow: Left turn signal and brake light (one light)
Green: Right turn signal and brake light (one light)

Note

As a precaution against rupture by vibration, all conductors should be of the stranded type and no single conductor smaller than No. 16 AWG shall be used. Multi-conductor cable circuits shall be no smaller than No. 18 AWG.

1.8.9 Trailer Storage

Boat storage is just a matter of common sense. If possible keep your boat and trailer under cover. If that is not feasible, keep your boat covered but vented (preventing mildew) by a good quality canvas or plastic boat cover. Put enough slats across the open areas of the boat to create a bulge or tent effect, so that rain cannot gather in depressions and leak into the boat. Small brackets are available in most good marine stores. Before going into storage with your boat and trailer, wash and wax both units. Wipe off upholstery with a clean damp cloth and vacuum the carpeting. Check bunks (pads) and make sure they fit properly and inspect them for damage. If the bunk (pad) carpet is damaged, it should be replaced before launching the boat to prevent possible hull damage. If the boat is to be stored for a lengthy period of time, drain the gasoline from the boat fuel tank, or preferably add *fuel stabilizer* to the gasoline. This saves the hassle of draining the fuel. If using a fuel stabilizer, run the motor for a period of time to insure that the stabilizer is throughout the fuel system. (Supply water to the outboard when running it.) Check the tire pressure on the trailer tires and check the wheel bearing grease condition. They may need to be repacked because of water leakage through a damaged seal.

1.9   Outboard Storage: Motors without Oil Injection

If the outboard is going to be put up for the off-season, the external metal, electrical parts of the powerhead should be cleaned and then sprayed lightly with a corrosion guard. The cover, exhaust housing, and lower unit can be waxed for protection over the winter months. In addition, the internal parts, such as the crankshaft, rods, and bearings, need to be protected with Mercury Storage Seal Rust Inhibitor or OMC Storage Fogging Oil—rust inhibitors. These products are used in conjunction with a flushette or test tank (30-gallon trash can) supplying the water.

 Safety

For safety reasons, *remove the propeller* if operating on a flushette. Remove the carburetor silencer; warm up the powerhead and spray the product directly into the carburetor throat until the exhaust smokes profusely. (Do not use WD40.) Disconnect the fuel line. The rust preventive fogs (bathes) the crankshaft, rods, and bearings with oil, *while the carburetors run out all the remaining gasoline,* so it won't turn to gum and varnish during the off-season (Figure 1–28).

When the powerhead stalls *because of the lack of gasoline,* apply the choke and try to restart, pulling out any remaining fuel from the carburetor circuits. If the carburetor(s) has a drain plug in the float bowl, remove it and drain any remaining fuel. Pull the spark plugs,

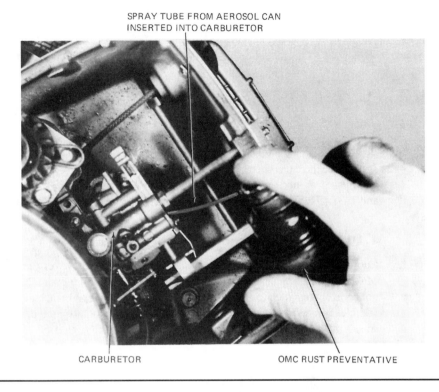

SPRAY TUBE FROM AEROSOL CAN
INSERTED INTO CARBURETOR

CARBURETOR   OMC RUST PREVENTATIVE

Figure 1–28 Fogging the powerhead with OMC rust preventative before storage.

ground the leads, and spray the fogging oil directly into the cylinders. Then turn the powerhead by hand spreading the oil around the cylinders. Reinstall the spark plugs and service the fuel filter by draining any remaining gasoline and cleaning the screen. Four-stroke outboards should have the oil and filter changed.

1.10 Outboard Storage: Engines with Oil Injection

Instead of running the motor out of fuel, you can use a Marine Fuel System Treatment and Stabilizer (Mercury). This keeps the fuel fresh for up to a year in storage. It prevents gum and varnish formation throughout the fuel system and helps protect internal fuel system components from corrosion. This also keeps the oil injection system operational. This means you don't have to drain that fixed fuel tank on larger boats. But, you must run the outboard at 1500 RPM for a minimum of five minutes to get the fuel stabilizer into the fuel pump and carburetors. When doing this, use a flushing attachment and *remove the propeller to prevent personal injury*! While the outboard is running, spray some fogging oil through the carburetor inlet(s) to coat the internal motor parts. Next, you should pull the spark plugs and ground the leads to protect the electronics. Then spray fogging oil into the cylinders and turn the motor over a couple of times to spread the fogging oil throughout the cylinders, pistons, and rings. Also spray the spark plug threads in the cylinder head and on the plugs.

 **Warning**

Always remove the propeller when treating the fuel system for storage.

1.10.1   Lower Unit

Just in case there has been seepage of water into the lower unit, the lubricant should be drained and fresh gear lubricant added to bring the oil level up to the vent hole (Figures 1–29, 1–30 and 1–31.) This new lubricant will protect the lower unit roller (ball) bearings and gears.

1.10.2   Other Components

Lubricate the throttle and shift linkage, along with the swivel pin and tilt mechanism. Make sure that any casting drain holes in the lower unit are open, to insure that water will not be in any cavities to cause corrosion, freeze, expand and crack the housing. Also check and clean out the speedometer pick-up tube with a small piece of wire. Disconnect the speedometer tube at the speedometer head and blow through it.

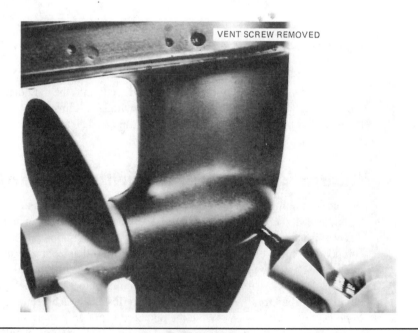

Figure 1–29   Filling lower unit with gear lubricant.

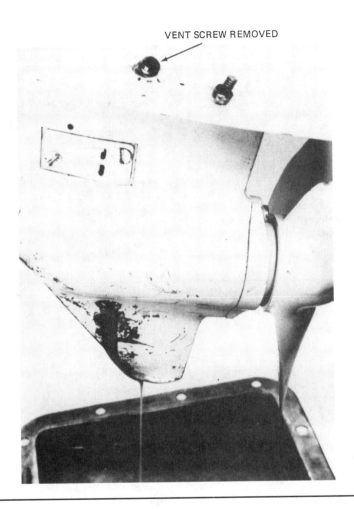

VENT SCREW REMOVED

Figure 1–30   Draining lower unit gear oil.

The propeller needs to come off of the propeller shaft, so corrosion will not cause it to seize up. Check the propeller hub for condition and the condition of the blades for chips and nicks. The blades can be dressed with a file but maintain the same leading and trailing edge contour. With the propeller off, apply non-corrosion grease (such as OMC Triple Guard Grease; Mercury Anti-corrosion, 2-4-C with Teflon or Special Lube 101) to the propeller shaft splines. These compounds resist moisture and will allow for easy removal of the propeller next season.

Next remove the battery and clean the exterior surfaces with bicarbonate of soda (baking soda). Then coat the terminals with a light coat of grease. Make sure that the electrolyte level is correct, and slow-charge the battery. The charge rate should be 6 for five or six hours or until fully charged, as indicated by a hydrometer reading (see Chapter 5).

Figure 1–31   Special gear lubricant for lower units continued. *(Courtesy of Outboard Marine Corporation and Mercury Marine)*

Store the battery in a cool dry place and fully recharge it every 60 days to keep the battery from sulfating.

 Safety

Keep the battery out of the reach of children.

Always use slow-charge. Store the outboard in the upright position; it can be left on the boat or put on a rack. If it is left outside, protect it with some type of tied-down weather-proof tarp. If you have fuel in fixed tanks, add a fuel stabilizer and run the outboard for at least five minutes to insure the stabilizer gets into the total fuel system.

If the outboard is stored, the boat is probably also going to be stored. Lubricate the shift, steering, and speed control cables with anti-corrosive grease. During the off-season go out several times and work the cables through full travel. Also, lubricate the cables once or twice during the off-season. Clean the boat, remove the transom drain plug, tilt the boat stern down, and install a tie-down tarp. Also protect the outboard motor with a tie-down tarp.

This is the best time to clean and pack the wheel bearings on the trailer. Even if Bearing Buddy II is installed, water may have entered the hub and deteriorated some of the grease. Make an operational check of the electrical system, and repair as necessary. The trailer paint, rollers, bolsters, and other contact points should also be repaired as necessary. Now you will be ready for the next boating season.

1.11   Preseason Check

The preseason check involves a lot of items and identifies problems if the outboard, boat, and trailer were not put in storage properly. The preseason check assumes that all preventive maintenance was done.

The spark plugs should be cleaned before the powerhead is started. Make sure that the boots and spark plug leads are in good shape. Check the oil level in the lower unit by adding oil to the unit until it runs out the vent. Check the oil level in the oil injection system if so equipped. If the exterior finish has deteriorated over the winter clean it by using automotive wax. Slow-charge the battery and test the electrolyte for battery state of charge using a hydrometer. It is a good idea to take the battery into a dealer or service center and have the battery load-tested. This test will determine if the battery capacity is up to specifications (see Chapter 5). Install the battery and cables applying anti-corrosive grease to the terminals. With the water hose and flushette attached or the motor in a water tank, start the powerhead. Blue smoke will emit from the exhaust due to the oil applied for storage protection.

Once the exhaust smoke clears, stop the powerhead and reclean the spark plugs or install new spark plugs. Lubricate the steering, shift, and speed-control cables. Operate them without the powerhead running, to their maximum travel, checking for ease of operation. With the powerhead at fast idle, check for water-pump operation and for overheating of the powerhead. (An impeller should be installed each spring. Be careful to install the correct type.) If the motor was removed from the boat, make sure that you use the special locknuts and lockbolts that were set aside earlier.

 **Warning**

Do not substitute any of this special hardware. Many of these parts are *rated* to the horsepower of the outboard for safety reasons.

Test the power trim if so equipped. Fill the power trim reservoir according to the service manual instructions.

You are not ready for the water until you inspect the boat hull and transom for deterioration and all safety gear required by the Coast Guard is inspected, installed and ready for use. A spare propeller, shear pin, cotter pin, and propeller nut, tools for the propeller exchange, and tools for installing extra spark plugs should be aboard. Remember that the correct preparation brings pleasure to boating.

1.12   Corrosion—What Can Help

A destructive, seemingly uncontrollable condition that develops on the outboard in the marine environment is galvanic action (corrosion) and/or electrolysis. Untold numbers of repairs to lower unit assemblies, exhaust housings, and powerheads have been made because of direct chemical attack or electrochemical reaction in these assemblies. Corrosion is a silent destroyer of the outboard casting and fasteners, as shown in Figures 1–32 and 1–33. The oxidation of aluminum leaves a white powdery coating on the surface. *Galvanic action (corrosion)* is an electrical process where atoms of one metal are carried in *electrolyte* solution (salt or brackish water) and deposited on the surface of a dissimilar metal. A protective sealant should have been appled to the bolts in Figure 1–32. Broken bolts can be removed by applying heat to the casting around the bolts.

BROKEN BOLTS SEIZED IN CASTING

WHITE POWDER (CORROSION)
EVIDENT IN BOLT HOLE

Figure 1-32   Corrosion damage to bolts.

The **electrolyte** is the water in which we boat, from brackish lake water to ocean water. You might wonder why unlike metals used in manufacturing of the outboard are not changed by the manufacturers? It is not quite that easy. A quick reference to the galvanic series chart shows the relative position of various metals and alloys. In general, depending upon the electrolyte, the farther apart on the chart, the greater the degree of attack. The less noble metal corrodes sacrificially and protects the more noble material. The chart shows that aluminum and steel are close together and therefore the least transfer of metal will occur.

This galvanic process can be used to an advantage. Add a piece of less noble material, usually zinc, to the outboard casting and it will corrode sacrificially. The trim tab attached to the anti-ventilation plate on the lower unit is often made of zinc, and is used as an anode to be sacrificed. (Not all trim tabs are made of zinc.) In other models an anode may be bolted to the lower unit bearing housing (see Figure 1–17). These anodes protect the lower unit. While the zinc trim tab or anode may have to be replaced from time to time, it saves the more expensive parts.

Another method used to protect the castings and the bolts that hold them, is to coat the bolts, studs, shift shaft splines, etc., with (OMC or Mercury) gasket sealing compound. Such a coating with non-hardening gasket compound insulates the dissimilar metals (steel and aluminum) from each other, effectively reducing the transfer of the metal. This com-

Figure 1–33   Corrosion damage to lower unit.

pound is water-resistant, remaining for long periods of time when in the water. It is better than any "waterproof" grease that is available.

Paint on the outboard also protects against corrosion, and should be maintained using matching paint from the dealer. The paint should not be applied to the casting where the zinc trim tab (or anode blocks) make contact, nor should the trim tab be painted. The zinc trim tab or zinc anode block needs to have direct contact with the water and clean contact to the casting for the best corrosion control.

Anode (trim tab or blocks) erosion or disintegration in salt or brackish water indicates that the anode is doing its job. The anodes should be inspected and/or replaced at intervals or corrosion of the outboard will increase. If an anode has been reduced to two-thirds its original size, it should be replaced.

To test for proper anode installation, use an ohmmeter on the lowest scale. Place one lead on the anode and the other lead on the powerhead ground. There should be a zero reading. If not, remove anode and clean contact area to the lower unit and check the connections of the powerhead ground.

When mooring a boat in the marina, the trim tab may not be able to handle the corrosion problem. In that case you can install a cathode system. This is an electronic device that hooks up to the boat battery. It has two leads that go through the transom to two sensors, the anode and the cathode, which are mounted on the transom under the water line. Power needed to operate this cathode system is minimal and should be no problem to a good deep-cycle battery. The cathode system should be checked annually where the boat is regularly moored, per manufacturer's instructions. This system is inexpensive in comparison to potential repairs and possible replacement of a lower unit gear housing.

Sometimes there are problems at the marinas with onshore electrical systems that run electrical cables in the water, out onto the docks. This can cause man-made accelerated electrolysis, which will overpower the zinc trim tab/blocks and/or the cathode system. In this situation, a galvanic insulator needs to be installed in the power line coming into the boat at the mooring. This insulator will control and halt the formation of corrosion caused by an onshore electrical condition.

Mercury Marine and OMC both market a rust inhibitor/corrosion guard in an aerosol can. It can be sprayed on the external parts of the powerhead to protect it from the atmosphere and water spray. Other parts around the steering wheel, swivel, transom bracket, etc., can also be sprayed. Other products available through local hardware stores are not recommended because they do not effectively control salts for long periods of time. This marine rust inhibitor can also be used inside the powerhead to control conditions when the motor is not going to be run for some time. Apply a coat of automobile wax at regular intervals to protect the smooth outer surfaces of the outboard.

If paint is to be applied to the hull of a metal boat, avoid paint that has copper in it. This will increase corrosion problems. Purchase an anti-fouling marine paint that contains tributyl-tin adipate (TBTA). (If TBTA is prohibited by law in your area, use copper-base paint.)

1.13   **Recovered Submerged Motors**

Outboards that take the dive do so under one of three conditions: (1) submersion into saltwater, (2) submersion into fresh- or saltwater *while running*, or (3) submersion into freshwater.

 **Warning**

One of the best cures for this is not to let it happen!

Maintenance performed on the transom brackets, use of proper bolts through the transom, and use of a safety chain or rope on the small fishing motor can prevent the outboard from going overboard.

On some of the larger outboards the bolts are put through the transom and sealed. As the seasons go by, they may corrode and weaken, dropping the outboard motor when under the strain of full power or upon impact with a submerged object (Figure 1–34).

These bolts should be inspected and renewed if deteriorated, as part of a normal maintenance program. There are cases of installations where the nuts have corroded away and yet the bolt looked perfectly fine. It pays to take a periodic look at these mounting nuts and bolts. The motor mounting hardware should be of the type approved by the outboard manufacturer and installed to the manufacturer's specifications.

When the outboard is submerged in saltwater the situation is critical. Due to the corrosive effect of the saltwater, no attempt should be made to start the motor. The powerhead will have to be completely disassembled. Saltwater will attack the finished surfaces of the crankshaft, rods, cylinder walls, pistons, and rings and all of the other parts that are inside

Outboard Mounting System and Persons Safety

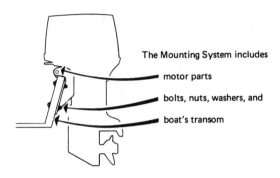

The Mounting System includes

→ motor parts

→ bolts, nuts, washers, and

→ boat's transom

<u>What's Important?</u>

Motor must <u>stay</u> in position on boat's transom

<u>What Could Happen?</u>

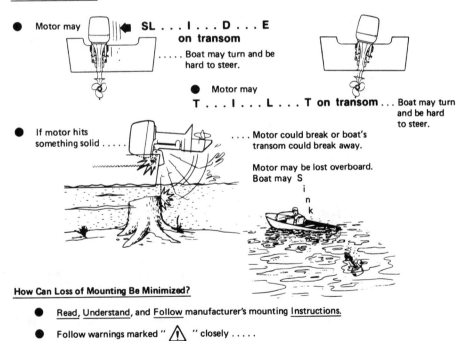

● Motor may   SL . . . I . . . D . . . E **on transom**

. . . . . Boat may turn and be hard to steer.

● Motor may

T . . . I . . . L . . . T **on transom** . . . Boat may turn and be hard to steer.

● If motor hits something solid . . . . .

. . . . Motor could break or boat's transom could break away.

Motor may be lost overboard. Boat may S i n k

<u>How Can Loss of Mounting Be Minimized?</u>

● <u>Read</u>, <u>Understand</u>, and <u>Follow</u> manufacturer's mounting <u>Instructions</u>.

● Follow warnings marked " ⚠ " closely . . . . .

Figure I–34   Outboard mounting system and persons safety (continued). *(Courtesy of Outboard Marine Corporation)*

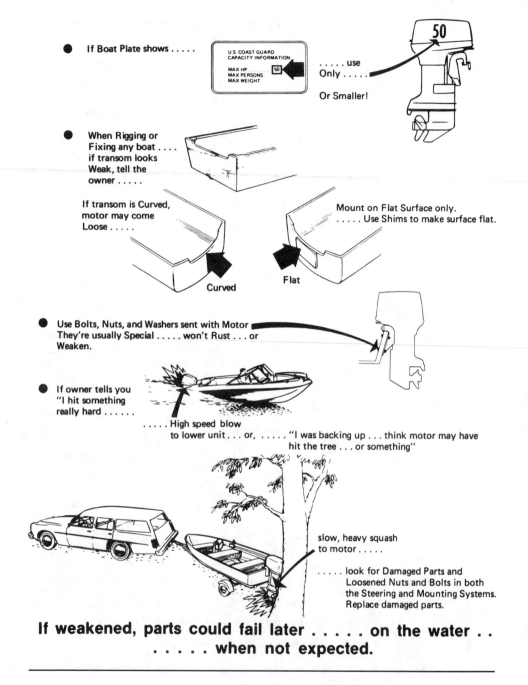

● If Boat Plate shows . . . . .

U.S COAST GUARD
CAPACITY INFORMATION

MAX HP 50
MAX PERSONS
MAX WEIGHT

. . . . . use
Only . . . . .

Or Smaller!

● When Rigging or Fixing any boat . . . . if transom looks Weak, tell the owner . . . . .

If transom is Curved, motor may come Loose . . . . .

Mount on Flat Surface only. . . . . . Use Shims to make surface flat.

Curved   Flat

● Use Bolts, Nuts, and Washers sent with Motor They're usually Special . . . . . won't Rust . . . or Weaken.

● If owner tells you "I hit something really hard . . . . .

. . . . . High speed blow to lower unit . . . or, ... "I was backing up . . . think motor may have hit the tree . . . or something"

slow, heavy squash to motor . . . . .

. . . . . look for Damaged Parts and Loosened Nuts and Bolts in both the Steering and Mounting Systems. Replace damaged parts.

If weakened, parts could fail later . . . . . on the water . . . . when not expected.

Figure 1–34   Outboard mounting system and persons safety *(continued)*. *(Courtesy of Outboard Marine Corporation)*

Figure 1–35 Keeping the powerhead submerged reduces oxidation (rust).

or outside of the powerhead. If you can make a quick recovery, and you can even get it started, the salt will still remain to do damage. Adding 50/1 oil in the fuel, even mixed to 24/1 with alcohol, still will not provide protection against the remaining salts! There is only one safe solution and that is to disassemble the powerhead and totally and thoroughly clean all the parts. If an outboard goes overboard and a marine dealer is not convenient for this disassembly work right after recovery, what do you do? Remove the cover and rinse the outboard totally with freshwater. If you are able to remove the powerhead from the adapter and exhaust housing, do so. Once removed, resubmerge it in a tub of freshwater. This is done so air cannot get to those finished surfaces and therefore to prevent oxidation (rusting). If you cannot take the powerhead off, do the next best thing—stand it upside-down in a tub of freshwater (Figure 1–35). Then take the outboard to your favorite dealer, where it can be submerged until it can be worked on.

There are additional problems if the outboard goes under when running. Water is drawn through the carburetor, into the crankcase, and then into the cylinder. This causes a hydro-static lock, and may possibly bend the connecting rod (Figure 1–36). Also, an oil line, oil tank, or bleed lines may have been damaged and water possibly entered these systems.

No attempt should be made to start the outboard after recovery, as sprung or bent parts could cause permanent damage. Keep the motor submerged and take it to your local dealer for total disassembly and repair of the powerhead, fuel, and electrical systems.

There are several ways to restart an outboard that was lost overboard in freshwater. First of all, evaluate the water condition. Was the water clear and clean? Was the water dirty with sand or silt? If the water had sand or silt in it, then you need to follow the in-structions given under saltwater submersion. If the outboard *wasn't running* and went under in clean freshwater, with recovery within 12 hours, you can pull the plugs and pour the water from the cylinders. Pour alcohol into the cylinders and carburetor and then rotate

Figure 1–36 A bent rod may be caused by a hydrostatic shock.

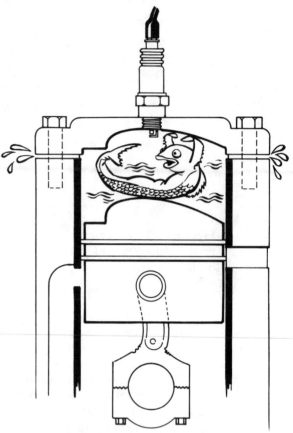

the flywheel. Squirt 50/1 oil into the cylinders while turning over the crankshaft using a rope start. Clean out the carburetor, fuel pump, fuel lines, oil container, oil lines (on oil-injection models), and bleed lines. You will also need to dry off the electrical parts, opening each and every connector, reassembling with *dielectric* grease to control corrosion. When there is a starter, it will have to be disassembled and cleaned, and the bearings will have to be lubricated.

If the boat has been swamped or capsized and then recovered, additional work will have to be done to controls and the electrical system. The shifting, steering, and throttle speed cables will need to be cleaned, lubricated, and tested for freedom of operation. The control head will need to be disassembled, cleaned, and lubricated by a qualified mechanic. The instrumentation at the helm will also have to be serviced, and generally speaking, may have to be replaced. The electrical wiring will need to be inspected for frayed, broken connections, damaged insulation, and mounting to the hull. If the boat has been roughed up a bit by the waves or swells, the hull's condition must be examined for seaworthiness.

KNOW THESE PRINCIPLES OF OPERATION

- Measuring transom height and mounting of the larger outboard.
- Steering system operation and safety.
- Installation and handling of the low horsepower outboard.
- Maintenance and points of lubrication.
- Performance of operational checks.
- Trailering the boat.
- Measuring trailer tongue weight.
- Making a pre-trip check of the trailer.
- Making a preseason check on the outboard.
- Corrosion control
- Procedures for submerged outboard recovery.

REVIEW QUESTIONS

1. What does the transom height determine?
   a. the distance from the bottom of the keel to the top of the transom
   b. how high or low the outboard sets on the transom
   c. the measurement between the skeg and the stern brackets
2. Performance of the boat will be best when the anti-ventilation plate is positioned so it is
   a. tilted up raising the stern of the boat.
   b. tilted down to increase the propeller left-right imbalance.
   c. running parallel to the surface of the water, and even or slightly above the bottom of the boat.
3. What happens at the steering wheel must happen at the
   a. power trim without undue resistance.
   b. outboard without grabbing or seizing.
   c. Both a and b are correct.
4. The total steering system should be checked and lubricated every _____ days of operation when operating in saltwater.
   a. 30
   b. 60
   c. 90
5. When the boat pulls to the starboard and trim is correct, the trim tab adjustment should be made to the_____.
   a. port
   b. center position
   c. starboard
6. When transporting the outboard after a fishing trip, it should be positioned
   a. with the powerhead up to allow for water drainage.
   b. with the propeller higher than the powerhead.
   c. on the left side with the exhaust outlet higher than the powerhead.

7.   When lifting a larger outboard, what device is used?
     a.   sling
     b.   universal puller with a lifting ring
     c.   Both a and b are correct.

8.   The heads of bolts are marked to tell the mechanic what the _____ strength and grade of the bolt is.
     a.   pitch
     b.   thread
     c.   tensile

9.   When bolts and nuts are installed on the lower unit, they are coated with
     a.   OMC gasket sealing compound.
     b.   Mercury perfect seal.
     c.   Both a and b are correct.

10. What seasonal service is performed on the propeller shaft?
     a.   check for alignment
     b.   replace the propeller nut
     c.   clean the splines and apply non-corrosive grease or gasket sealing compound to the splines

11. What causes the trim tab to corrode?
     a.   galvanic corrosion
     b.   electrolysis
     c.   Both a and b are correct

12. At wide-open throttle the powerhead should be running in the
     a.   upper one-quarter of the operating range.
     b.   lower one-quarter of the operating range.
     c.   just above maximum operating range.

13. To raise the operating range RPM of the powerhead you should
     a.   increase pitch.
     b.   reduce pitch.
     c.   install the same size propeller but with a cup.

14. How often is the lower unit oil drained and flushed?
     a.   every 25 hours in saltwater
     b.   every 30 hours in freshwater
     c.   every 50 hours in saltwater

15. When greasing the Bearing Buddy II on the boat trailer, you are still able to move the piston by pushing on the grease gun hose or tube. This would indicate that
     a.   the bearing has adequate grease.
     b.   the bearing needs more grease.
     c.   the retainer ring is broken.

16. The trailer wiring color code for ground is white. The ground wire is run through the
     a.   frame of the trailer bypassing the connector.
     b.   wiring harness and is grounded at the tow-vehicle frame and the trailer frame.
     c.   hitch of the trailer.

17. When placing the outboard into storage, the carburetors are run dry and at the same time
    a. water is supplied to the water inlet.
    b. rust-preventive oil is injected into the powerhead through the carburetors.
    c. Both a and b are correct, plus checking the lower unit oil and giving the outboard a complete lubrication per service manual instructions.
18. When the battery is placed in storage it is
    a. slow-charged until fully charged at 1.260 SG.
    b. recharged every 60 days while in storage.
    c. Both a and b are correct, plus cleaning with bicarbonate of soda.
19. The battery state of charge is checked with a_____.
    a. voltmeter
    b. load tester
    c. hydrometer
20. What type of metal is used as a sacrificial anode when installed in blocks or used as the basic metal of the trim tab?
    a. copper
    b. zinc
    c. brass
21. An outboard breaks away from the transom when running and goes under in saltwater and is recovered quickly; you would need
    a. to cleanup the fuel system and put it back into service.
    b. to have the powerhead disassembled, cleaned and reassembled.
    c. the electrical system dried out, plus doing a and b.
22. Until repair work can be performed on the powerhead which has been submerged in saltwater, the unit should
    a. be injected with oil in each cylinder.
    b. have the plugs removed and the water removed from each cylinder.
    c. remain submerged in a tub of freshwater.
23. An anode should be replaced when it is
    a. three-quarters gone.
    b. two-thirds gone.
    c. one-third gone.
24. To test the continuity of the anode (trim tab) circuit, set the ohmmeter on low scale and the ohmmeter leads are placed
    a. on the trim tab and the exhaust housing.
    b. on the powerhead and lower unit.
    c. on the trim tab and powerhead ground.

CHAPTER 2

Two-Stroke and Four-Stroke Principles of Operation

Section I: Two-Stroke Principles of Operation

Objectives

After studying the first section of this chapter, you will know:

- Principles of two-stroke operation.
- Design of direct charge powerhead.
- Design of power porting powerhead.
- Design of loop scavenging powerhead.

2.1   Two-Stroke Operation

A *two-stroke* outboard can be an efficient operating unit. Many people have had a bad experience with the two-stroke powerhead either because of a lack of understanding or possibly the lack of maintenance and operational skills. Very often we consider ourselves to be good mechanics, because of the maintenance skills we have developed over the years on cars. Some of those skills, however, do not transfer to the two-stroke powerhead (engine) and new skills are needed. This section of the chapter will discuss the operation of the two-stroke powerhead as it relates to the outboard and other small engines.

The two-stroke powerhead is like a pump. A pump has one port for the fluid to enter and one port for the fluid to be discharged. In the same way, the two-stroke powerhead receives fuel and exhausts spent gases. The two-stroke refers to one revolution of the crankshaft, which forces the piston outwards within the cylinder and then withdraws the piston back to the bottom of the cylinder. Simply said, the cycle involves intake, compression, power, and exhaust. With each revolution the fuel is taken in and compressed, the ignition spark ignites the fuel developing power (heat), and the spent gases are released into the exhaust.

2.1.1   Intake/Compression Stroke

Let's take a look at what is happening on the intake/compression stroke (Figure 2–1).

Remember that the engine operates in the same way as a pump; one inlet is the carburetor. Atmospheric pressure at sea level is 14.7 PSI. A negative pressure (vacuum) is created in the crankcase by the outward movement of the piston, moving toward the closed end of the cylinder. This causes air to rush into the crankcase through the carburetor and reed valves, picking up the fuel/oil mix from the carburetor circuits. This happens through the venturi action at high speed and idle circuit at low speed. This fills the crankcase of the two-stroke powerhead with fuel mix, and is a partial intake stroke. The 50/1 oil used has the tendency to fall out of the air stream since it is in mist form. It coats the cylinder wall, and replenishes the oil on the moving parts of the engine: the crankshaft, bearings, pistons, and rings. This 50/1 TC-W3 oil is very adequate lubrication for maximum engine RPM, provided that the oil is properly mixed with fuel and is in the correct ratio.

Think for a moment about the action of drinking with a straw. The straw provides a means of carrying the drink from the glass to our mouth; our lips, mouth, and lungs do the rest. But what happens if the straw has a hole? Nothing! Why? The air coming into the

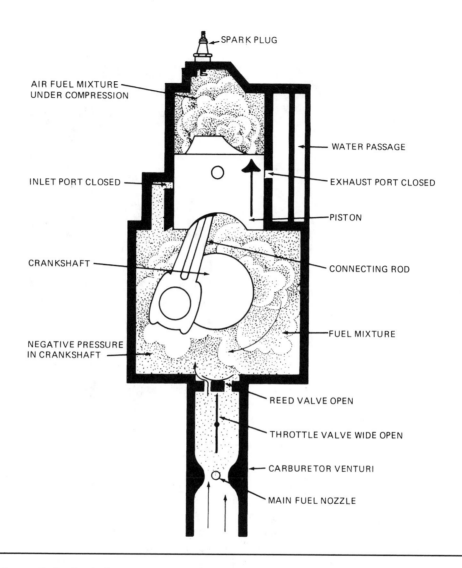

Figure 2–1   Intake/compression stroke as viewed from the flywheel end of the crankshaft.

straw effectively leaves the drink in the glass. The remedy, of course, is to plug the hole in the straw and continue drinking. Apply that analogy to the intake stroke of the two-stroke powerhead and the importance of the crankshaft seals, crankshaft seal rings, gaskets, and O-rings used to seal the crankcase of the engine. Any leak in the crankcase other than the carburetor airhorn will cause air to come in, and fuel leanness will develop. Because lean mixtures burn hotter than normal, problems can ensue: e.g., burned pistons and scored cylinder walls. There is also the possibility of damaged bearings since a crankcase leak ef-

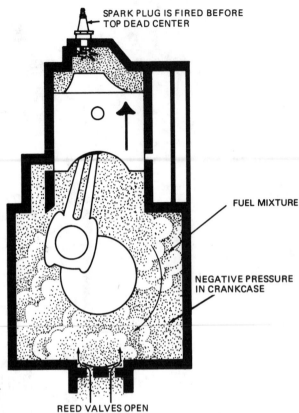

SPARK PLUG IS FIRED BEFORE
TOP DEAD CENTER

FUEL MIXTURE

NEGATIVE PRESSURE
IN CRANKCASE

REED VALVES OPEN

Figure 2–2 Timed ignition as viewed from the flywheel end of the crankshaft.

fectively leans out lubrication in the crankcase. It is, therefore, absolutely necessary to be sure that the crankcase is positively sealed during any repair work that is done. Also as an operator, watch for signs of oil leaks between the crankcase half and cylinder assembly or for oil on ignition or stator parts under the flywheel. Oil leaking out also means air leaking in, and the two-stroke powerhead cannot tolerate this condition.

Now let us look closer at the compression stroke (see Figure 2–2). Realize that the intake stroke and compression stroke are happening at the same time. A partial intake stroke is taking place below the piston in the crankcase, and the compression stroke is taking place in the cylinder beyond the piston head. Fuel that has entered the cylinder through the intake ports at the end of the previous power/exhaust stroke from applied positive crankcase pressure has been deflected by the piston and port design to the spark-plug end of the cylinder. (Some engines do not use the deflector-dome type of piston but instead use a crowned, or flat piston head.) This directed flow of fresh fuel purges the cylinder of exhaust gases. As the piston moves outward, the intake and exhaust ports are closed by the piston, and the fuel is trapped and compressed in the cylinder to a predetermined ratio in the combustion chamber. Heat is added to this fuel charge by means of compression and engine heat, making the fuel charge very volatile. This fuel is unable to escape the cylinder because of the sealing effect of the piston rings, gaskets, and, design permitting, a cylinder head. The fuel charge is then ready to be ignited from the ignition system via a timed spark at the spark plug. Remember that two-cycle oil (TC-W3) is in this fuel so lubrication of the upper cylinder wall and all piston rings is accomplished. This nearly eliminates cylinder wear, which is a major cause for overhaul in a four-stroke engine. We are not trying to control oil consumption in the two-stroke, so there are no oil rings, just two or three compression rings.

2.1.2   Power/Exhaust Stroke

The spark at the spark plug actually occurs a few degrees before top dead center of the compression stroke (Figure 2–2). This gives the fuel time to burn, giving up its heat, and develops maximum pressure by a given point, after top dead center. Maximum pressure on the piston in the power stroke is at a predetermined point in the rotation of the crankshaft (Figure 2–3).

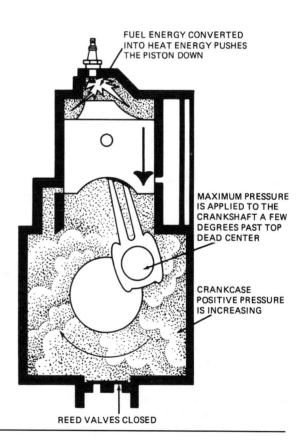

FUEL ENERGY CONVERTED INTO HEAT ENERGY PUSHES THE PISTON DOWN

MAXIMUM PRESSURE IS APPLIED TO THE CRANKSHAFT A FEW DEGREES PAST TOP DEAD CENTER

CRANKCASE POSITIVE PRESSURE IS INCREASING

REED VALVES CLOSED

Figure 2–3   Power applied to the crankshaft as viewed from the flywheel end of the crankshaft.

The angle on the crankshaft throw is thus correct to receive power and complement the rotation of the crankshaft. If this power (maximum pressure) came into the crankshaft when the crankshaft throw was in the more straight up-and-down position, the pressure of the expanding gases would be applied directly onto the piston, rod, main bearings, and cylinder assembly. This, of course, would not produce an efficient or smooth-running engine, and is one reason why it is important to follow timing specifications. In this way, the engineering of the engine can be accomplished and horsepower specifications met. The burn of the fuel within the cylinder is considered to be a controlled burn. It is a progression of the flame front across the combustion chamber. If we were to burn a field like the farmers do in the Pacific Northwest, the fire is started at one point in the field and it progresses across the field, controlled all the way. This analogy describes the burn within the cylinder. (See "Normal and Abnormal Combustion" in Chapter 10.)

Several factors can throw this burn off or out of control: incorrectly set timing, fuel of wrong octane, overheating because of an improperly operating cooling system, or glowing carbon deposits in the combustion chamber. Preignition or detonation may occur if any of these factors are present. The end result is damage to the piston and piston rings with the resulting loss of compression. As the pressure continues to move the piston down the cylinder, the exhaust port is opened and the intake port is next to open. As soon as the exhaust port opens, the pressure from the hot expanding gas formed by the burning fuel is released into the exhaust system (Figure 2–4). Here the exhaust is cooled by water coming

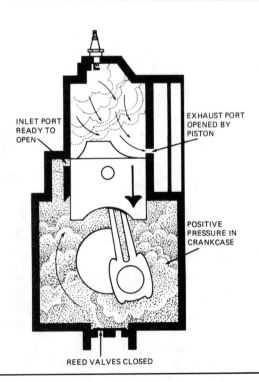

INLET PORT
READY TO
OPEN

EXHAUST PORT
OPENED BY
PISTON

POSITIVE
PRESSURE IN
CRANKCASE

REED VALVES CLOSED

Figure 2–4 Exhaust of spent combustion gases as viewed from the flywheel end of the crankshaft.

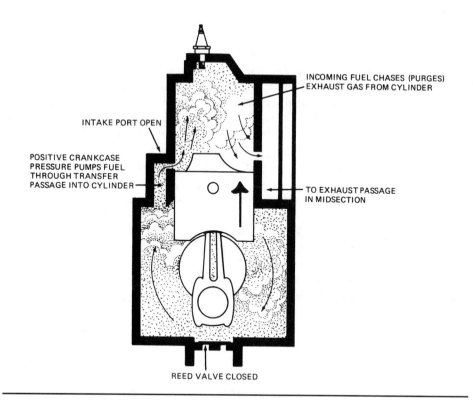

INCOMING FUEL CHASES (PURGES)
EXHAUST GAS FROM CYLINDER

INTAKE PORT OPEN

POSITIVE CRANKCASE
PRESSURE PUMPS FUEL
THROUGH TRANSFER
PASSAGE INTO CYLINDER

TO EXHAUST PASSAGE
IN MIDSECTION

REED VALVE CLOSED

Figure 2–5   Intake of fuel/air mixture as viewed from the flywheel end of the crankshaft.

from the powerhead, so the exhaust housing is not overheated. Exhaust gas goes through the exhaust housing and the propeller hub, where it is emitted into the cone area behind the propeller, filling a void. It is silenced as it goes into the water.

As the piston travels down the cylinder the spring-loaded reed valves are closed, blocking fuel from leaving the crankcase. The volume or square-inch area of the crankcase becomes smaller. The fuel/air mix in the crankcase brought in during the intake stroke is thus compressed. This pressurized fuel is then ready to be pumped through the transfer passage into the cylinder, just as the intake port(s) is opened. The fuel in vapor form is then deflected out to the spark plug end of the cylinder by the piston head deflector (on a direct-charge piston), and at the same time pushes the spent exhaust gases out of the cylinder (Figure 2–5).

The incoming fuel attempts to chase the exhaust gas flowing out through the exhaust port. Because of exhaust tuning and the closing of the intake and exhaust ports by the piston, the fuel mix is trapped in the cylinder. The compression stroke is then ready to begin all over again. It took one revolution of the crankshaft to complete the intake/compression and power/exhaust strokes.

2.2   **Methods of Porting**

There are different methods of porting the incoming fuel into the cylinder from the crankcase. They are commonly referred to as direct charge, power porting, and loop charge induction. These methods deal with the way the fuel mixture enters the cylinder. Direct charge is the standard method, using the conventional transfer passage, porting, and a deflector-dome piston. (Some outboards may not use the deflector dome.) Intake and exhaust ports are located across from each other and the deflector on the piston is used to direct the flow of incoming fuel upward. As the fuel comes back down, it travels across the cylinder heading for the exhaust port, causing a cross flow through the cylinder, and purging exhaust gases from the cylinder (see Figure 2–6).

Power porting (Mercury) adds to this by adding a hole in the piston skirt, a beveled edge in the piston crown directly in front of the deflector, and a "third port" (booster port) in the cylinder wall, as shown in Figure 2–7. This allows additional fuel to enter the cylinder, increasing horsepower without adding weight.

In the *loop charge induction* method, no piston deflectors are required and the piston head is basically flat. The piston skirt has three holes that align with the three intake passages (ports) in the cylinder (Figure 2–8).

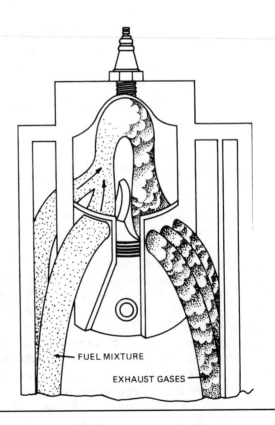

Figure 2–6   Direct charge design.

FUEL MIXTURE

EXHAUST GASES

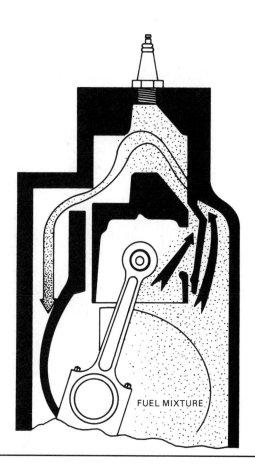

Figure 2–7   Power porting design shows hole in piston skirt, modified piston head, and transfer passage for additional fuel source, as viewed from the flywheel end of the crankshaft.

FUEL MIXTURE

These intake passages are tapered upward and angled towards each other to direct the streams of incoming fuel mixture to the top of the cylinder. These fuel streams coming into the cylinder converge ahead of the piston causing a looping or swirling of the fuel mixture in the cylinder. Such a flow pattern forces the spent exhaust gases, which are ahead of the fuel, down into the exhaust ports resulting in a complete purge of the exhaust gases and a greater in-filling of fuel mixture for the next compression stroke. This looping of the fuel mixture more completely fills the cylinder and less fresh fuel charge is required to expel the burned exhaust gases. This system is more fuel efficient than either the cross-flow, direct charge or power porting systems. Again, horsepower is added to the powerhead without additional parts or weight.

A *flywheel* is added to the crankshaft to smooth out the power pulses. It is light in comparison to the automotive-type flywheel and carries momentum once it is spinning. Aluminum is used on many late models, allowing a fast acceleration yet retaining some effects of inertia. This helps smooth out the power pulses by carrying the pistons through those phases when power is not being produced. The flywheel and flywheel hub may be used to

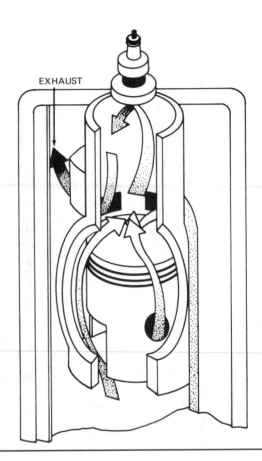

EXHAUST

Figure 2–8   Loop charge
induction (scavenging) design
shows arrows indicating fuel
from three ports, deflecting
upward.

mount magnets for the ignition and charging systems, as well as a ring gear for the starter
drive gear to index and rotate the crankshaft.

This in-depth study of two-stroke operation equips the serious mechanic for troubleshooting. Realizing what is happening during these strokes will help you in making the
correct diagnosis, and therefore in making effective repairs.

KNOW THESE PRINCIPLES OF OPERATION

- The intake and compression stroke.
- The power and exhaust stroke.
- Design of direct charge powerhead.
- Design of power porting powerhead.
- Design of loop charge powerhead.

Now let's look at the operation of four-stroke motors.

Section II: Four-Stroke Principles of Operation

Objectives

After studying the second section of this chapter, you will know:

- Principles of four-stroke operation.
- Single overhead cam design.
- The pressurized lubrication system.
- Classification and function of motor oil.
- The cooling system.

2.3   Four-Stroke Operation

The *four-stroke* outboard engine is a water-cooled overhead valve engine. It runs smooth, clean, and efficient. There is no mixing of the gas and oil. It is much quieter than the two-stroke outboard and very fuel efficient. (see Figure 2–9)

The four-stroke outboard engine is mounted with the crankshaft in the vertical position and the pistons move horizontally. The crankcase is bolted to the side of the cylinder assembly and the engine is mounted above the oil sump and exhaust housing. The four strokes indicate one complete power cycle to accomplish work. It requires two revolutions

Figure 2–9   OMC four-stroke engine.

of the crankshaft, and therefore four strokes of the piston to complete the intake, compression, power, and exhaust strokes. The strokes of the four-stroke differ from the two-stroke in the way fuel is brought into and exhausted from the cylinder. The four-stroke outboard engine uses a belt driven overhead *camshaft* timed to the crankshaft to precisely open and permit closing of the valves. The camshaft lobes contact and operate directly on the rocker arms.

2.3.1   Intake Stroke

When the piston moves down the cylinder, a vacuum is created in the cylinder. The camshaft opens the intake valve and fuel enters the cylinder (see Figure 2–10a). The cross-fire valve (Yamaha) fuel charging system helps move the fuel into the cylinder. This fills the voided low pressure space above the piston. The heat of a previous compression and power stroke helps to vaporize the incoming fuel. As the piston nears the bottom of the intake stroke, the intake valve closes because of camshaft and valve spring action. This effectively seals the cylinder, closing off the intake manifold and carburetor opening. The fuel is now trapped in the cylinder and the compression stroke is ready to begin.

2.3.2   Compression

As the piston and crankshaft move through bottom dead center, the next stroke of the piston highly compresses the fuel-air mixture against the closed intake, exhaust valves and the cylinder-head combustion chamber (see Figure 2–10b). This adds heat, making the fuel extremely volatile. At a few specified degrees before top dead center, the spark for ignition is introduced, igniting the fuel. Igniting the fuel just before top dead center allows time for combustion to expand and develop maximum pressure on the piston head. This pressure is transferred to the connecting rod and on to the crankshaft rod journal at a specified angle of crankshaft rotation.

2.3.3   Power Stroke

As the flame front advances, developing pressure in the combustion chamber, the piston is forced into the cylinder towards the crankshaft (see Figure 2–10c). This rotates the crankshaft, thus work can be done at the driveshaft end of the crankshaft. Horsepower developed here is the force required for propeller rotation. Just before bottom dead center, the exhaust valve opens, releasing spent exhaust gases and the piston helps push them into the exhaust system (see Figure 2–10d).

In only one stroke out of four, the piston delivers power to the crankshaft. Through the other three strokes this is reversed; the crankshaft acts on the piston/rod, pushing it into and pulling it back through the cylinder. In order to keep the crankshaft turning more steadily between power strokes, a flywheel is attached to the upper tapered end. This is simply a heavy metal wheel that carries momentum once it is spinning. This adds to crankshaft momentum, spinning the crankshaft more smoothly. In multi-cylinder engines, the next power

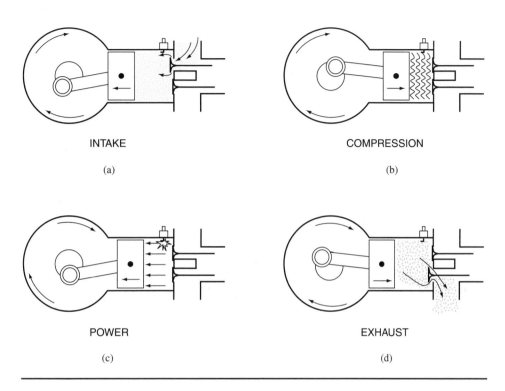

INTAKE

(a)

COMPRESSION

(b)

POWER

(c)

EXHAUST

(d)

Figure 2–10   Four-stroke cycle.

stroke in the firing order pushes another piston into a cylinder, assisting crankshaft rotation and momentum.

2.3.4   Exhaust Stroke

As the camshaft lobe moves the rocker arm, the exhaust valve is opened and combustion pressure is released into the exhaust housing (see Figure 2–10d). The piston moves into the cylinder pushing exhaust gases out of the cylinder. At this point the exhaust system reduces the noise and exhaust is directed through the propeller hub into the water. If the engine has been properly tuned and adjusted to factory specifications, the hydrocarbons (carbon monoxide) released in the exhaust gases are at acceptable levels. A few degrees before top dead center (**TDC**) of the exhaust stroke, the intake valve opens and for a few degrees, both intake and exhaust valves are open. As the exhaust valve seats, the intake valve remains open. This starts the next intake stroke. The engine has completed two revolutions (four strokes of the piston) of the crankshaft, to complete one combustion cycle and provide one power stroke, in which it has developed power to rotate the driveshaft, propeller shaft, and the propeller. Figure 2–11 shows the four-stroke operating cycle.

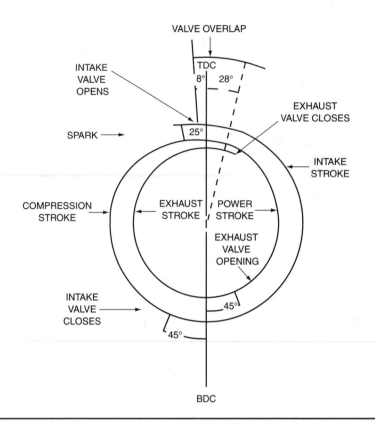

Figure 2–11   Diagram of four-stroke operating cycle.

In the powerhead shown in Figure 2–11, the intake valve opens at 8 degrees before top dead center (**BTDC**) and stays open for 233 degrees. The compression stroke continues for 135 degrees. Ignition occurs at 25 degrees BTDC, and the power stroke continues for 135 degrees. Then the exhaust valve opens for 235 degrees. The valve overlap continues for 36 degrees, and then the next cycle begins. In Figures 2–12 and 2–13 the valve springs, rocker arm, combustion chamber, and camshaft sprocket are shown.

2.4   Single Overhead Cam Design

2.4.1   Cylinder Block

The engine cylinder assembly (block) is the main engine structure and other parts are installed in it or bolted to it. Cylinders are bored horizontally in the block to guide the piston and rings. A plateaued finish is used on the cylinder wall and leaves a surface that distributes oil. It also acts as a bearing surface on which an oil film can form to lubricate the rings

CAMSHAFT PULLEY   VALVE SPRING
                                                  AND RETAINER

Figure 2–12 OMC cylinder
head, showing valves, rocker
arms, and oil pump location.

OIL PUMP   ROCKER ARM

and piston skirt. The main bearings support the crankshaft between the cylinder assembly and
the crankcase. The cylinder head is bolted to the side of the vertical block engine. The block is
a rigid casting and it can maintain correct alignment of all engine parts. A water jacket surrounds the cylinders in the block. It routes and carries water pumped from the water pump
mounted on the lower unit to cool the engine. Oil passageways are drilled in the block directing motor oil to the overhead camshaft, main bearings, and rods. The oil pump is driven by
the lower end of the camshaft and is of the gerotor gear design which supplies 35–45 PSI at
5000 RPM (OMC 15 HP). The cylinder side of the block and the crankcase close the engine,
and oil is in a sump in the upper part of the exhaust housing. To change the oil, it is withdrawn
with a suction tool or drained through a screw hole in the exhaust housing.

2.4.2   Crankshaft

The reciprocating motion of the piston is converted to rotary motion at the orbiting
crankpin moving around the center line of the crankshaft. Opposite the crankpin, which

COMBUSTION   CAMSHAFT
CHAMBER   PULLEY

Figure 2–13 OMC four-stroke cylinder head showing valves and combustion chamber.

INTAKE   EXHAUST
VALVE   VALVE

mounts the connecting rod, are the counterweights. Counterweights offset the weight of the crankpin (throws) and connecting rods providing crankshaft balance. The difference from the crankshaft centerline to crankpin journal center is one-half the stroke of the engine. The crankshaft is mounted to the block using insert-type bearings secured by the crankcase. The crankshaft main bearing journals are polished and in true alignment with each other. The crankshaft is constructed of cast steel; it mounts the connecting rods and flywheel and is splined to receive the lower unit driveshaft (Figure 2–14).

2.4.3 Bearings

The crankshaft main *bearings* used in the four-stroke engine are of the precision insert-split type. The bearings are steel-backed and use softer laminated materials, such as alloys of copper, aluminum, babbitt, and lead. A film of motor oil carries the load and keeps the

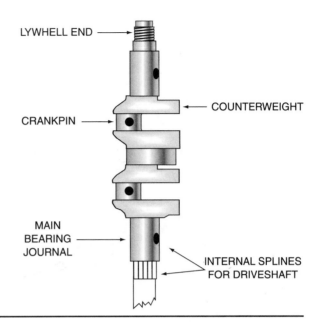

LYWHELL END

CRANKPIN

COUNTERWEIGHT

MAIN
BEARING
JOURNAL

INTERNAL SPLINES
FOR DRIVESHAFT

Figure 2–14   Four-stroke
crankshaft.

bearing material away from the bearing journal. A main bearing insert may have a special thrust flange on it, or a thrust washer to control crankshaft end play. The bearing inserts have grooves or holes drilled through them to facilitate the flow of motor oil through the bearing. Bearing halves use locating lugs (tangs) to align the main bearing into their bores during assembly. Filtered motor oil supplied under pressure from a camshaft gear-driven oil pump maintains an oil film between bearings and journals.

Crush height, which is the bearing protruding slightly beyond the bore, is used to insure that a radial pressure is applied against the bore permitting good heat transfer. In essence, the bearing insert is being squeezed into the main bearing bores as the crankcase is installed (Figure 2–15).

2.4.4   **Connecting Rod**

The connecting rod is used to connect the piston to the crankshaft. As combustion pressure is applied to the piston head it is transferred to the piston pin and then through the connecting rod to the crankpin on the crankshaft. The offset location of the crankpin applies a rotational force to the crankshaft. The small end of the connecting rod is attached to the piston pin, where it oscillates because of movement of the big end of the rod, mounted to the crankshaft crankpin. The connecting rod is made of aluminum and has a plain bearing (Figure 2–16).

2.4.5   **Piston**

A special alloy aluminum piston moves within the cylinder by the force of combustion. The head of the piston is flat. An orientation mark is on the piston head which is used for

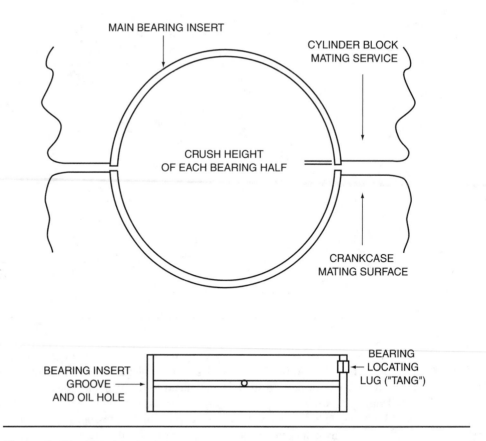

MAIN BEARING INSERT

CYLINDER BLOCK
MATING SERVICE

CRUSH HEIGHT
OF EACH BEARING HALF

CRANKCASE
MATING SURFACE

BEARING INSERT
GROOVE ⟶
AND OIL HOLE

BEARING
⟵ LOCATING
LUG ("TANG")

Figure 2–15   Main bearing inserts.

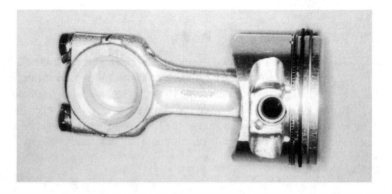

Figure 2–16   OMC four-stroke piston and rod assembly.

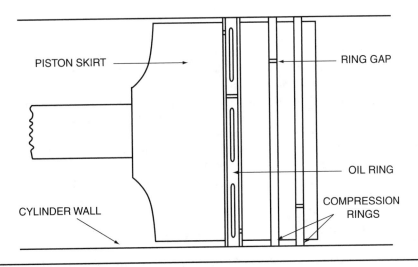

PISTON SKIRT

RING GAP

OIL RING

COMPRESSION
RINGS

CYLINDER WALL

Figure 2–17   Piston and rings.

proper installation. The piston has three ring grooves to mount two compression rings and one oil control ring. The rings seal against the cylinder wall and piston, preventing combustion gases and oil from escaping around the piston. Machined slots are in the piston behind the oil control ring to permit oil collected from the cylinder wall to return to the crankcase and sump. The piston pin mounts in the heavier piston bosses and is held in position by a press fit or with snap rings. The piston head is round and is smaller than the piston skirt. This is because the head faces the higher temperatures of combustion and expands more. The piston skirt is cam ground (oval shaped), therefore keeping the piston skirt (thrust area) in close contact with the cylinder wall when cold or hot. As the piston heats up from combustion it expands along the piston pin, making the piston skirt more round. This allows the piston to fit the cylinder better when cold and at operating temperatures (Figure 2–17).

2.4.6   Piston Rings

Piston rings provide a moving seal between the cylinder and piston. To perform satisfactorily, rings are engineered to conform to the cylinder wall and maintain adequate tension while maintaining their sealing action at required operating speeds and temperatures. There are two *compression rings* near the head (top) of the piston and one *oil ring* assembly under them. A wear resistant surface (chrome) is applied to each compression ring face. The compression rings seal the piston to the cylinder wall against the forces of combustion and transfer combustion heat. They serve a dual role. They must seal and conduct heat, and also assist the oil ring in controlling oil (Figure 2–17).

When the engine is running, oil that is thrown off the bearings is scraped from the cylinder wall by the oil ring assembly. For an oil ring to be effective, it must prevent excessive amounts of oil from passing between the face of the ring and cylinder. The rings also keep oil from passing through the ring gap and passing around behind the ring. The oil ring has a slotted segmented (spacer) ring along with two steel rails—one on top of the spacer and one below it. This keeps the steel rails separated and slots in the piston ring groove behind the oil ring allow the oil to return back to the sump. The oil ring is to maintain enough lubricating oil on the cylinder wall throughout the entire length of piston travel and thus limit piston ring wear and cylinder wear. It also controls the thickness of oil film so that satisfactory oil control is realized.

The ring gap is a space between the ring ends. This space provides for heat expansion and allows the ring to be spread open for installation. Ring tension/expansion makes the ring slightly larger than the cylinder and it spreads out (pushes) against the cylinder wall when installed.

2.4.7   Overhead Valve Camshaft

The camshaft journals are mounted in bearings in the cylinder head (see Figure 2–12). The camshaft is mounted vertically and is used to open the engine valves in a precise order. There is one machined and polished lobe for each engine valve. Mounted to the upper end of the camshaft is a sprocket turned by a timing belt, which is driven by the vertical crankshaft and sprocket. The camshaft sprocket is twice as big as the crankshaft sprocket. Therefore, it turns at one-half the speed of the crankshaft. The valves are precisely timed to open as the cam lobe toe moves against the rocker arm. This action corresponds to the four-stroke cycle by opening the intake valve on the intake stroke and opening the exhaust valve on the exhaust stroke. The camshaft lobes hold these valves open for a precise amount of degrees, permitting fuel to enter and exhaust gases to escape. Valve springs close and seat the valves as the camshaft lobe rotates. The lower end of the camshaft turns the gerotor-type oil pump.

2.4.8   Valves

The **valve guides** are mounted in the cylinder head and guide the valves as they open and close the combustion chamber (see Figures 2–12 and 2–13). A valve seal is installed on the valve stem or guide to prevent oil from entering the combustion chamber. Machined **valve seats** in the combustion chamber (head) provide a contact area for the valves to seal against combustion pressures. With camshaft lobe action, the valves open the combustion chamber for incoming fuel or spent exhaust gases. **Valve spring** action closes and seats the valves, sealing the combustion chamber as the camshaft lobe moves away. This permits: (1) the intake of fuel through an opening and closing of the intake valve; (2) a sealing of the combustion chamber; (3) a compression stroke which heats and pressurizes the fuel air mixture; (4) the spark of ignition that ignites the fuel before top dead center developing combustion pressure for power; then (5) a camshaft lobe and rocker arm open the exhaust valve, exhausting the spent exhaust gases into the exhaust housing. As the camshaft and crankshaft continue rotating, the cycle starts all over again.

Because of wear the valves need to be adjusted periodically. This is done on a cold engine. You will find specifications and procedures in the service manual.

2.4.9   Gaskets and Oil Seals

Gaskets are used to seal against compression, motor oil, fuel, water, and vacuum. They are made of varying materials depending upon what is being sealed. Some materials used are RTV silicon, fiber material, paper, neoprene rubber, cork, or thin steel. The mating surfaces being sealed should be inspected for scratches, nicks, grooves, and warpage. Light alloy metals can be damaged by a putty knife during excessive scraping, therefore use a chemical gasket remover to quickly remove old gaskets. As the parts are reassembled and bolted together, the gasket is reformed as it is compressed, sealing any minor imperfections.

Generally speaking, a gasket is replaced because it is leaking or a major repair is being made. Follow this sequence when replacing a gasket:

1.   Be careful when cleaning and washing the mating surfaces and check each surface for flatness and corrosion.
2.   Compare your new gasket to the bolt pattern and mating surfaces.
3.   Use sealer as specified.
4.   Apply anti-corrosion grease to bolts. If you are replacing any bolts, use the same grade as indicated in parts or the service manual.
5.   Align the part and start all bolts by hand, tightening them in sequence.
6.   Apply *torque* as instructed in the service manual.
7.   Operate the engine and check for leaks.

Oil seals are used to seal around the rotating crankshaft and are installed in the cylinder/crankcase assembly. (Also they are used in the lower unit.) Seals are made of an outer metal shell, inner metal shell (in some applications), steel garter spring (some may not use a spring) and an auxiliary dirt lip. A synthetic bonded sealing member is held against the shaft with an exact uniform pressure by the garter spring.

When installing a seal, the lip of the seal is always pointed towards the lubricant and a sealer is used between the outer metal shell and the cylinder/crankcase. Never strike a new seal directly with a hammer. Pre-lube the seal lip with lubricant to be retained. Seals have two vital jobs: (1) they retain oil needed to protect bearings; (2) they prevent entrance of dirt and water into the engine.

2.5   Motor Oil Classification

Motor oil is classified by two organizations: the Automotive Petroleum Institute (API) and the Society of American Engineers (SAE). When you purchase motor oil you read the viscosity on the label of the oil container, "SAE 30 motor oil." On the back of the container, it tells you that it is approved for or exceeds API service class SJ/CD. "What does all this mean?" "SAE 30" indicates the viscosity (thickness) of the motor oil at 212° F (100° C). The viscosity number *30* refers to its flow resistance. The Society of Automotive Engineers (*SAE*) has established a numbering system used for both single- and multi-grade oils. Single-grade oils have a single number, such as "SAE 30." Multi-grade oils have a dual designation, such as "SAE 10W-40." (It acts like 10W when cold and 40 when hot.) A *W* suffix indicates that the various grades meet certain viscosity requirements at winter temperatures

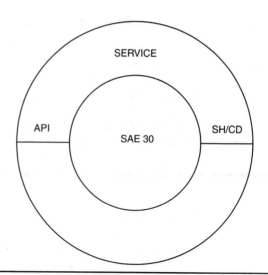

Figure 2–18 A typical label on a container of motor oil contains information on its viscosity and service classification.

from 23° F (-5° C) down to -31° F (-35° C) as well as minimum viscosity requirements at 212° F (100° C). Grades without the *W* such as SAE 30 oil meet only the requirements at 212° F (100°C). The numbers in use today are SAE 5W, 10W, 20W, 30, 40, and 50. The thick, slow-flowing oils have high numbers and the thin, free flowing oils have low numbers. The Automotive Petroleum Institute (API) indicates the service classification using a designation such as *SH*. This gives the engine manufacturers maintenance service application for their engines. Manufacturers may recommend oil service class SG, SH or SJ. You can use a higher rating, but *never* a lower rating. These are all detergent oils. The higher ratings will maintain a lubricating film under higher temperatures and loads. The ratings tell you that this classification of oil will keep the engine cleaner and retard the formation of gum and varnish deposits and provide adequate lubrication (Figure 2–18).

SAE 10W-30 SG or SH is a recommended oil for use in the Mercury and OMC four-stroke outboards. Multi-grade motor oil is recommended for all four-stroke outboards if temperatures under 40° F (4° C) are expected. Check your service manual for the recommended oil for your outboard.

2.6   **Function of Motor Oil**

Motor oil has five purposes. It cools, cushions, cleans, seals, and lubricates. Lubrication, one of the primary purposes, relieves friction—the resistance to motion when one dry surface rubs against another. We consider the crankpin and main bearing journals smooth surfaces. However, under microscopic examination there are high and low spots. The best machining in the world cannot remove these minute high points. If allowed to rub together, scoring and metal transfer would take place. This would destroy the parts while increasing clearances. Motor oil under pressure is pumped into the bearings by a camshaft-driven oil pump. This keeps the bearings and journals separated with a film of oil, relieving friction (Figure 2–19).

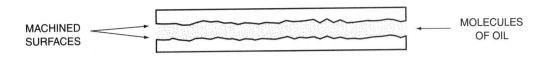

Figure 2–19   Motor oil separates and lubricates machined surfaces.

Figure 2–20   OMC four-stroke gerotor oil pump which is driven by the camshaft (shown with cover removed).

Oil is drawn from the oil sump and pumped through an oil filter into oil galleries (passageways) in the cylinder assembly, to crankshaft main bearings, rod and camshaft bearings, and rocker arms. Oil coming out of the crankshaft and main and rod bearings splashes or squirts onto the cylinder wall, pistons, and rings. The oil in the bearings is under pressure and separates the moving parts with a thin oil film (Figures 2–19 and 2–20). The OMC, 15 HP engine operates on 35–45 PSI at 5000–6000 RPM.

 Note

Oil pressure may be monitored on some models and a light will turn on or a warning horn will sound if low oil pressure develops. An oil flow diagram is shown in Figure 2–21.

OIL FLOW DIAGRAM

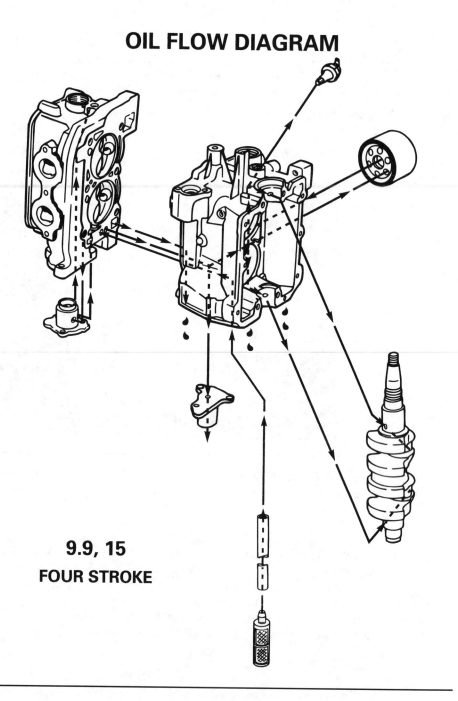

9.9, 15

FOUR STROKE

Figure 2–21   Oil flow diagram for a four-stroke engine. *(Courtesy of Outboard Marine Corporation)*

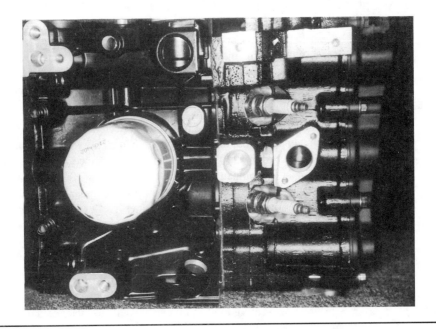

Figure 2–22   Oil filter on a 15 HP OMC. Carburetor and spark plug wires are removed.

Foreign particles created by engine wear get in the engine oil and are filtered out at the full-flow oil filter (OMC 15 HP). All oil passes through the filter before going to the bearings. This prevents the particles from getting into the bearings and between the piston and cylinder wall. If the filter is not changed regularly and it becomes clogged, a filter bypass valve will open and dirty oil will continue to be pumped to the bearings. This will prevent engine bearings from running dry and causing major engine damage. Though the dirty oil will cause engine wear, the engine will continue to run and get you back to the harbor. When the engine is shut down, an anti-drain back filter valve prevents oil from draining out of the filter back into the crankcase. The oil and filter should be changed at 100 hours on regular oil or 200 hours on OMC oil or once a season (see Figure 2–22).

On smaller 8 HP (OMC) motors the oil filter is serviceable and is rinsed with solvent and dried with compressed air. The filter is located on the oil pick-up inside the oil sump.

2.7   Changing Motor Oil

Why should you change the motor oil and when?

- Because crankcase oils become loaded (like dishwater) with undesirable materials which should be drained away.
- Because oil additives are used up in doing their job of helping the oil lubricate, clean, and cool.

- Because even the best-quality oils can be used too long and become contaminated past the point of proper engine protection.
- Because motor life and performance are adversely affected by neglect of oil and filter change.
- Because reasonable and regular oil changes are a "best buy" as insurance against untimely breakdowns and costly repairs.

Change the oil according to the instructions from the outboard manufacturer.

To effectively do its job, the specially compounded Evinrude/Johnson Ultra 4-stroke oil has to be changed every 200 hours of operation or once a season. Regular off-the-shelf motor oil from an oil company has to be changed every 100 hours of operation. Check the oil level with the outboard in the *vertical* position before each launch.

To change the oil, supply water to the water inlet or use an adapter and let the engine warm up at idle speed. Warm engine oil drains more quickly and completely and contains more suspended sludge than cold oil.

Motor oil is used to lubricate the engine because it does the following:

- Reduces friction, preventing excessive wear.
- Picks up heat and carries it to a cooler oil sump.
- Cushions the bearing load and prevents metal to metal contact.
- Acts as a cleaning agent.
- Retards acid, varnish, and gum formations.
- Minimizes engine power loss.
- Controls rust and corrosion.
- Forms a gas-tight seal between the piston rings and cylinder.

No wonder the engine oil must be changed regularly. It gets dirty doing its job and if not changed regularly, the above elements and conditions will not be controlled thereby shortening engine life.

2.8   Oil Pressure Test

An oil pressure test can be made to determine operating pressure. By removing the oil pressure sending unit, a pressure gauge can be screwed in. Oil pressure varies depending upon RPM. Check the service manual for your engine's make and model. Oil pressure on a Johnson 15 HP should be 10 PSI at 2000 RPM within 10 seconds of start up. As the engine is warmed up and is accelerated to between 5000 and 6000 RPM, pressure should be 35–45 PSI.

If specifications are not met there may be possible problems with the oil level—a clogged filter, a clogged inlet screen, or a defective oil pump or pressure regulator.

2.9   Cooling System

Water pumps are mounted on the lower units of four-strokes as they are on two-stroke outboards. The water pump impeller is driven by the driveshaft and water is brought in

WATER FLOW DIAGRAM

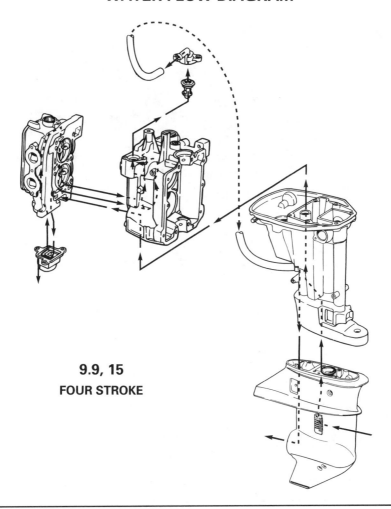

9.9, 15

FOUR STROKE

Figure 2–23 Water flow diagram on a four-stroke engine. *(Courtesy of Outboard Marine Corporation)*

through the screened pickup on the side of the lower unit. The water pump is of the same basic design as two-stroke outboard pumps and service requirements are the same.

Pressurized water flow is from the pump through the water tube/exhaust housing where it is heated; then into the cylinder block, cylinder head, and through a pressure release/thermostat to the midsection; then it exits through the lower unit along with the exhaust gases through the propeller housing. See Figure 2–23 for an OMC diagram of water flow.

Figure 2–24   Thermostat and poppet valve.

The thermostat is used to bring engine temperature up to a minimum specification of 125 degrees. If the engine is accelerated to a high RPM when cold, the thermostat spring will permit water to bypass the thermostat or a poppet valve will open (Figure 2–24). This allows for increased water flow through the engine.

Engine temperature for both four-stroke and two-stroke powerheads is checked in the same way. That is, with the use of Thermomelt Stiks. Use Thermomelt Stiks that are 125 and 163 degrees. The Stiks are applied like crayon marks to the thermostat pockets. The 125-degree mark should melt and the 163-degree mark should not melt. If the 125-degree mark does not melt, check the system for causes of overcooling. If the 163-degree mark melts, check the cooling system for causes of overheating.

If the outboard has been running in brackish or saltwater, the cooling system must be flushed.

 **Warning**

Before flushing the engine *remove* the propeller to prevent personal injury.

Apply the recommended lubricant to the propeller shaft splines upon reinstalling the propeller. Run the engine on a flushing attachment or in a 30-gallon trash can for a minimum of five minutes at 1500 RPM. If the outboard is going into storage, run the carburetor out of gasoline or add fuel stabilizer to the fuel before you start flushing the engine. Fuel stabilizer must be in the carburetor, so run the engine enough to get stabilizer into the carburetor float bowl. If the outboard is going into storage, spray some fogging oil through the carburetor opening just before you shut the engine down. This will lubricate the internal parts of the engine and prevent rusting. After shut down, pull the spark plugs and spray fogging oil into the cylinders and rotate the engine to spread the oil throughout the cylinders.

2.10   **Temperature Switch**

If the engine temperature rises to an unsafe operating level (240° F, 116° C), a temperature switch will activate a horn and light (if so equipped) at the helm. This activation can be caused by something minor like a plastic bag or weeds over the water inlet or something serious like a failed water pump impeller.

To test the electrical warning system, connect a jumper wire from the temperature switch to the engine ground. With the key on, the horn should sound and the indicator light should be lit (if so equipped).

KNOW THESE PRINCIPLES OF OPERATION

- Four-stroke operation.
- Operation of the camshaft and valves.
- Main bearings and purpose of crush height.
- Pressurized lubrication system.
- Function and classification of motor oil.
- How the engine is cooled.

REVIEW QUESTIONS

Chapter 2, Section 1, Principles of Two-Stroke Operation

1.   The two-stroke powerhead is considered to be a _____.
2.   When is negative pressure present in the crankcase?
3.   What oil type is used in the two-stroke outboard?
4.   What opens the reed valves?
5.   When is there positive pressure in the crankcase?
6.   Where does the exhaust leave the outboard?
7.   How are the internal parts of the powerhead lubricated?
8.   What happens to the fuel mixture when the crankcase seal or O-ring is leaking?
9.   Is properly mixed 50/1 two-cycle oil very adequate for maximum RPM?
10. On the intake stroke fuel comes into the crankcase. What is happening in the cylinder at the same time?
11. The deflector dome on the piston is used to _____.
12. With the ports closed, what seals the fuel charge and combustion in the cylinder?
13. Why is there very little wear on the upper cylinder wall?
14. When is the spark plug fired?
15. When is maximum pressure developed on the piston?
16. If there is preignition, what part can be damaged?

17.   What can throw the combustion burn off and out of control?
18.   When the piston is moving into the crankcase, which port is opened first?
19.   In a running powerhead, when are the reed valves closed?
20.   When is a positive pressure applied to the upper crankcase seal?
21.   How is fuel mixture put into the cylinder?
22.   How are the exhaust gases purged from the cylinder?
23.   It takes one revolution of the crankshaft to complete the _____.
24.   A flywheel is added to the crankshaft to _____.
25.   The flywheel is also used to _____.
26.   What are the three different methods used to port the incoming fuel into the cylinder?
27.   What is special about the design of a power-ported piston?
28.   What closes the reed valves?
29.   What is the main purpose of power porting?
30.   When the engine is power ported, there is a port in the piston skirt and in the _____ _____ .

Chapter 2, Section II, Principles of Four-Stroke Operation

1.   The crankshaft of a four-cycle outboard is mounted
     a.   horizontally.
     b.   vertically.
2.   The crankcase is bolted to the
     a.   bottom of the cylinder assembly.
     b.   side of the cylinder assembly.
     c.   flywheel end of the engine.
3.   It takes _____ revolution(s) of the crankshaft to complete one power cycle.
     a.   one
     b.   two
4.   The four-stroke camshaft is
     a.   gear driven and timed to the crankshaft.
     b.   belt driven and timed to the crankshaft.
5.   To open the intake or exhaust valve, the camshaft lobe contacts the
     a.   valve stem.
     b.   valve spring retainer.
     c.   rocker arm.
6.   Fuel enters the combustion chamber when the piston moves toward the crankcase and the _____ valve is open.
     a.   intake
     b.   exhaust
7.   Which valve is open on the compression stroke?
     a.   intake
     b.   exhaust
     c.   neither valve is open

8. When is the ignition spark delivered for igniting the fuel mixture?
   a. after top dead center
   b. before top dead center
   c. at top dead center
9. What pushes the piston down on the power stroke?
   a. crankshaft
   b. burning fuel
   c. combustion pressure
   d. Both b and c are correct.
10. What stroke follows the power stroke?
    a. intake
    b. compression
    c. exhaust
11. The intake valve opens near the end of the exhaust stroke.
    a. true
    b. false
12. The main and most rigid structure of the engine is the
    a. cylinder head.
    b. cylinder assembly.
    c. crankcase.
13. The main bearings are supported by the cylinder assembly and the crankcase.
    a. true
    b. false
14. The water jacket routes water around the cylinders.
    a. true
    b. false
15. The engine oil sump is located
    a. in the crankcase.
    b. under the engine in the upper portion of the exhaust housing.
    c. at the flywheel end of the engine.
16. The crankpin revolves (orbits) around the center line of the crankshaft.
    a. true
    b. false
17. What type of main bearings are used in the four-stroke?
    a. ball
    b. needle
    c. insert
18. The lower unit drive shaft splines fit (mesh) into the lower end of the crank-shaft.
    a. true
    b. false
19. Main bearings use a lobe (tang) to align the bearing inserts during assembly.
    a. true
    b. false

20. Grooves in the main bearing are to facilitate oil flow around the crankshaft main journal and bearing.
    a. true
    b. false
21. Bearing crush height is used to
    a. permit a good heat transfer.
    b. to squeeze (push) the insert tightly into the bore.
    c. Both a and b are correct.
22. The small end of the connecting rod is mounted to the
    a. crankpin.
    b. piston pin.
    c. piston boss.
23. Which piston ring is mounted closest to the piston skirt?
    a. compression ring
    b. oil ring
24. When the cam ground piston heats up, it expands
    a. through the piston skirt.
    b. through the heavier piston pin bosses.
    c. along the ring grooves, to seal the rings to the piston.
25. Piston rings and motor oil provide a moving seal against the cylinder wall to prevent/reduce combustion leakage.
    a. true
    b. false
26. What conducts heat away from the piston?
    a. water flow in the water jacket
    b. cylinder wall
    c. piston rings
    d. All of the above are correct.
27. The camshaft pulley is twice as big as the crankshaft pulley.
    a. true
    b. false
28. The exhaust and intake valves are opened by the same lobe on the camshaft.
    a. true
    b. false
29. The valve guide supports and aligns the valve stem.
    a. true
    b. false
30. Valve springs are used to seat and hold the valves closed.
    a. true
    b. false
31. Light alloy metal parts that the gaskets seal against are easily damaged when being cleaned.
    a. true
    b. false

32. Seals are used on rotating shafts to
    a. seal in motor oil.
    b. seal out water on the lower unit.
    c. Both a and b are correct.
33. SAE 30 (motor oil) refers to flow resistance.
    a. true
    b. false
34. The *W* listed in 10W-40 motor oil indicates that it meets requirements for summer operation.
    a. true
    b. false
35. Engine manufacturer's maintenance service applications for motor oil are listed as
    a. SAE 30.
    b. 10W-40.
    c. SG/SH.
36. It is best to change motor oil when the engine is cold.
    a. true
    b. false
37. At the bottom of your answer sheet list five purposes for using motor oil.
38. The primary function of motor oil is to relieve friction.
    a. true
    b. false
39. What will happen if a full-flow oil filter becomes clogged?
    a. the engine will shut down
    b. a warning horn will sound
    c. a bypass valve will open
40. Oil pressure varies with engine RPM.
    a. true
    b. false
41. Water used to cool the engine initially comes in through the
    a. water pump inlet.
    b. lower unit inlet and screen.
    c. exhaust housing.
42. If the outboard is accelerated before the thermostat opens, what happens to permit increased water flow?
    a. water pump impellers flex, permitting continued low volume water flow.
    b. increased water pressure lifts the spring loaded thermostat off its seat for increased water flow.
    c. increased water pressure opens a poppet valve.
    d. Either b or c are correct depending upon the model.
43. How is engine operating temperature checked?
    a. by measuring water temperature as it exits at the exhaust
    b. Thermomelt Stiks are applied to pockets near the thermostat.
    c. Thermomelt Stiks are applied to the cylinder assembly.

44. An outboard run in the ocean should be flushed for a minimum of 5 minutes at 1500 RPM with freshwater supplied and the propeller removed for safety.
    a.   true
    b.   false
45. An outboard that is not going to be used for 30 days or more should have a fuel stabilizer added to the fuel. Then it should be run long enough to get the stabilizer into the carburetor.
    a.   true
    b.   false
46. When engine operating temperature gets too high, a temperature switch will activate and sound a horn (OMC 15 HP).
    a.   true
    b.   false

CHAPTER 3

Two-Stroke and Four-Stroke Powerhead Repair

Section I: Two-Stroke Powerhead Repair

Objectives

After studying the first section of this chapter, you will know:

- The two major parts of the cylinder assembly.
- Methods used in sealing each crankcase.
- How to make cylinder measurements.
- The need for a crosshatch pattern of the cylinder wall.
- Piston nomenclature.
- Measuring the piston size.
- Conditions that occur on the piston.
- Ring gap measurement.
- Connecting rod service.
- Crankshaft service.
- Oil injection.
- Cooling systems.

3.1   Cylinder Assembly and Sealing

The cylinder assembly is made of aluminum and has cast iron/steel cylinders. It is the major part of the powerhead and care must be given to this part when service work is performed. Mishandling or improper service procedures performed on this assembly may make scrap out of an otherwise good casting. The cylinder assembly and other major castings on the outboard are expensive and need to be cared for accordingly.

There are two parts to the cylinder assembly—the cylinder block and the crankcase half. These two castings are married together and line-bored to receive the crankshaft bearings, reed blocks, and sealing rings (design permitting). After this machining they are treated as one casting. Remember that anything done to the mating surfaces during service work will change the inner bore diameter for the main bearings, reed blocks, and *sealing rings* (design permitting), and possibly prevent the block and crankcase mating surfaces from sealing (coming together). The only service work allowed on the mating surfaces is a lapping operation to remove nicks from the surface.

 **Warning**

Do not lower the basic surface!

Carefully guard this surface when other service work is being performed. There are different sealing materials used to seal the mating surfaces; with sealing strips use Permatex #2 Form-A-Gasket. When sealing strips are not used, use Loctite Master Gasket Sealer or OMC Gel-Seal. Remember that the two-stroke powerhead operates like a pump, with one inlet and one outlet for each crankcase cylinder. Special sealing features must be designed into the cylinder assembly to seal each individual cylinder crankcase onto itself in a multi-

OMC LABYRINTH SEAL SHOWN IN CYLINDER ASSEMBLY

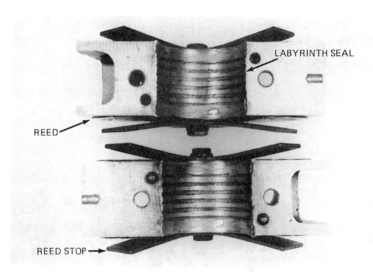

DAMAGED MERCURY LABYRINTH SEAL
SHOWN WITHIN REED BLOCK

Figure 3–1   Labyrinth seals.

cylinder powerhead. Each crankcase and inlet manifold must be completely sealed both for vacuum and pressure (see Figure 3–1).

One way of sealing this internally is with a labyrinth seal, which is located between two adjacent cylinders next to the crankshaft. It may be of aluminum or brass, formed in the assembly and machined with small circular (annular) grooves running very close to a machined area on the crankshaft. The tolerance is so close that fuel residue puddling in the seal effectively completes the seal between the cylinder block and crankcase halves against the crankshaft. Crankcase pressures are therefore retained to each individual cylinder. No repair of the labyrinth seal is made. If damage occurs to the seal, the main bearings allow the crankshaft to run out and rub.

Another method of internal sealing between crankcases is with *seal rings*. These rings are installed in grooves in the crankshaft. When the crankshaft is installed, the sealing rings mate up to and seal against the web in the cylinder block crankcase halves and crankshaft (Figure 3–2).

Sealing rings of different thicknesses are available for service work. The side tolerance is close, so puddled fuel residue will effectively complete the seal between crankcase and crankshaft.

To seal the upper and bottom crankcases around the crankshaft, O-rings are installed around the end caps, and neoprene seals are installed inside the cap and seal against the crankshaft (Figure 3–3).

Follow the service instructions carefully on how to position the seal(s) in a bearing cap. One seal may have the lip up and the other may be placed with the lip down, or both may be placed lip down.

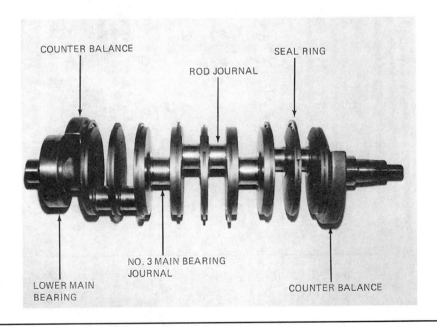

Figure 3–2   V-6 crankshaft. *(Courtesy of Outboard Marine Corporation)*

3.1.1   Troubleshooting the Cylinder Assembly

Because the two-stroke powerhead is a pump, the crankcase *must be sealed* against positive pressure created on the power/exhaust stroke and the vacuum created when the piston moves toward the spark plug end of the cylinder on the intake/compression stroke. If there are air leaks into the crankcase, insufficient fuel will be brought into the crankcase and into the cylinder causing abnormal combustion. If there is a very small leak, the powerhead will run poorly, because the fuel mixture will be lean, and cylinder temperatures will be hotter than normal. Air leaks are possible around any seal, O-ring, cylinder block mating surface, or gasket. Always replace O-rings, gaskets, and seals when service work is performed. If the powerhead is running, soapy water can be sprayed onto the suspected sealing areas. If bubbles develop, there is a leak at that point. Oil around sealing points and on ignition/stator parts under the flywheel indicates a crankcase leak. The base of the powerhead and lower crankshaft seal is impossible to check on an installed powerhead. This is always a questionable area (see Figure 3–3). When every test and system has been checked out and the bottom cylinder seems to be affecting performance, then go for the lower seal. You can prove whether the lower seal is leaking by pressure-testing the crankcase after the powerhead is removed.

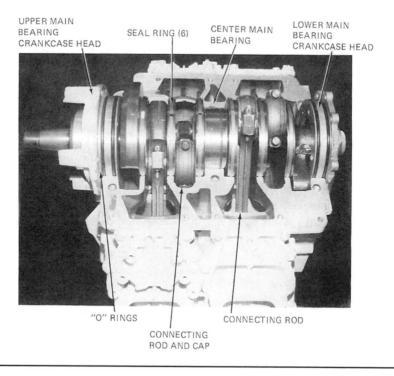

Figure 3–3   V-4 cylinder assembly with crankshaft installed.

To pressure-test the crankcase, make adapters to fit the carburetor mounting studs. Into one adapter, place an air fitting which will accept a hand air pump or tap the adapter for the Stephens pump, which is used for checking lower units. With the powerhead on the bench, place a thick piece of rubber over the exhaust outlet and back it up with a thick steel plate. Hold the plate in position with bolts or a powerhead stand adapter and compress the rubber to effect a seal on the exhaust outlet, leaving water passages open. Using the hand pump, pressurize the crankcase(s) to five pounds of pressure. Spray soapy water around the lower seal area and other sealed areas watching for bubbles, which indicate the leaking point. Also, pull a vacuum to stress the seals in the opposite direction and watch for a pressure drop. After repairs or overhauls are completed, use this pressure test to prove sealing of the total crankcase. By turning the powerhead upside down and filling the water jacket with water, any cracks or corrosion pin holes, which may have been causing overheating, will become evident if bubbles show up in the water when a positive pressure is applied to the crankcase. There are adapter plates available from some manufacturers to seal the base of the powerhead for pressurizing the water jacket. This type of testing takes a little time, and is not for every job, but for that problem-plagued powerhead, it may be necessary.

Every time the cylinder head is removed, it should be checked for warping. Do this with a straight edge, a surface block, or automotive window glass. If it is warped, surface it by using emery paper in a figure-eight motion on a surface block until the surface is true. Check the head for cracks and damage to the bolt holes caused by galvanic corrosion. On models that do not use a cylinder head, check the cylinder dome for holes or cracks caused by overheating and preignition. The spark plug threads may also be damaged by over-torquing of the spark plug.

Quite often the small bolts around the crankcase sealing area are seized by corrosion. If white powder is evident around the bolts, stop! This calls for an experienced, trained technician! Galvanic corrosion is probably seizing the shank of the bolt, and possibly the threads as well. Putting a wrench on them may just twist the head off, creating one big job! Know the strength of the bolt and stop before it breaks. If it breaks, don't reach for the easy out. It won't work! A good service tool for these seized bolts is localized heat from a torch. Heat the aluminum casting (not the bolt), which will expand faster than steel. This releases the bolt from the corrosive grip by creating clearance between the bolt, the corrosion, and the aluminum casting. Be careful because too much heat will melt the casting! Many bolts can be released in this way, preventing drilling out the total bolt and heli-coiling the hole, or tapping the hole oversize. For more detailed information see Chapter 1, "Corrosion," Figures 1–17 and 1–32.

Sometimes we attempt to remove the powerhead and it won't come free from the adapter (lower motor cover) on the lower unit. The gasket may hold the powerhead. Rock the powerhead back and forth or give it a gentle nudge with a pry bar. If the gasket breaks loose and the powerhead still will not come free, then the drive shaft is seized to the crankshaft and the splines. STOP! It is time for an experienced, trained technician. For more detailed information on this problem see Chapter 11, "Troubleshooting the Lower Unit."

3.2   Cylinder

The cylinder is actually a container with one end closed and the other movable. The purpose of the cylinder is to help lock in combustion gases. It provides a guided path for the piston to travel in. It provides a lubricated surface for the piston rings to seal against and transfer heat into the cooling system. These functions are carried out through all RPM ranges (see Figure 3–4). To function properly the cylinder has to have a true machined surface and must have the proper finish installed on it to retain the lubricant. The roundness of the cylinder and the straightness of the cylinder wall should be looked at first. Six micrometer readings should be taken to determine the cylinder condition. Start at the bottom using an inside micrometer or telescoping gauge and an outside micrometer. By starting at the bottom, below ring travel, cylinder bore diameter can be determined. You will know immediately, therefore, if the powerhead is standard or has been bored oversized. Sometimes you may find a mark on the piston head which will alert you to an oversized bore. There is generally no wear on the cylinder in this location. Take a second reading straight up from the first reading in the area of the ports and you will find that the used cylinder is larger here. A third reading is taken straight up from the second reading and within one-half inch of the top end of the cylinder. These three readings are taken in a straight line up the cylinder. Next turn the micrometer 90 degrees and take three more readings in a straight line up the cylinder, as shown in Figures 3–5 and 3–6.

Figure 3–4   Cutaway of cylinder with piston.

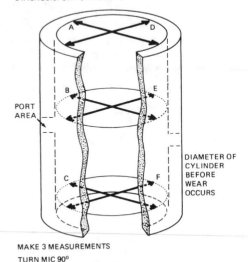

EXAMPLE OF MEASUREMENTS

CYL. MEASUREMENTS	AT 90 DEGREES CYL. MEASUREMENTS	OUT OF ROUND
A 3.002	D 3.001	.001
B 3.003	E 3.002	.001
C 3.001	F 3.000	.001
TAPER .002	TAPER .002	

DIAGNOSIS: OK FOR RE-RING

PORT AREA

DIAMETER OF CYLINDER BEFORE WEAR OCCURS

MAKE 3 MEASUREMENTS
TURN MIC 90°
MAKE 3 MORE MEASUREMENTS

Figure 3–5   Cylinder measurement.

You will then have enough information to accurately assess the cylinder condition to determine if rings can be put in and if the cylinder has possibly been bored oversized. Does it meet service tolerances? While measuring the cylinder you should also notice if there is a pattern on the cylinder wall, and if there is any scoring, scuffing, or scratches. Once all information is at hand, it is decision time. What is going to be done to the cylinder wall to bring it into specifications? If it is scored, then reboring, resleeving, or a short block are the choices. If the cylinder measurements do not exceed the factory wear tolerances, the cylinder can be deglazed using a ridged or flex hone. Did you notice a pattern? This pattern is used to condition the surface of the cylinder wall for oil retention and ring seating. As the piston rings move in and out on the wall, a glaze develops. The hone is used to remove this glaze and reestablish the **crosshatch** (basket weave) pattern (see Figure 3–7). This pattern and therefore the finish is established by using a medium-grit (220) honing stone or flex hone. There are about 75,000 cutting points per square inch of stone surface which generate millions of tiny superimposed crisscross grooves in the cylinder wall during the honing process. The resultant diamond shaped crosshatch pattern has a satin-looking finish and makes an ideal surface for good retention of the two-cycle oil on the cylinder wall.

Micrometer Readings and Measuring Cylinder Bore

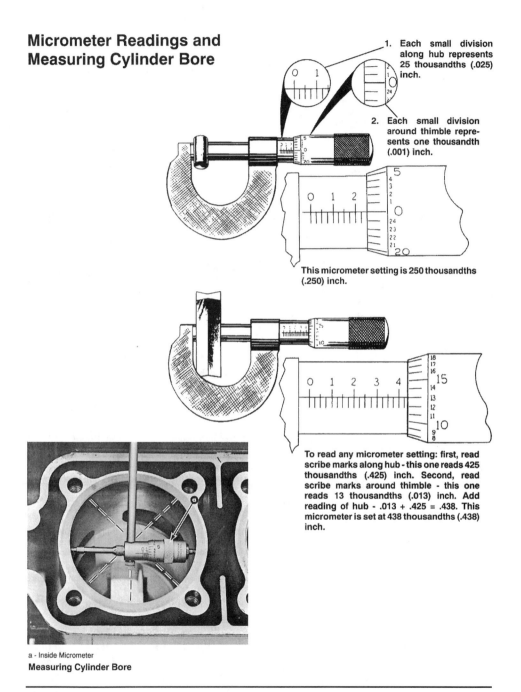

1. Each small division along hub represents 25 thousandths (.025) inch.

2. Each small division around thimble represents one thousandth (.001) inch.

This micrometer setting is 250 thousandths (.250) inch.

To read any micrometer setting: first, read scribe marks along hub - this one reads 425 thousandths (.425) inch. Second, read scribe marks around thimble - this one reads 13 thousandths (.013) inch. Add reading of hub - .013 + .425 = .438. This micrometer is set at 438 thousandths (.438) inch.

a - Inside Micrometer
Measuring Cylinder Bore

Figure 3–6   Micrometer readings and measuring cylinder bore. *(Courtesy of Brunswick Corporation)*

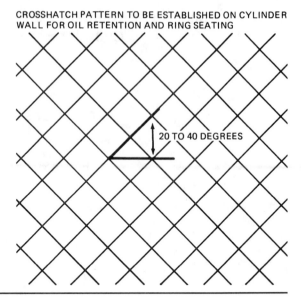

CROSSHATCH PATTERN TO BE ESTABLISHED ON CYLINDER WALL FOR OIL RETENTION AND RING SEATING

20 TO 40 DEGREES

Figure 3–7   Cylinder cross-hatch pattern.

Various manuals state that the crosshatch pattern should include an angle of between 22 and 60 degrees. There is nothing magic about the angle, but there should be one similar to what the factory used, which is around 45 to 60 degrees (Figure 3–7). Too steep or too flat a pattern is not acceptable and is not desirable for good ring seating. Since the honing tool reverses reciprocation, the pattern in the cylinder is really one of many different angles, all crisscrossing each other, with every single abrasive grain tracing a curve on the cylinder wall (Figure 3–7).

This pattern installed in a new cylinder or on one within wear tolerances, is the secret for longevity of the cylinder and of the piston rings. The pattern allows two-cycle oil to flow under the piston ring bearing surface, and prevents a metal-to-metal contact between cylinder wall, piston skirt, and piston rings. The satin-finished diamond crosshatch pattern is necessary to prevent early break-in scuffing and to seat the rings correctly. Perhaps you have noticed something in the owner's manual about break-in of the powerhead. The manufacturer is suggesting that powerhead RPM be limited for the first few hours to allow for some break-in time of the critical surfaces and parts. The cylinder and rings are two of these critical areas. The manufacturer may also suggest increasing the oil mix for the first tank of gasoline. The amount and viscosity of the oil are very important. The viscosity is controlled in the TC-W3 two-cycle oil used. Notice that the container is not marked for viscosity (weight) like four-cycle oil. It is all formulated to do the best lubricating job for the cylinder and bearings at wide-open throttle. (See "Lubrication" in this chapter.)

In theory, one-tenth of a micron of oil lubrication or film between two rubbing surfaces will prevent them from touching. Use of the proper two-cycle TC-W3 oil mixed in proper proportions and a good cylinder wall surface pattern prevents ring scuffing upon start-up. The oil and pattern continue to provide a good surface for the rings to seal against because of the retention of the two-stroke oil under the ring bearing surface. Because this powerhead is of two-stroke design, the oil is replenished on the cylinder wall above and below the rings after each power stroke. Therefore, there is very little wear on a two-stroke cylinder wall. In the two-cycle powerhead we are not concerned with oil consumption, but rather ring sealing. We do see a few problems on cylinder walls because of overheating, improper oil mixing, poor cooling system circulation, timing and synchronization, and lean mixtures in the fuel system. Remember that lean mixtures involve not only fuel, but oil and therefore lubrication throughout the powerhead. Being a hot area, the cylinder will quite naturally score when the lubrication film can no longer support the ring bearing surface or piston skirt.

After the cylinder hone operation is complete, one *very* important job remains. The grit that was developed in the machining/honing operations must be cleaned up. This is where many mechanics fall short! They may cut short the clean-up operation, not thinking of the damage that the shortcut will cause. Grit left in the powerhead will find its way into the bearings and piston rings, and become embedded in the piston skirt, effectively grinding (sanding) away at these precision parts. Relate this to emery paper applied to a piece of steel, or steel against a grinding stone. The effect is removal of material from the steel. Grit left in the powerhead will "kill" the powerhead in a few short hours of operation.

Many people try to wipe down the cylinder with an oily rag and others try solvent from a wash tank.

 Note

These methods *do not remove the grit.* Cleaning must be thorough to be sure that all abrasive grit has been removed from the cylinders.

It is important to use a scrub brush with hot soapy water or a special cleaning compound designated for this purpose. Remember that aluminum is not safe with all compounds. After the cylinder is thought to be clean, test it with a white towel or cloth. Put it through the cylinder and wipe, rubbing hard. If any gray color shows on the towel, the cylinder still has abrasive grit in it. Re-scrub it, and use the white-towel test again. When the cylinder passes the test, apply 50/1 oil on the cylinder walls immediately, for they will begin to rust quickly.

Some cylinders are chrome-plated. Service work on these cylinders must be treated differently. Measuring the cylinder is the same; however, the repair work is different from that discussed for cast iron cylinders. If there are any aluminum deposits left from a scored piston, you can try to remove them with muriatic acid (found in toilet bowl cleaner). Be careful to confine the muriatic acid to the aluminum deposits in the cylinder. Only *very light* honing should be done in the chrome cylinder. The cylinder bore is inspected for scoring, wearing through, and flaking of the chrome. If any of these conditions are found, replace the cylinder assembly and matched crankcase unit. You might consider a short block to replace the damaged unit. Porosity in the chrome of the chrome cylinder will not hurt, and probably was there when new.

3.3   Piston

The piston is the movable end of the cylinder. The cylinder provides a guided path for the piston allowing approximately a .005-inch clearance between the piston skirt and cylinder

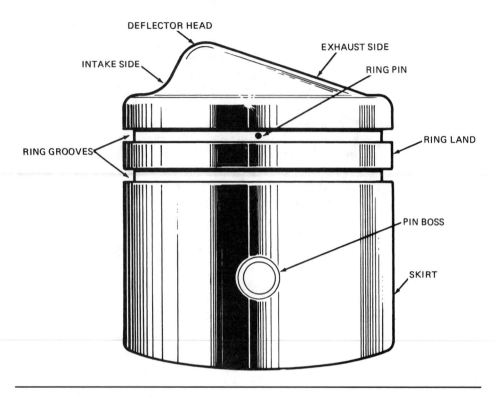

Figure 3–8   Piston identification (deflector head).

wall. This clearance allows for piston expansion and controls piston rock within the cylinder. The deflector head piston design is such that the head can direct incoming fuel toward the outer end (spark plug end) of the cylinder. This design is called a deflector type of piston head (Figure 3–8).

Note also that the opposite side of the piston head is sloped to direct the exhaust gas to the exhaust port in the cylinder wall. The deflector dome directs the incoming fuel outward to the spark plug end of the cylinder, partially cooling the cylinder and sparkplug tip. It also purges the spent exhaust gases from the cylinder. Actually, the incoming fuel charge is chasing out the exhaust gases from the cylinder. Not all piston designs are of the deflector head type. Other pistons have a small convex crown on the piston head (Figure 3–9) or they may be flat.

In this case port design aids in directing the incoming fuel outward. The piston head bears the brunt of the combustion force and heat. Most of the heat is transferred from the piston head through the rings to the cylinder wall and then on to the cooling system (Figure 3–10).

A hole is placed in the side of the standard piston. This is the ***piston*** or ***pin boss*** and it is used to mount the piston pin. The combustion is transferred to the piston pin and connecting rod bearing, then on to the crankshaft where it is converted to rotary motion. The pin is fitted to the piston bosses. The piston pin is the inner bearing race for the bearing

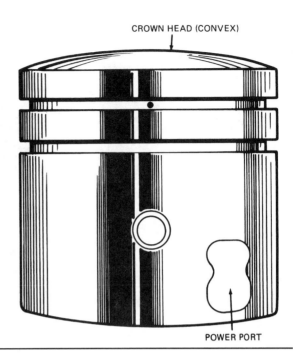

CROWN HEAD (CONVEX)

POWER PORT

Figure 3–9   Piston identification (crown head).

mounted in the small end of the connecting rod. This transfers the combustion pressures into the connecting rod and allows the rod to swing with a pendulum-like action. The piston has machined ring grooves in which the rings are installed. They are carried along with the piston as it travels in and out on the cylinder wall. There is one small pin in each ring groove to prevent the ring from rotating into a port in the cylinder wall. The piston skirt is the bearing area for thrust, and rides on the cylinder-wall oil film. The side thrust of the piston is dependent upon pin and connecting-rod position. If the pin is in the center of the piston, then there will be more thrust. If the pin is offset a few thousandths of an inch from the center of the piston, there will be less thrust. Notice the piston skirt after the piston has been in service. One piston skirt is brighter and shows more signs of wear than the opposite skirt. This is the major thrust side of that piston (see Figure 3–11).

The thrust is caused by the pendulum action of the rod following the crankshaft rotation, which pulls the leg (rod) out from under the piston. The combustion pressure therefore pushes and thrusts the piston skirt against the cylinder wall. Some heat is also transferred at this point. The other skirt receives only minor pressure. Some pistons have small grooves (machining rings) circling the skirts. This helps retain oil in the critical area between the piston skirt and the cylinder wall. Other piston skirts are smooth.

The piston design can be round or it can be cam ground. The cam ground design allows for expansion of the piston in a controlled manner. As the piston heats up, expansion takes place and the piston moves out along the piston pin, becoming more round as it warms up. Other designs may expand at 15 degrees to the pin (see Figure 3–12). The piston may also

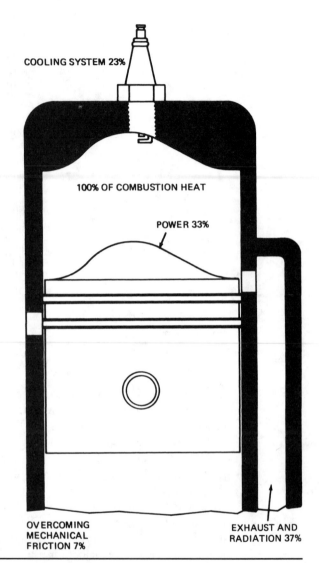

Figure 3–10 Approximate heat distribution of the powerhead (as viewed from flywheel).

be barrel-shaped as well as cam ground. This barrel shape allows the piston to rock. Very minor rocking motion helps to keep the rings free.

To measure the piston, an outside micrometer is placed on the skirts. All pistons in a powerhead should read the same unless one cylinder has been bored oversized. Oversized pistons can be put in one or more cylinders. (Cam-ground pistons may be installed in the top and bottom cylinder, 50-55-60 OMC). Check the service manual for placement of the micrometer and specifications when piston measurements are taken. (Figure 3–13).

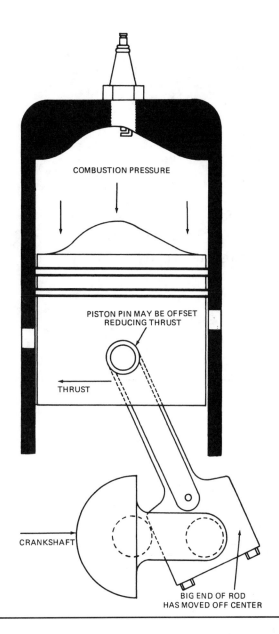

Figure 3–11 Piston major thrust side as viewed from the flywheel.

Piston pins may be secured into both piston bosses. All have retainers (locks), and in addition the majority use a press fit to secure the pin. Some models use a slip fit. Check the service manual for special instructions on the method of retainer (lock) installation. OMC makes some of their pistons with one tight and one loose piston boss. This aids in removal of the pin without collapsing the piston. With this design of piston, always press on the pin

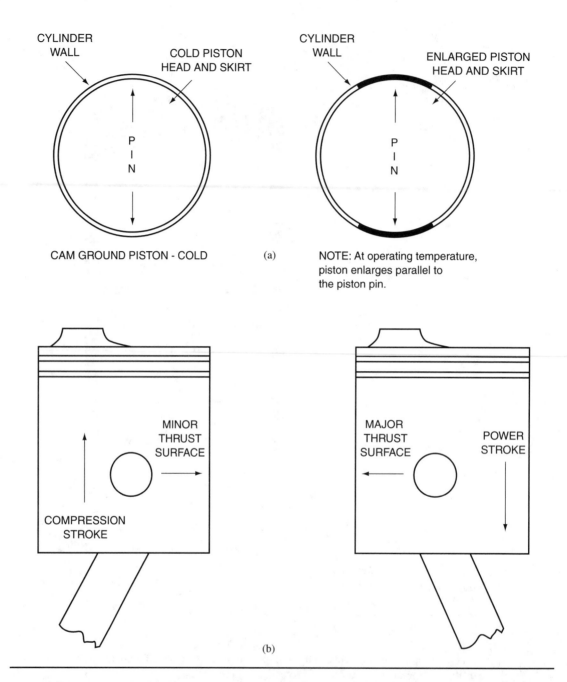

Figure 3–12 Diagram of the top view of a cam ground piston. Expansion occurs as the piston warms to operating temperature (a). The minor and major thrust surfaces shift (b) as the crankshaft rotates, moving the connecting rod left to right.

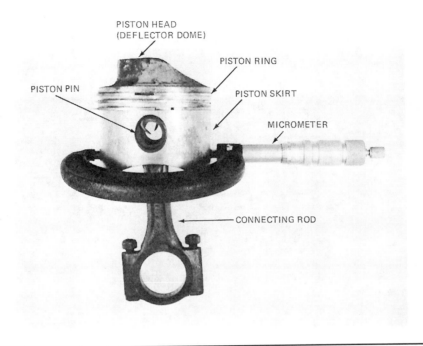

Figure 3–13   Measuring the piston skirt.

from the loose boss side. The piston is marked on the inside of the piston skirt with the word *loose* to identify the loose boss. Always press with the loose side up and press the pin all the way through and out. When reinstalling, press with the loose side up. In all pressing operations set the piston in a cradle block to support the piston. Some pistons require heating with a torch lamp to expand the piston bosses so the pin can be pressed out without collapsing the piston. Follow the manufacturer's special instructions carefully so no damage to the piston will occur. Other pistons may just have a slip fit. Check the service manual carefully. When installing the connecting rod, the long slopping side of the piston must be installed toward the exhaust side of the cylinder assembly (Direct Charge pistons), and if there is an oil hole in the connecting rod, position the oil hole up. Some crown-type, power-ported piston designs are marked with the word *up*, and this side is placed in the up position, meaning toward the flywheel end of the crankshaft. Also, this type has an extra window in the piston skirt for fuel transfer through the power port (see Figure 3–9).

3.3.1   Troubleshooting the Piston

The piston needs to be inspected for damage. Check the head for erosion caused by excessive heat, lean mixtures, and misadjusted timing/synchronization. Examine the ring land area to see if it is flat and not rounded over. Also look for burned through areas caused by preignition. Check the skirt for scoring caused by a break-through of the oil film, excessive cylinder wall temperatures, incorrect timing/synchronization, or inadequate lubrication (Figures 3–14 through 3–19).

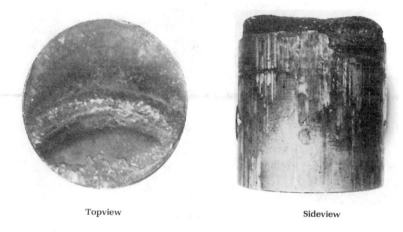

Topview   Sideview

Figure 3–14   Preignition: erosion of deflector and ring land damage.

Topview   Sideview

Figure 3–15   Damage caused by preignition.

Topview   Sideview

Figure 3–16   Carbon stuck ring, preignition, shearing action on the deflector and head, residue on the head and water in cylinder.

Figure 3–17   Hotter plug
installed, causing preignition,
detonation.

Topview

If the piston looks reasonably good after cleaning, take a close look at the ring lands. Wear may develop on the bottom of the ring lands. This wear is usually uneven, causing the ring to push on the higher areas, and loading the ring unevenly when inertia is greatest. Such uneven support of the ring will cause ring breakage and the piston will need to be replaced.

Figure 3–18 Water entered cylinder, causing corrosion to develop between piston and cylinder wall.

Sideview

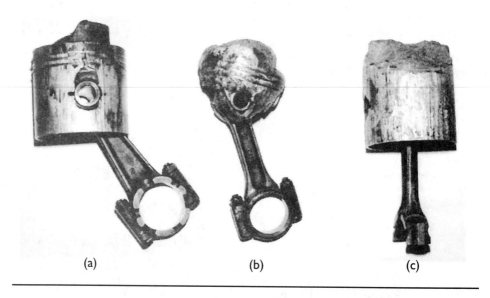

(a)   (b)   (c)

Figure 3–19 Piston retainer came loose (a); piston skirt broke operation continued to return to port (b); detonation with deflector erosion with ring and skirt damage (c).

When installing a new piston ring, make sure the groove has been cleaned of carbon then measure the ring side clearance against specifications. Also check to see if the ring pins are there and that they have not loosened. Measure the skirt to see if the piston is collapsed. Compare the readings against other pistons if the factory service manual does not have a specification for the piston diameter. Check piston pin retainer grooves for evidence of the retainer (locks) moving as they may have been distorted. Always replace the retainers once they have been removed. If there is evidence of wear in any of these areas, the piston should be replaced. Look at the piston pin for wear in the bearing area. Rust marks caused by water moisture will leave a needle bearing imprint. Chatter marks on the pin indicate that the piston pin should be replaced. If these marks are not too heavy, they may possibly be cleaned up with 320 emery paper for loose needle bearings, or with crocus cloth for caged bearings.

3.4   Piston Ring

The piston ring is a seal itself, just as other seals are used on the crankshaft and lower unit. To perform correctly, the ring must conform to the cylinder wall and maintain adequate pressure to insure its sealing action at required operating RPMs and temperatures. There are different designs used throughout the outboard industry (see Figure 3–20). A given manufacturer will select a ring design that will meet the operating requirements of its powerhead. This may be a standard ring or perhaps a pressure back (keystone) ring or a combination of rings.

3.5   Ring Travel

Let us put ring travel within the cylinder into perspective. In general, a single-cylinder, small-bore outboard running at 3600 RPM and consuming one gallon of fuel would need 10,000 gallons of air (equal to a $12 \times 12 \times 8$-foot room), with the air going into the carburetor at a rate of 24 MPH. The piston would be moving at a rate of 7,200 feet per minute, which means that the rings are also moving on the cylinder wall at the same rate. With these approximates in mind, it is easy to see why good surface for ring travel is important and why ring design selection for the powerhead is essential.

The function of the piston rings include sealing the combustion gases so they cannot pass between the piston and the cylinder wall into the crankcases upsetting the pulse, and maintaining an oil film in conjunction with the cylinder wall finish throughout the ring travel area. This prevents boundary lubrication on the ring bearing areas. The rings also transfer heat picked up by the piston head during combustion. This heat is transferred into the cylinder wall and thus to the cooling system. There are either two or three rings per piston that perform these functions. Since this powerhead is of the two-stroke design, we do not have an oil control ring. The rings are all of the compression type. This means that they are for sealing the clearance between piston and cylinder wall. They are not allowed to rotate on the piston as four-stroke piston rings may. They are prevented from rotating by a pin in the piston ring groove (Figure 3–8). (Some small horsepower powerheads do not have pins in the ring groove.)

If the ring was allowed to turn on larger bore powerheads, a ring end could snap into the cylinder port and become broken. The ring ends are specially machined to compensate for the pin. As the rings warm up in a running powerhead, they expand, thereby requiring a specific end gap between the ring ends for expansion. This ring gap decreases upon warm-up, effectively limiting blow-by gases (from the combustion process) from going into the crankcase. The rings ride in a piston ring groove with minimal side clearance, which gives them support as they move in and out on the cylinder wall. With this support, combustion gas pressure and oil effectively seal the piston ring against the ring land and the cylinder wall. As long as the oil mix is correct and temperatures remain where they should, the rings will provide service for many hours of operation.

3.5.1   Troubleshooting the Piston Ring

One of the first indications of ring trouble is the loss of compression and performance. When compression has been lost or lowered because of the ring not sealing, the ring is either stuck with carbon, gum, or varnish, or it is broken. Improper oil mixing and stale gasoline provide the carbon, gum, and varnish that cause the rings to stick. Low octane fuel, misadjustments in the *timing/synchronization*, and lean fuel mixtures can damage the ring land and the ring because of high temperatures. Running the outboard out of water for even a few seconds can have a damaging effect on the rings, pistons, cylinder walls, and water pump.

Outboard motors use a selection of ring designs. Examine the rings and refer to Figure 3–20. To determine if the rings fit the the cylinder and piston, two measurements are taken—ring gap and ring side clearance. The ring is pushed into the cylinder bore using the piston skirt (or flat head piston), so it will be square. Position each ring, one at a time, at the bottom of the cylinder (the smallest diameter), and measure the expansion space between the ring ends. This is known as the *ring gap measurement* (Figure 3–20). Check against the manufacturer's specifications and refer to ring instructions on the package for proper installation. Next, position each ring in the piston ring groove and measure side clearance between the ring and the piston ring land (Figure 3–21). Check this measurement against specifications. Use the feeler gauge for these measurements. You should also check the ring land condition.

After the ring gap measurements are taken, install the rings according to the instructions. When the cylinder bore has passed the white-cloth test (rubbing a white rag in the cylinder), dunk the assembled piston and rings in a can of 50/1 oil to thoroughly saturate the assembly. This oil lubricates the assembly for initial start-up, and will prevent scuffing of the rings on the cylinder wall.

 **Warning**

Now that you are ready to install the piston assembly, *do not use an automotive-type ring compressor!* It will not work and will allow the ring to turn over the pin and become broken.

Ring compressors are available from the powerhead manufacturer, or use ice cream sticks (a blunt instrument that won't damage the ring), keeping the ring ends over the pins.

Use these blunt instruments to push in on each ring to close the gap and push the piston down one ring *with your hand*. Proceed to the next ring and repeat the process until all the rings are installed. You will probably need some helping hands to do this. With the piston and rings installed, push the piston down until the rings are visible in the ports. Using a

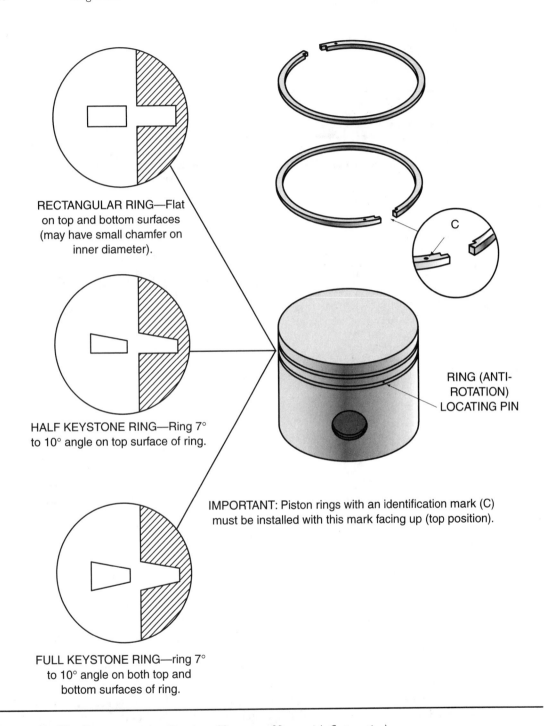

RECTANGULAR RING—Flat on top and bottom surfaces (may have small chamfer on inner diameter).

HALF KEYSTONE RING—Ring 7° to 10° angle on top surface of ring.

FULL KEYSTONE RING—ring 7° to 10° angle on both top and bottom surfaces of ring.

RING (ANTI-ROTATION) LOCATING PIN

IMPORTANT: Piston rings with an identification mark (C) must be installed with this mark facing up (top position).

Figure 3–20   Piston ring identification. *(Courtesy of Brunswick Corporation)*

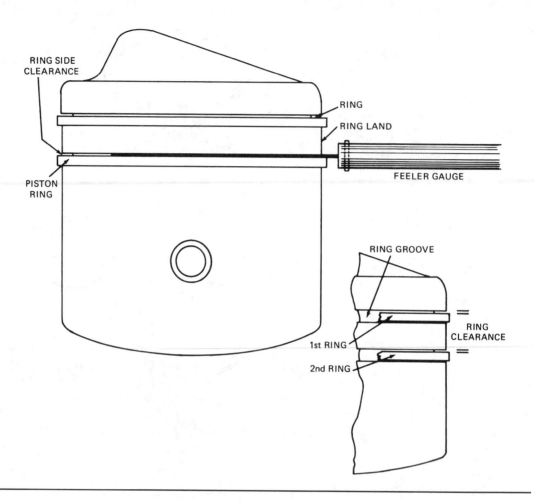

Figure 3–21   Measuring standard ring side clearance with clearance exaggerated.

blunt instrument, push on each ring through the port and see if it springs in and out. If it does, the ring is not broken. If it does not spring in and out, the ring was broken during installation and must be replaced.

For diagnostic work pull the bypass cover (transfer passage cover not on all engines). This will expose the intake ports through which the rings can be seen, and a blunt instrument can be used to see if the rings are stuck or broken. You will also be able to check the piston skirt and see if there is damage. To expose the exhaust side of the piston, the exhaust cover (not on all engines) has to be removed. On Mercury powerheads there is no satisfactory way around the piston ring compressor tool. You must use Mercury's ring compressor, which is available through the manufacturer, or take this phase of the repair to the Mercury dealer.

3.6   **Connecting Rods**

The connecting rod transfers the combustion pressure from the piston pin to the crankshaft, changing the horizontal motion into rotary motion. In doing so, the connecting rod swings back and forth on the piston pin like a pendulum, while it is moving in and out. It goes out by combustion pressure and goes in by flywheel momentum and/or other engine power strokes on a multi-cylinder engine. The connecting rod can be of aluminum on smaller horsepower fishing outboards or of steel on larger horsepower models.

If the rod is made of aluminum, it may be of plain bearing design. Another design uses a steel liner with needle bearings in the large end and a pressed-in needle bearing in the small end (Figure 3–22).

On the large end of the rod, a removable cap allows for attachment of the rod to the crankshaft. The parting surfaces of the rod and cap need some consideration. They have a machined parting surface (line) and the cap and rod both have match marks for rod cap alignment to the rod. The cap can be put on backwards with the misalignment of inner bore. Be very sure that the match marks, which are dots, casting protrusions, knob marking, or etched marks are matched together to ensure inner-bore alignment of the large end of the rod (see Figures 3–23 and 3–24).

To obtain this alignment the rod bolts have to be torqued to specification so the parting surfaces and rod material will be properly stressed. The rod bolts are special bolts which can take the specified torque.

The steel rod is a bearing race at both the large and small ends of the rod. It is hardened to withstand the rolling pressures applied from the loose or caged needle bearings. These rods also have match marks that have to be aligned. The large end has the cap which has to be absolutely aligned upon reassembly. There is no bearing insert and the bearings run directly on the hardened rod and cap and over parting lines. On one style of OMC connecting rod the parting line is a fracture. This means that the rod cap has been broken from the rod

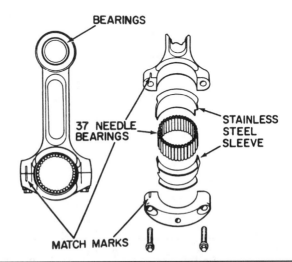

Figure 3–22   Aluminum connecting rod with sleeve and needle bearings. *(Courtesy of Tecumseh Products Company)*

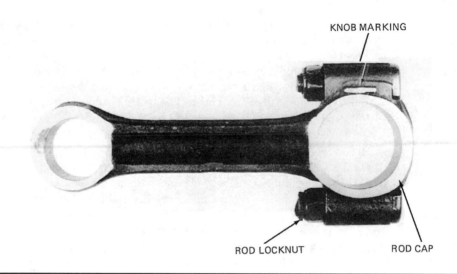

Figure 3–23 Knob marking on Mercury connecting rod.

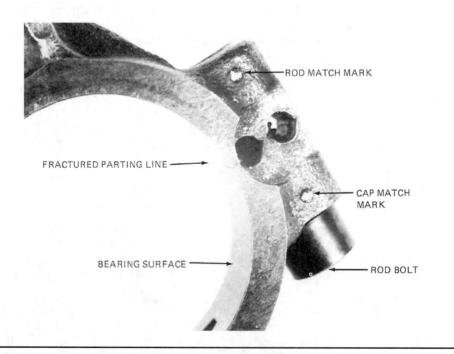

Figure 3–24 Match marks on OMC connnecting rod.

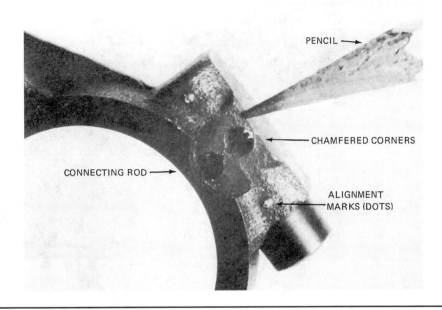

Figure 3–25   OMC fractured connecting rod.

during manufacture, after the large end inner bore was machined. Upon reassembly, the inner bore has to be put back in its pre-broken position. To aid the mechanic in the re-assembly, there are chamfered marks machined on the corners of the large end of the rod. All four chamfers must be aligned. To determine alignment a lead pencil or scribe is run over the parting lines at the chamfers (all four) to see if a step can be felt (Figure 3–25).

If a step is evident, then the rod cap is not in alignment to the rod. The needle bearings on the inner bore would be forced to run over the step, damaging the rod and needle bearings. This would cause early failure of the bearing. When the cap is in alignment, you will feel no step, it will be smooth. Therefore, alignment of the chamfers on this style of steel rod is very important. (An alignment tool is available for newer 40 through 50 HP models.)

 Note

Do not torque the rod bolts down until rod cap alignment is made.

Keep the bolts finger-tight and shift the cap back and forth until alignment is made. Then torque the rod bolts in two stages to specifications. If the rod bolts are torqued before alignment of the chamfers is obtained, the high points of the fractured parting line will be damaged, and true alignment will be nearly impossible to achieve. Again, the rod bolts are special so don't substitute them (Figure 3–26).

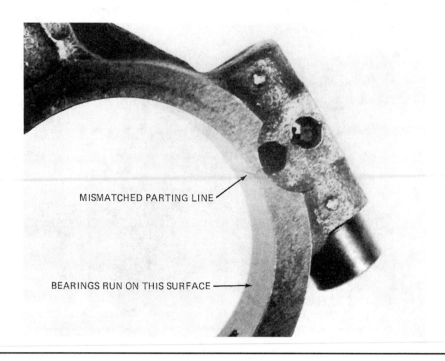

MISMATCHED PARTING LINE

BEARINGS RUN ON THIS SURFACE

Figure 3–26 Misalignment of rod and cap.

Another type of steel rod has machined parting lines, making it easier to refit the cap to the rod. This type of rod relies upon precision drilling and mounting of the rod bolts passing through the cap. Shoulders on these bolts and etch marks precisely align the cap to the rod. Self-locking rod nuts are used to secure the cap to the rod (see Figure 3–27). It is recommended that once the rod nuts have been used, you should not reuse them. (Install new ones!) The plastic material used for locking has been damaged and may not hold the nuts. This could result in major damage to the powerhead. The rod is a hardened bearing race and must be aligned and handled like a bearing. Remember this key point when working on these steel rods!

All caps and rods are machined as one piece. Therefore the caps cannot be interchanged from one rod to another. Upon disassembly, place the cap back on the rod and mark the rod cylinder number immediately so there will be no mix-up. Do the same for the needle bearings—they must stay with a given rod once run on that bearing surface.

Let us consider rod design for aiding lubrication. Some of the rods have a trough design in the shank area. Oil holes may be drilled into the bearing area at both ends of this trough (see Figure 3–28).

Oil mist that falls out of the fuel will settle into the rod trough and collect. As the rod moves in and out of the cylinder, the oil is sloshed back and forth in the trough and out the oil holes into the rod or pin bearings. This provides sufficient lubrication for these bearings. When the rod is equipped with oil holes, the holes have to be placed in the upward position toward the tapered end of the crankshaft when reassembled.

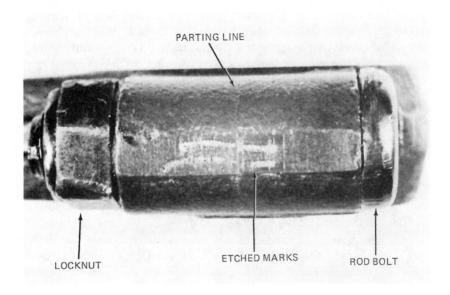

Figure 3–27   Etched alignment marks, Mercury connecting rod.

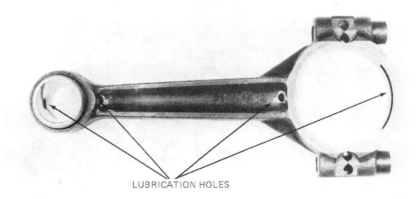

Figure 3–28   Lubrication holes in OMC connecting rod.

3.6.1   Troubleshooting the Rods

Damage to the aluminum rod can be caused by lack of lubrication, and will result in galling of the plain bearing, with seizing to the crankshaft. A break may occur at the large end. Improper torque to the rod bolts results in damage to the parting line area and threads and damage to the large end. Over speeding of the powerhead may also cause the upper shank area of the rod to stretch and break near the piston pin.

Steel rods are inspected in the bearing areas, much like you would inspect a roller bearing. Look for scoring, pit marks, chatter marks, rust, and color change (Figure 3–29). A blue color indicates overheating of the bearing surface. Minor rust marks or scoring may be cleaned up using crocus cloth for caged needle bearings or 320 emery paper for loose needle bearings. A piece of round stock, cut with a slot in one end to accept a small piece of emery paper and mounted in a drill motor, can be used to clean up the rod ends (see Figures 3–30 and 3–31). This is a clean-up operation, not an attempt to remove material from the rod.

The rod also needs to be checked to see if it is bent or twisted. To do this, remove the piston and place the rod on a surface plate or a piece of flat glass (automotive window). Using a flashlight behind the rod and looking from the front of the rod, check for any light which can be seen under the rod ends. If light can be seen shining under the rod ends, the rod is bent and it must be replaced. You can also use a .002 feeler gauge (Figure 3–31). See if it will *start* under the machined area of the rod. If it will, the rod is bent. Examine the rod bolts and studs for damage and replace the nuts where used. Always reinstall the rod back on the same crankpin journal from which it was removed. The needle bearings, rod bearing surface, and crankpin journal are all mated to each other once the powerhead has been run.

(a)

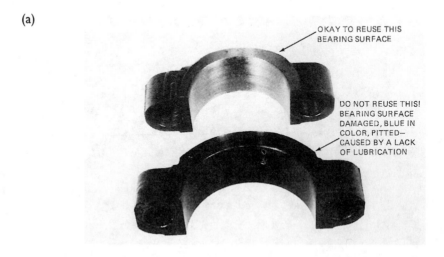

Figure 3–29   Rod and cap damaged by lack of lubrication and overheating (a); water-damaged rod and main needle bearings (b) (continued on next page).

(b)

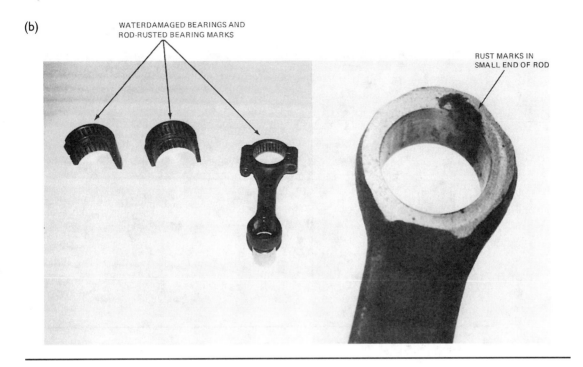

WATERDAMAGED BEARINGS AND
ROD-RUSTED BEARING MARKS

RUST MARKS IN
SMALL END OF ROD

Figure 3–29   Continued.

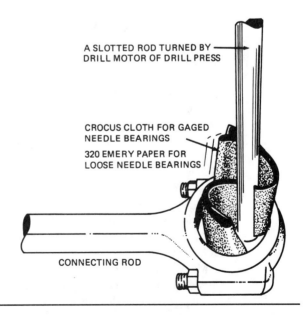

A SLOTTED ROD TURNED BY
DRILL MOTOR OF DRILL PRESS

CROCUS CLOTH FOR GAGED
NEEDLE BEARINGS

320 EMERY PAPER FOR
LOOSE NEEDLE BEARINGS

CONNECTING ROD

Figure 3–30   Cleaning connecting rod bearing surface (crankshaft end shown).

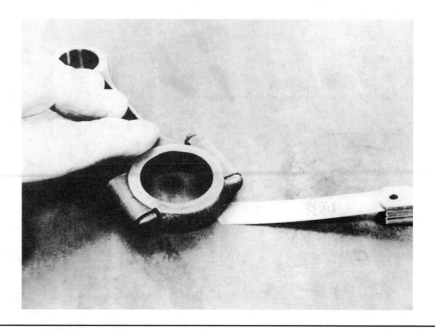

Figure 3–31   Check for bent or twisted rod.

3.7   Rod and Main Bearings

Needle *bearings* are used to carry the load which is applied to the piston and rod. This load is developed in the combustion process and the bearings reduce the friction between the crankshaft and the connecting rod. They roll with little effort and at times have been referred to as anti-friction bearings, since they reduce friction by reducing the surface area that is in contact with the crankshaft and the connecting rod. Needle bearings are of two types, loose and caged. When loose bearings are used, there can be up to 32 loose bearings floating between the crankpin and the connecting rod. These bearings are aided in rolling by the movement of the crankpin journal and the connecting rod pendulum action. The surface installed on the journal and rod encourages needle rotation because of its relative roughness. If the journal and rod surface were polished with crocus cloth, the loose needle bearings would have the tendency to scoot, wearing both surfaces. So, journals of rods which use the loose needle bearings are cleaned up using 320 grit emery paper.

Caged needle bearings use a reduced number of needles, and the needles are kept separated and are encouraged to roll by the cage. The cage also controls end movement of the bearings. Because of the cage, the journal and rod surfaces can be smoother, so these surfaces are polished with crocus cloth.

Main bearings are used to mount and control the axial movement of the crankshaft. They are either ball, needle, or split race needle bearings. The split race needle bearings are held together with a ring and are sandwiched between the crankcase and cylinder assembly.

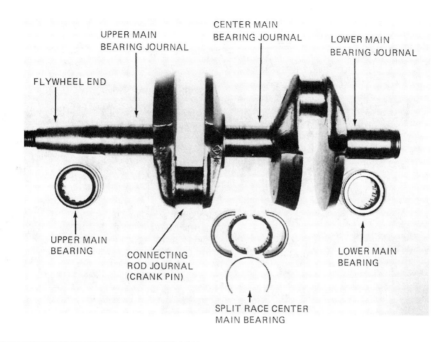

UPPER MAIN
BEARING JOURNAL

CENTER MAIN
BEARING JOURNAL

LOWER MAIN
BEARING JOURNAL

FLYWHEEL END

UPPER MAIN
BEARING

CONNECTING
ROD JOURNAL
(CRANK PIN)

LOWER MAIN
BEARING

SPLIT RACE CENTER
MAIN BEARING

Figure 3–32   Crankshaft mounted in needle bearings.

The split race bearings are commonly used as center main bearings, as this is the only type of bearing that can be easily installed in this location. The ball bearings may be mounted as top or bottom mains on the crankshaft (Figures 3–32 and 3–33).

The bearing is made up of three parts—the inner race, needle, and the outer race. In most industrial applications, the outer or inner race of a needle bearing assembly is held in a fixed position by a housing or shaft. The connecting rod needle bearings in the outboard powerhead have the same basic parts, but differ in that both inner and outer races are in motion. The outer race—the connecting rod—is swinging like a clock pendulum. The inner race—the crankshaft crankpin—is rotating and the needle bearing is floating/rolling between the two races.

3.7.1   Troubleshooting the Bearings

When the powerhead is disassembled and inspection of the parts is made, then by necessity, along with examining the needle bearings, the crankshaft main bearing journal, rod journal, and connecting rod bearing surfaces are also examined. The surfaces of all three of these parts can impart information to the technician, and the examination will determine if the parts are reusable. The surfaces are examined for scoring, pitting, chatter marks, rust marks, spalling, and discoloration from overheating of the bearing surfaces (see Figure 3–34). Minor scoring or pitting and rust marks may be cleaned up and the surfaces brought

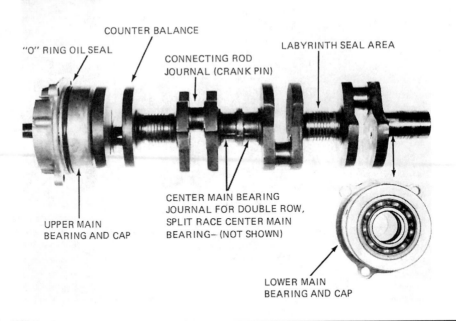

COUNTER BALANCE

"O" RING OIL SEAL

CONNECTING ROD
JOURNAL (CRANK PIN)

LABYRINTH SEAL AREA

CENTER MAIN BEARING
JOURNAL FOR DOUBLE ROW,
SPLIT RACE CENTER MAIN
BEARING– (NOT SHOWN)

UPPER MAIN
BEARING AND CAP

LOWER MAIN
BEARING AND CAP

Figure 3–33 Crankshaft mounted with ball bearings and center main needle bearing.

back to a satisfactory condition. This is done using crocus cloth for caged needle bearings and 320 grit emery paper for loose needle bearings. This is not a metal-removing process, rather just a clean-up of the surfaces.

Needle bearings are used as main bearings and are inspected for the same conditions as listed earlier. There are no oversize bearings available for rod or main bearings. Because of the hardness (Rockwell C60-63) of the crankshaft, which in reality is a bearing race, it should not be turned or welded up in order to bring it back to standard size. The welding process may stress the metallurgical properties of the crankshaft, developing cracks.

The caged rod bearings and split race main bearings are inspected for the same condition as loose needle bearings, plus the cage is examined for wear, cracks and breaks.

Ball bearings are used for top and bottom main bearings in some powerheads. These may be pressed onto the crankshaft or pressed into the end cap. To examine these bearings, wash, dry, oil, and check them on the crankshaft or in the bearing cap. Turn the bearing by hand, and feel if there is any roughness or catching. Try to wobble the bearing by grasping the outer race (inner race), checking for looseness of the bearing. Replace the bearing if any of these conditions are found. If the bearing is pressed off (out), the bearing will probably be damaged, and should be replaced.

Consult the service manual when installing a bearing. Check to see if the bearing should be installed with the bearing numbers up or down, and if the snap ring should be placed up or down when installing the split race center main bearing.

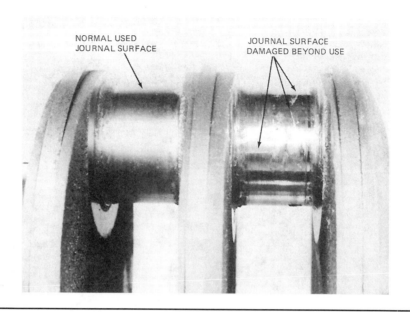

NORMAL USED
JOURNAL SURFACE

JOURNAL SURFACE
DAMAGED BEYOND USE

Figure 3–34   Damaged crankshaft connecting rod journal.

If new or used bearings are contaminated with grit or dirt particles at the time of installation, abrasion will naturally follow. Many bearing failures are due to the introduction of foreign material into the internal parts of the bearing during assembly. Misalignment of the rod cap, torque of the rod bolts, and lack of proper lubrication also cause failures. Bearing failure is usually detected by a gradual rise in operating noise, excessive looseness (axial) in the bearing, and shaft deflection. Keep the work area clean and use OMC needle bearing grease or Mercury multipurpose grease to hold the bearings in place. This grease will dissipate quickly as the fuel mixture comes in contact with it. Do not use an automotive wheel bearing or chassis grease as this will cause damage to the bearings. Oil the ball bearings with two-cycle oil upon installation. Remember to keep them clean.

3.8   **The Crankshaft**

The crankshaft is used to convert horizontal motion received from the mounted connecting rod into rotary motion, which turns the drive shaft. It mounts the flywheel, which imparts a momentum to smooth out pulses between power strokes. The crankshaft also provides sealing surfaces for the labyrinth seals to hold oil against and a groove in which sealing rings are installed to seal pressures into each crankcase. Mounted main bearings control the axial movement of the crankshaft as it accomplishes these functions. The crankshaft bearing journals are case-hardened to be able to withstand the stresses applied by the floating needle bearings used for connecting rod and main bearings. In essence, the crankshaft journals are the inner bearing surfaces for the needle bearings (Figure 3–35).

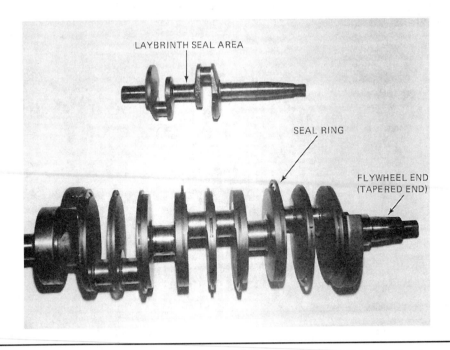

Figure 3–35   V-6 and two-cylinder crankshafts.

3.8.1   Troubleshooting the Crankshaft

Pressure from the power stroke applied to the crankshaft rod journal by the needle bearings has a tendency to wear the journal on one side. During crankshaft inspection, the journals should also be measured with a micrometer to determine if they are round and straight. They should be inspected for scoring, pitting rust marks, chatter marks, and discoloration caused by heat. Check the sealing surfaces for grooves worn in by the upper and lower crankshaft seals. Take a look at the splined area which receives the drive shaft. Inspect the side of the splines for wear. This wear can be caused by lack of lubrication or improper lubricant applied during a seasonal service. An exhaust housing/lower unit that has received a sudden impact can be warped, and this can also cause spline damage in the crankshaft.

The crankshaft cannot be repaired because of the case hardening and the possibility of changing the metallurgical properties of the material during the welding and machining operation. Also, there are no oversized bearings available. Repairs are limited to cleaning up the journal surface with 320 grit emery paper when loose needle bearings are run on the journal. Where caged bearings are run, the journal may be polished with crocus cloth. For more detailed information read the section on "Rod and Main Bearings" within this chapter.

The tapered end of the crankshaft has a spline or a keyway and key which times (positions) the flywheel to crankshaft. Inspect the spline, key, and keyway for damage. The crankshaft taper should be clean and free of scoring, rust, and lubrication. This taper must match the flywheel hub. If someone has hit the flywheel with a heavy hammer or has used an improper puller to remove the flywheel, the flywheel and hub may be warped. Place the flywheel on the tapered end of the crankshaft and check the fit. If there is any rocking, indicating a distorted hub, replace the flywheel. Always use a puller that pulls from the bolt pattern or threaded inner hub of the flywheel.

 Warning

Never use a puller on the outside of the flywheel! (see Figures 3–36 and 3–37).

The taper is used to lock the flywheel hub to the crankshaft. When mounting the flywheel to the crankshaft, the taper on the crankshaft and in the flywheel hub must be cleaned with fast evaporating solvent. *No lubrication should be done on the crankshaft taper or flywheel hub*. The flywheel nut must be torqued to specification to obtain a press fit between the flywheel hub and the crankshaft taper. If the nut is not brought to specifications, the flywheel may spin on the crankshaft causing major damage.

The flywheel key is for alignment purposes and sets the flywheel's relative position to the crankshaft. Check the key for partial shearing on the side. If there is any indication of shearing, replace the key. Also check the keyway in the flywheel and crankshaft for damage. If there is damage which will allow incorrect positioning of the flywheel, the powerhead timing will be off (Figure 3–38). If damage did occur, try lapping the flywheel hub to the crankshaft mating surface. The crankshaft/flywheel key is to be installed parallel to the center line of the crankshaft.

Figure 3–36   Universal puller installed on OMC flywheel.

Figure 3–37   Puller installed on Mercury flywheel.

DAMAGE TO TAPERED
END OF CRANKSHAFT

DAMAGE TO FLYWHEEL HUB

Figure 3–38   Flywheel spun on crankshaft at high speed—caused by under-torque of flywheel nut.

3.9   Lubrication

Recommendations for lubrication of the powerhead are basically the same from all manufacturers—either use the manufacturer's recommendation or, when not available, use a NMMA-certified TC-W3 oil from the after-market. This means the National Marine Manufacturers Association engineering committees have developed procedures to certify the oil. Companies using those procedures then certify that their oil is compounded for two-cycle water-cooled (TC-W3) outboard motors. The National Marine Manufacturers Association is the umbrella name for organizations dedicated to expanding, improving, and servicing the boating industry. Look on your oil container and make sure it states that the oil is NMMA, TC-W3. This classification of oil is "ashless," designed specifically for the water-cooled two-stroke outboard motor. Such oil will maintain lubrication for high-power output, protecting cylinders, pistons and those anti-friction needle roller bearings on the crankshaft. It will minimize carbon build-up on the piston head, and control varnish build-up in the crankcase, keeping the piston rings free so they can seal. This TC-W3 oil also has rust-preventive properties, which will prevent or reduce rusting of the needle bearings, crankshaft, and connecting rods. Rusting of these vital parts can begin upon cooldown of the powerhead. Protection against rusting is necessary for powerhead longevity. NMMA, TC-W3 oil is blended to mix easily, and will not separate easily from the gasoline when in temporary storage. However, it will not prevent gasoline from going stale. Use fuel stabilizer products to maintain gasoline in fixed tanks throughout the winter.

There is a two-cycle oil (TC) for air-cooled engines available. This oil is for air-cooled outboards, chain saws, and motorcycles. It should not be substituted for NMMA, TC-W3 oil. TC-W3 oil has a higher viscosity content, called "Hi Vis." In other words, it has a stronger oil base. TC oil uses a "cut-back oil" thinned by solvents, therefore having a weaker oil base.

Once inside the crankcase of the powerhead, the oil turns into a mist and is subject to positive and negative pressures. In mist form, the oil bathes and lubricates the cylinder walls, piston, crankshaft, connecting rod, and bearings. It is extremely important to measure your oil correctly for the proper ratio of oil and gasoline. No guesswork here! Insufficient oil will cause boundary lubrication (a rupture of the oil film), which will cause galling of the piston skirts and cylinder walls, and damage to the bearings and crankshaft. Many people tend to make the fuel mix rich with oil. Most of the time the outboard is operating below maximum RPM (e.g., when trolling or in controlled speed limits), and is actually running with too much oil in the crankcase for the lower RPMs. When an inaccurate measure of oil is added and you think that adding more oil will be better, problems are created because too much oil is falling out of the air stream and is really over-flooding the moving parts of the crankcase. This creates a mixture in the combustion chamber that is harder to ignite. Among the problems that can develop are carbon build-up in the exhaust ports and combustion chamber, fouled plugs, low-speed miss, and an outboard that won't idle smoothly. The designated fuel/oil ratio (50/1) for most powerheads will supply adequate lubrication for maximum sustained operating RPM without powerhead damage. So, when filling your tank or adding to it, measure fuel/oil exactly, and you won't have to worry about lubrication or lubrication-related problems. Many larger late-model outboards use an oil injection system, which varies the oil ratio to match RPM.

 **Warning**

If filling a portable fuel tank, remove it from the boat or truck and place it on the ground to discharge any static electricity build-up. Keep fuel nozzle in contact with fuel tank! An arc from static electricity can cause an explosion and/or fire.

Automotive oils of multi-grade should not be used. Straight-grade automotive oil can be used in a pinch. Remember that automotive oil really isn't designed to be burned, as is the case with two-cycle oil. It leaves unwanted deposits in the combustion chamber. You really can't beat today's two-cycle TC-W3 oil for water-cooled two-stroke outboards. It gives excellent lubrication if properly mixed, and adds many hours of operating time to your outboard powerhead (Figure 3–39).

3.10   The Economixer (OMC)

The OMC economixer was developed to fill the need for an accurate oil mix and to offset problems that arise when the outboard is placed on a larger boat using fixed fuel tanks, as well as the Sea Drive unit. Using microprocessor technology, a variable ratio oil injection system was developed. This electronic oil injection system is able to vary the oil mix from 50/1 at wide open throttle to 150/1 at idle. Oil is injected into the fuel line with each revolution of the crankshaft based on throttle opening (See Figures 3–40 and 3–41).

The pump is married to the electronic control module and gauge, which is a status indicator. The status indicator has a yellow warning light, a flashing red light and a horn to get the attention of the operator when there is a problem. The yellow light indicates low oil level. When the yellow light comes on, there is approximately one-third of a tank of oil left. When the flashing red light, horn, and yellow light are on, the powerhead will stop in 30 seconds from when the warning began. This indicates that there is no oil. The powerhead can be restarted repeatedly, but will stop after each 30-second run. The flashing red light and horn are indications that the system is not delivering sufficient oil. Again, as in the previous example, the powerhead will stop every 30 seconds, and while it can be restarted, it will run for only another 30 seconds. This shutdown is designed, of course, to save the powerhead.

The microprocessor is programmable for various horsepowers, from V-4 up to V-6. Programming is done on top of the microprocessor by positioning the wires and socket on the pins according to the letter codes for a given horsepower.

The microprocessor must receive information from four systems on the powerhead. B-plus (battery) voltage indicates to the economixer that it is time to test and get ready to operate. The CD ignition system tells the economixer that the powerhead is running. A pulse from the alternator charging system (tach signal) tells the economixer exactly how fast the powerhead is running. A rheostat on the upper carburetor linkage gives information on throttle opening, and therefore powerhead load. This determines how much oil is to be pumped. Powerhead load is related to the propeller pitch, hull condition, passenger load, and other factors.

Because the B-plus signal is electrical, there is a possibility that it can fail. In this case,

CRANKSHAFT & PISTON

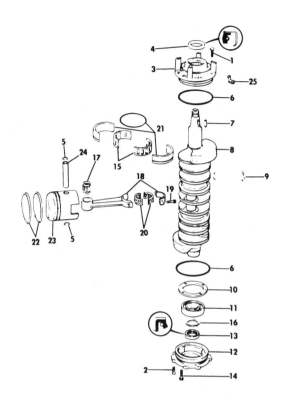

OP0922C

832241

Ref. No.	P/N	Name of Part	Qty.	Ref.No.	P/N	Name of Part	Qty.
1	306049	SCREW, Crankcase head upper	4	14	317763	SCREW, Plate	4
2	308742	SCREW, Crankcase head	8	15	387041	RETAINER & NEEDLE SET	2
3	387432	CRANKCASE HEAD AND BRG.		16	308213	RING, Retaining, bearing	1
		ASSY.	1	17	388595	BEARING, Wrist pin	6
4	322575	. SEAL, Upper	1	18	393754	CONNECTING ROD ASSY.	6
5	317831	RETAINING RING	12	19	308186	. SCREW, Rod cap	2
6	305123	O-RING	4	20	390824	RETAINER SET	6
7	307480	KEY, Flywheel	1	21	320499	SPLIT SLEEVE, Center bearing	2
*8	388306	CRANKSHAFT ASSY.	1	22	390509	RING SET, Std. 2.6 litre	6
9	319244	. SEAL RING, One stripe	AR	22	390510	RING SET, .030, 2.6 litre	6
10	317762	PLATE, Lower bearing	1	23	394428	PISTON, Std. 2.6 litre	6
11	385503	BEARING, Ball	1	23	394507	PISTON, .030 2.6 litre	6
12	385502	CRANKCASE HEAD ASSY.	1	24	327861	WRIST PIN	6
13	321668	. SEAL	1	25	326439	CLAMP, Retainer to crankcase head	4

· **Note** Lap fit the flywheel taper to crankshaft taper with valve lapping compound. Use OMC Ultralock P/N 388517 on taper. Use OMC Nutlock P/N 384849 on crankshaft thread. Use a new locknut whenever service requires removal of flywheel.

Figure 3–39   Crankshaft and piston (continued on next page). *(Courtesy of Outboard Marine Corporation)*

CYLINDER & CRANKCASE

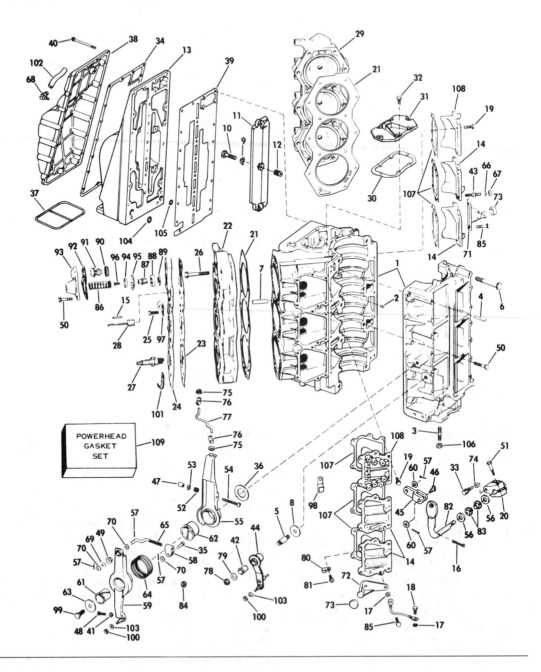

Figure 3-39 Cylinder and crankshaft (continued on next page).

CYLINDER & CRANKCASE

OP0923C

Ref. No.	P/N	Name of Part	Qty.	Ref. No.	P/N	Name of Part	Qty.
1	394533	CYLINDER C'CASE ASSY., 2.6 litre	1	52	308993	SPRING, Idle adjust	1
2	304178	. PIN, Dowel bearing	2	53	203688	NUT, Idle adjust	1
3	314616	. STUD	2,4	54	328065	SCREW, Idle adjust	1
4	305070	. PIN, Taper	2	55	320836	LEVER, Spark advance	1
5	320808	. STUD, Shift lever	1	56	202026	WASHER, Shift lever	2
6	316988	. SCREW, Lockplate cyl. to c'case	8	57	305650	COTTER PIN	5
7	322411	. DEFLECTOR, Water passage	6	58	310524	YOKE, Throttle cam	1
8	320744	. WASHER, Stud	1	59	389821	LEVER, Throttle	1
9	309577	. LOCKPLATE	2	60	302474	WASHER, Link, Shift arm	2
10	324565	. SCREW	6	61	320811	BUSHING	1
11	329775	. FILLER BLOCK	2	62	320812	BUSHING, Spark lever	1
12	327323	. INSERT	6	63	306124	WASHER, Throttle lever	1
13	329379	EXH. COVER, Inner, 2.6 litre	1	64	321270	SPRING, Throttle lever	1
14	392860	COVER, Bypass	4	65	320951	LINK, Throttle lever	1
15	303806	TUBING	2	66	303480	LOCKWASHER, Bolt double end	1
16	120356	COTTER PIN, Shift lever	1	67	301472	NUT, Double end bolt, bypass cover	1
17	303480	LOCKWASHER, Ground strap	4	68	321886	ELBOW, Indicator	1
18	376349	STRAP, Ground	1	69	204757	SPRING WASHER	1
19	306049	SCREW, Bypass cover	31	70	302474	WASHER, Link	3
20	326644	LEVER, Shift rod	1	71	319217	SUPPORT BRACE, Port	1
21	329774	GASKET, Cylinder head, 2.6 litre	2	72	319218	SUPPORT BRACE, Starboard	1
22	325639	CYL. HEAD, Stbd., 2.6 litre	1	73	313581	BUMPER, Support brace	2
23	321028	GASKET, Cover	2	74	302290	WASHER, Pin, Lever	1
24	320840	COVER, Cylinder head	2	75	313700	SOCKET, Linkage	2
25	306409	SCREW, Cylinder head cover	40	76	313703	BALL, Socket linkage	2
26	307267	SCREW, Cylinder head	28	77	320950	ROD, Spark advance	1
27	384771	SPARK PLUG, Champion UL-77V	6	78	315077	NUT, Shift lever	1
28	386686	TEMPERATURE SWITCH	2	79	309335	WASHER, Nut, shift lever	1
29	325638	CYL. HEAD, Port 2.6 litre	1	80	309322	CLAMP, Stator leads	1
30	322858	GASKET, Water passage cover	1	81	308762	SCREW, Clamp	1
31	323896	COVER, Water passage	1	82	324362	SHIFT LEVER	1
32	306049	SCREW, Water passage cover	6	83	312708	BUSHING, Shift lever	2
33	327590	PIN, Shift rod lever	1	84	301472	NUT, Rod, lever to cam	1
34	323217	GASKET, Exhaust cover, 2.6 litre	1	85	306418	SCREW, Support bracket to bypass cover	4
35	310525	PIN, Throttle cam to yoke	1	86	317245	SPRING, Pressure release valve	2
36	320813	WASHER, Spark advance lever	1	87	321027	VALVE, Pressure release	2
37	323625	SEAL, Exhaust manifold, 2.6 litre	1	88	320880	SEAT, Valve	2
38	323213	COVER, Exhaust outer, 2.6 litre	1	89	321222	WASHER, Valve	2
39	330291	GASKET, Inner cover, 2.6 litre	1	90	310058	GROMMET, Thermostat	2
40	323218	SCREW, Exh manf to cyl., 5" long	6	91	378065	THERMOSTAT ASSY.	2
40	909533	SCREW, Exh manf to cyl., 1" long	2	92	321184	GASKET, Thermostat cover	2
40	319126	SCREW, Exh manf to cyl., 1-3/8" long	4	93	321013	COVER, Thermostat	2
40	323626	SCREW, Exh manf to cyl., 1-7/8" long	4	94	321220	RETAINER, Spring valve	2
40	317972	SCREW, Exh manf to cyl., 4-1/4" long	6	95	321221	SEAL, Pressure release valve	2
40	324424	SCREW, Exh manf to cyl., 2-5/8" long	4	96	306643	SCREW, Seal to valve	2
41	204046	NUT, Throttle stop	1	97	309322	CLAMP, Coil leads	8
42	308532	BUSHING, Shift lever	2	98	311339	CLAMP, Leads to bypass cover	4
43	322068	BOLT, Double end, bypass cover	1	99	306834	BOLT, Throttle lever retainer	1
44	389438	SHIFT ARM ASSY.	1	100	328064	NUT, Casing guide to throttle arm & shift lever	2
45	324363	LINK, Shift arm	1	101	310439	CLAMP	4
46	324364	PIN, Shift arm link	1	102	327720	HOSE, Elbow to nipple	1
47	318660	CAP	1	103	306015	WASHER, Shift & throttle lever	2
48	316602	SCREW, Throttle stop	1	104	329380	O-RING, Manifold	1
49	314125	BUSHING, Link	1	105	329381	O-RING, Manifold	1
50	306778	SCREW	20	106	315204	NUT, Powerhead to adapter	2,4
51	323480	SCREW, Lever shift rod	1	107	307133	GASKET, Bypass cover	6
				108	392861	COVER, Bypass, upper	2
				109	394855	POWERHEAD GASKET SET, (Cylinder head gasket not incl)	1

Figure 3–39   Cylinder and crankshaft (continued).

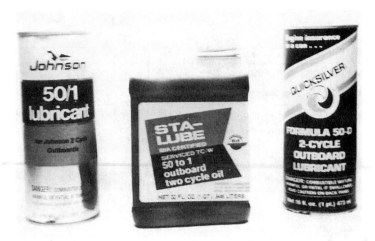

Figure 3–40   Fifty/one BIA TC-W outboard lubricant.

the ignition system would be interrupted. If the ground circuit fails, there is a back-up ground wire. Two wires at different locations on the powerhead almost eliminate the possibility of ground failure.

As the oil is injected into the fuel line, a check valve is necessary to prevent the gasoline from being pumped into the oil tank. The oil line required must be a minimum of three feet and a maximum of eight feet. This establishes the minimum and maximum lag time before an oil ratio change will be brought to the powerhead.

(a)

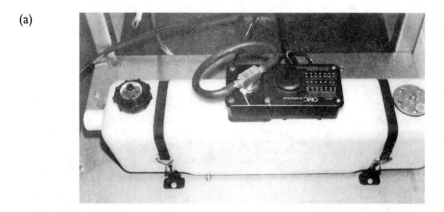

Figure 3–41   OMC economixer (a) components of economixer (b) (continued on next page).

(b)

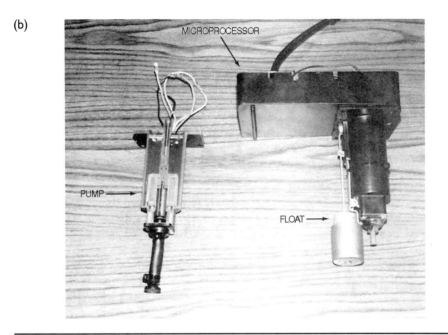

Figure 3–41   Continued.

A boat equipped with two outboards has two economixers. They may be installed together in the larger seven-gallon oil tank or in separate smaller $3^1/_2$-gallon oil tanks for convenience.

Because the economixer is sensitive to pulse signals, the early models may require the use of resistor spark plugs. Later models may have shielded ignition wires to protect the microprocessor from RFI signals, which can cause a momentary shutdown on the economixer.

3.10.1   Troubleshooting the Economixer

When the powerhead is started, the red warning light will flash and the horn will sound as a test sequence. This is a normal condition.

When there is a problem in the system, the status indicator will inform the operator with a steady yellow light, a red flashing light, and the horn or alarm sound, and the powerhead will stop. The problem is generally found in the outboard electrical system or electrical leads, leading to the economixer. An ohmmeter and a peak voltmeter can be used to check the system. The rheostat on the carburetor indicates throttle position, and is checked with an ohmmeter. The B-plus signal is checked with a voltmeter for 12V. As the powerhead is running, the peak voltmeter can be used to check for 300V coming from the power pack, measured on the economixer side. It is also used to check the tachometer signal coming from the alternator. If the four signals coming from the powerhead meet specifications, then the problem is in the leads, and they can be checked for continuity with an ohmmeter. If the leads are okay, then, through the process of elimination, you have proven the economixer at fault. A Stevens Tester is available to test the economixer.

Detailed procedures are given in the service manual for the four problems indicated on the status indicator. These procedures should be used, if checking of the outboard systems does not reveal the problem.

3.11   Oil Injection System (Mercury)

The reason for oil injection is to deliver a precise amount of TC-W3 oil to match powerhead demand (RPM), thus doing away with over-lubrication at lower RPMs, which leads to plug fouling. Oil injection is accomplished by using an oil pump, which is gear driven from the crankshaft. See Figure 3–42 for the Mercury V-6 oil pump drive system. On some smaller outboards (Merc 40 HP) a gear driven oil pump is used to move oil from the vented under cowl oil tank through a 2 PSI check valve (which keeps fuel out of the oil line) and into the fuel line before the diaphragm fuel pump. Oil and fuel are mixed as they flow through the fuel pump and fuel filter and on into the carburetor float bowls. If the oil tank runs low on oil, the warning module and low oil sensor will activate an intermittent "beep" (horn) at the helm (Figure 3–43). If the engine overheats, the warning horn will sound a constant "beep."

On larger horsepower powerheads (Mercury 135–200 HP V-6) a variable oil ratio is accomplished with a linkage between the oil pump and the carburetor throttle shafts. Every movement of the carburetor throttle linkage is applied to a motion sensor and the oil ratio is changed.

As shown in Figure 3–44 a deck-mounted remote oil tank (A) supplies oil through a screened pickup tube to the under cowl oil tank (B). By using a check valve (C) , if the remote oil line becomes obstructed, or positive crankcase pressure is not applied to the air space above the oil in the remote oil tank, the check valve opens preventing a vacuum in the under cowl oil tank. The positive crankcase pressure is routed via the check valve (D) and line to apply pressure to the oil in the remote oil tank. This pressure forces oil from the remote oil tank to the under cowl oil tank. The gear driven variable ratio oil pump (E) takes oil from the under cowl oil tank and pumps it into the fuel pump (F), where the oil is premixed with the fuel before entering the carburetors. A 2 PSI check valve (G) prevents gasoline from being forced into oil lines. Powerhead RPM, therefore motion sensor movement to throttle linkage settings, determines the ratio of oil to be added to the fuel. An electrical sensor warning module (H) and a pump motion sensor (I) make up a warning system which will emit an intermittent "beep" (horn) which alerts the operator when the under cowl tank oil level is low.

3.11.1   Troubleshooting the Oil Injection System

At a low level of oil or if the oil pump shaft does not turn the required number of RPMs, when compared to ignition signals for engine RPM, the horn will sound. Should the remote oil tank be emptied, the under cowl oil tank ensures several minutes of operation at full throttle with correct oil ratios.

135 thru 200 HP V-6 Oil Pump Drive System

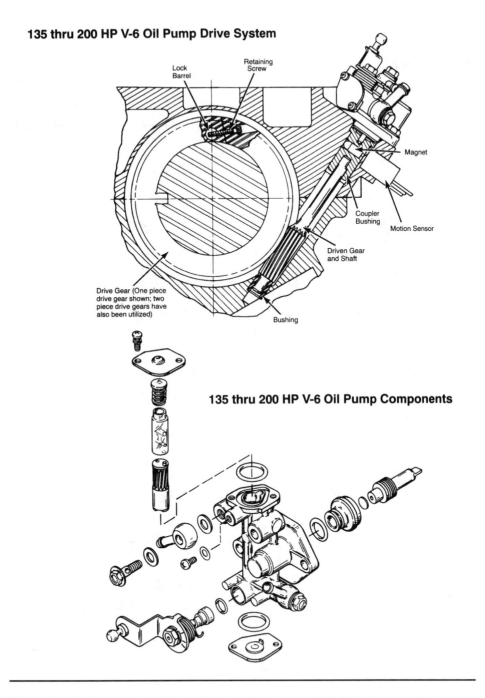

135 thru 200 HP V-6 Oil Pump Components

Figure 3–42 One hundred thirty-five thru two hundred HP V-6 oil pump drive system and components. *(Courtesy of Brunswick Corporation)*

40 HP (4 Cyl.) Oil Injection System

MERCURY

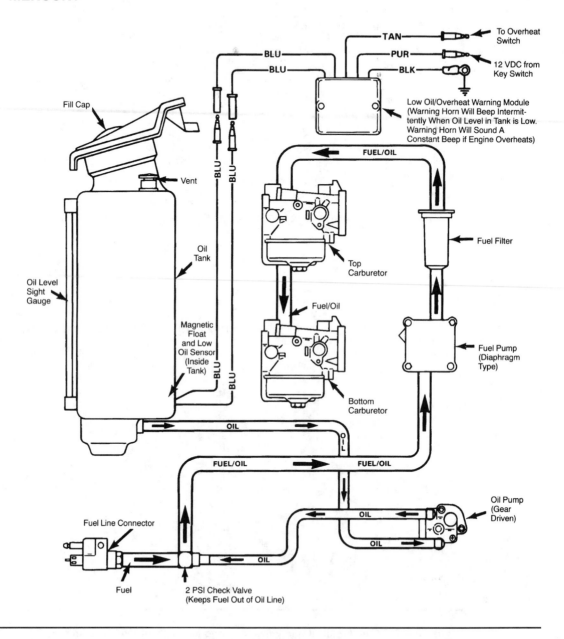

Figure 3–43   Forty HP (4 cylinder) oil injection system. *(Courtesy of Brunswick Corporation)*

135 thru 200 HP V-6 Oil Injection System (Check Valve Mounted on the Fuel Pump) [V-200 (w/Carbs.) System Shown]

MERCURY

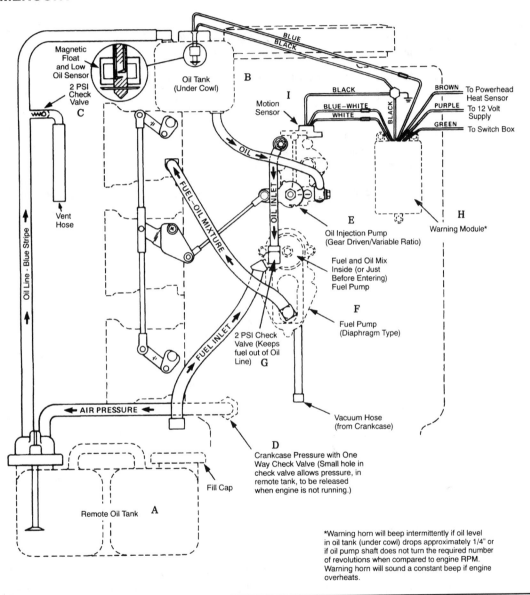

Figure 3–44   One hundred thirty-five thru two hundred HP V-6 oil injection system (check valve mounted on the fuel pump) [V-200 (with carbs.) system shown]. *(Courtesy of Brunswick Corporation)*

A low level of oil may be caused by the following:

- The quick disconnect on the remote tank is not fully engaged.
- The fill cap on the remote tank is not sealed.
- A blocked or ruptured air hose coming from the crankcase.
- A faulty pulse check valve.
- Oil pickup filter screen in remote tank is clogged.
- Hose with blue strip from remote tank is kinked.

If the ignition switch is turned to the On position and there is no "beep" the following may be causing the problem:

- Between the outboard and horn, the tan wire is open.
- The warning module needs to be checked for power and ground.
- The module is not working.
- The horn is not working.

If the warning horn sounds continuously whenever the ignition switch is on, check for the following:

- The overheat sensor.
- The low oil sensor.
- The pump motion sensor.
- Oil injection pump failure.
- Warning module malfunction.

 CAUTIONS

Use Merc Quicksilver oil only, as other oils may coagulate and clog the remote pickup tube filter screen.

Always securely tighten the remote oil tank fill cap, so there will be NO air pressure leak!

3.12 Cooling Systems

Between different manufacturers of outboard powerheads, there are air-cooled and various types of water-cooled cooling systems. There is also the combination of air/water-cooled outboards.

3.12.1 Air-Cooled System

The air-cooled or the air/water-cooled systems are found on the lower horsepower fishing motors. Fins on the spinning flywheel act as a fan and produce air movement. Shrouds and possibly a gas tank are used to direct air flow over the finned cylinder areas, preventing localized hot spots. Heat developed from the combustion process in the cylinder is transferred by conduction into the cylinder fins. By means of **convection**, the heat is carried away into the atmosphere (Figure 3–45).

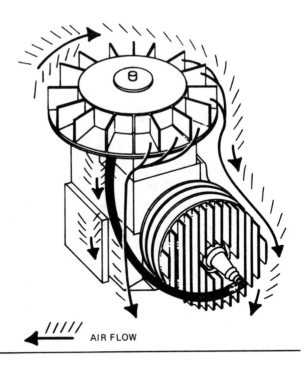

Figure 3–45   Air-cooled
powerhead.

AIR FLOW

The air-cooled system may have a water-cooled exhaust leg (housing). Water is pumped to near the top of the exhaust leg and is dropped into the powerhead exhaust. The water therefore cools the exhaust, keeping the exhaust leg at safe temperatures. There are relief holes in the leg to reduce exhaust back pressure for starting. After start-up, water spray can be felt coming from these holes indicating that the water pump is working.

3.12.1   Water-Cooled System

The water-cooled system is designed to maintain powerhead temperatures controlled by a thermostat, pressure relief poppet valve passage size and water pump RPM. There are variations of this system, with some older outboards using a cold water return to the pump. On most outboards the water inlet is located on the lower unit. From the inlet, water is pulled through a screen into the pump, and pushed up a water tube to the exhaust side of the powerhead. The incoming water passing near the exhaust is warmed and then circulated past the thermostat, or through the thermostat, controlling water temperature at idle in the water jacket passages. It then exits at the bottom of the powerhead into the exhaust. (Some models route the water to the cylinders first.) A tell-tale outlet is used on some models indicating water pump operation. Some will not show water pump operation until the thermostat opens. Water dropped into the exhaust stream cools the exhaust leg and the housing preventing damage to the casting (Figures 3–46 and 3–47).

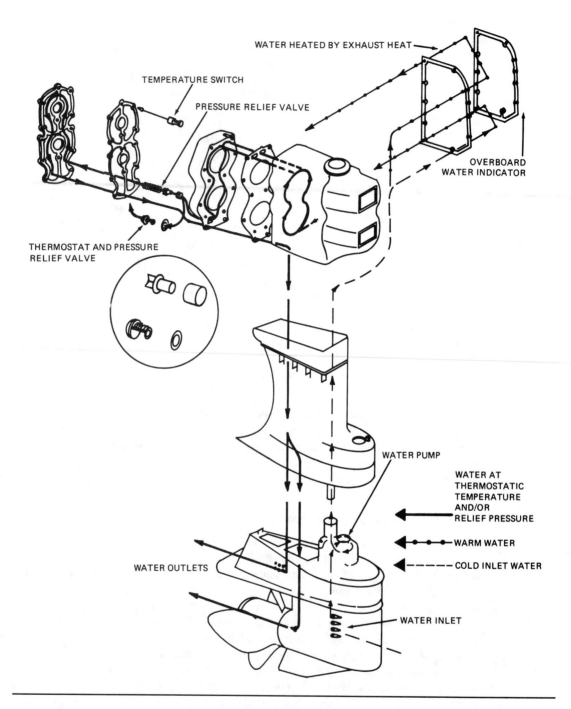

Figure 3–46   Waterflow diagram.

V-6 Water Flow (1989 and newer)

MERCURY

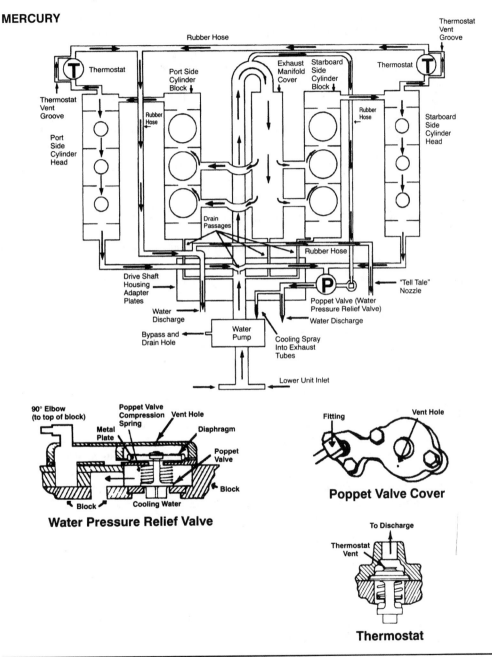

Figure 3–47   V-6 water flow (1989 and newer). *(Courtesy of Brunswick Corporation)*

3.12.3   Water Pump

The water pump utilizes a synthetic rubber impeller driven by the drive shaft, sealing between an offset housing and lower plate to create a pumping action (flexing) on the impeller blades. Mercury has two types of pumps, the volume type and the pressure type. The rubber impeller inside the pump maintains an approximate equal volume of water flow at most operating RPMs (Figure 3–48).

At low speeds the pump acts like a full displacement pump with the longer impeller blades following the contour of the pump housing. As pump RPMs increase, and because of resistance to the flow of water, the impellers bend back away from the pump housing and the pump acts like a centrifugal pump. If the impeller blades are short (Mercury) they remain in contact throughout the full RPM range supplying full pressure.

Warning

The outboard should never be run without water, *not even for a moment.*

As the dry impeller tips come in contact with the pump housing, the impeller will be damaged from the friction and heat. (Figures 3–49 and 3–53). On later models, the impeller comes in contact with a stainless steel cup inside a nylon housing and damage occurs to the impeller in seconds.

3.12.4   Thermostat and Pressure Relief

The thermostat controls the water temperature by regulating the flow of water through the water jacket passages at idle. At high RPM, water pressure is great enough to force the thermostat off its seat, compressing a spring and causing a release of pressure. There may be a separate poppet relief valve that starts to open at 4 PSI and is wide open at 15 PSI at higher RPM. Whether the thermostat or poppet valve is closed or open, water flows out with the exhaust (see Figure 3–50).

Routing of water through a water jacket may be accomplished by using water deflectors placed in the water jacket. These neoprene deflectors must be properly positioned between cylinders, as indicated in the service manual. They are to deflect the water flow and prevent it from going between the lower cylinders. They route the water flow upward and around the upper cylinders. If left out or improperly installed, overheating will occur (Figure 3–51). The heat developed in the combustion process, which is conducted through the cylinder walls, is picked up and carried away by circulating water, and flows out with the exhaust gases. Anodes may also be placed inside the water jacket. These protect against corrosion within the water jacket which is caused by brackish or saltwater operating conditions.

3.13   Overheat Warning System

If the powerhead overheats on a medium horsepower or larger powerhead, a warning is signaled by sounding a horn. The horn is hooked up through the ignition switch, and a tem-

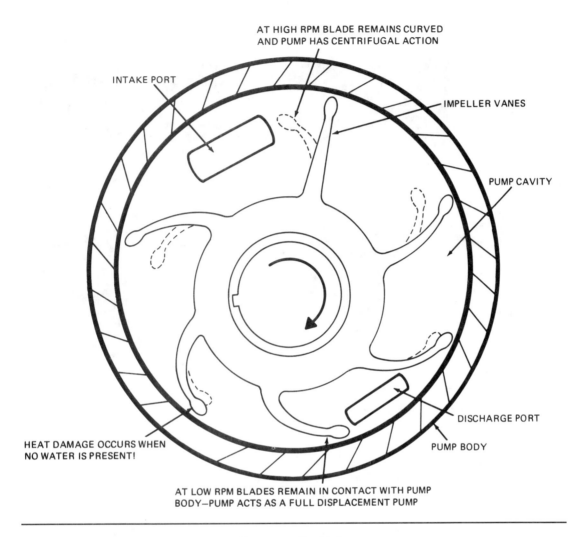

AT HIGH RPM BLADE REMAINS CURVED
AND PUMP HAS CENTRIFUGAL ACTION

INTAKE PORT

IMPELLER VANES

PUMP CAVITY

DISCHARGE PORT

HEAT DAMAGE OCCURS WHEN
NO WATER IS PRESENT!

PUMP BODY

AT LOW RPM BLADES REMAIN IN CONTACT WITH PUMP
BODY—PUMP ACTS AS A FULL DISPLACEMENT PUMP

Figure 3–48   Water pump with long rubber propeller blades.

perature-sending unit registering "overheat" completes the circuit to ground and sounds the horn. This should happen before major damage can occur. Reasons for overheating can be as simple as a plastic bag over the water inlet, or as serious as a leaking head gasket.

Another method of checking for overheating is by using Thermomelt Stiks. They come in two ratings—125 degrees and 163 degrees—for outboard use, and resemble a crayon. The Thermomelt Stik is applied to the cylinder head or thermostat pocket areas and is watched to see if it melts. At 900 RPM the 125-degree Stik should melt and remain glossy. This indicates a normal temperature. If the lower RPMs do not indicate an overheat condi-

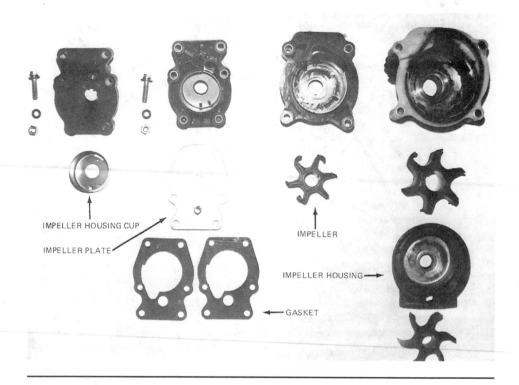

IMPELLER HOUSING CUP

IMPELLER PLATE

IMPELLER

IMPELLER HOUSING

GASKET

Figure 3–49   Damaged water pump components.

(a)

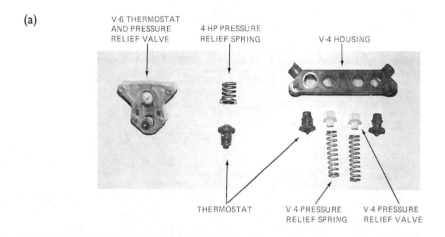

V-6 THERMOSTAT
AND PRESSURE
RELIEF VALVE

4 HP PRESSURE
RELIEF SPRING

V-4 HOUSING

THERMOSTAT

V-4 PRESSURE
RELIEF SPRING

V-4 PRESSURE
RELIEF VALVE

Figure 3–50   OMC V-6, a thermostat and relief for each cylinder head (a) OMC V-4, a thermostat and relief for each cylinder head (b) (continued on next page).

(b)

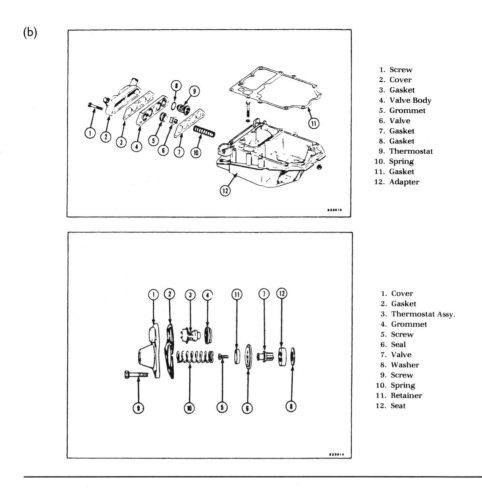

1. Screw
2. Cover
3. Gasket
4. Valve Body
5. Grommet
6. Valve
7. Gasket
8. Gasket
9. Thermostat
10. Spring
11. Gasket
12. Adapter

1. Cover
2. Gasket
3. Thermostat Assy.
4. Grommet
5. Screw
6. Seal
7. Valve
8. Washer
9. Screw
10. Spring
11. Retainer
12. Seat

Figure 3–50   **Continued.** *(Courtesy of Outboard Marine Corporation)*

tion, then increase RPM to 5000. Run the powerhead for 5 minutes and the 163-degree stik should remain dull and chalky. If it turns glossy, troubleshoot for overheating. The outboard should be run in a test tank using a test prop when performing these checks. However, the best way to check for overheating is while the boat is underway. (See Figure 3–52)

A digital pyrometer can also be used to check operating temperatures.

3.13.1   Troubleshooting the Cooling System

Overheating can occur with the air-cooled powerhead. One cause of overheat is a wrong oil mix. An improper amount of oil elevates temperatures, and can cause a carbon build-up in the combustion chamber, reducing heat transfer. An improper selection of propeller, which has too much pitch, will lug the powerhead down, overloading it. Adjustments of timing

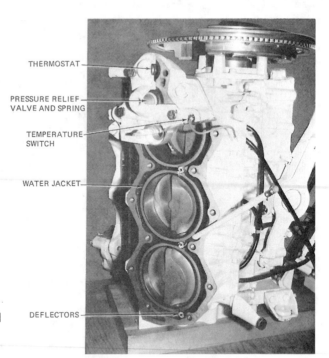

Figure 3–51 Deflectors located in water jacket; pencil pointing at deflector.

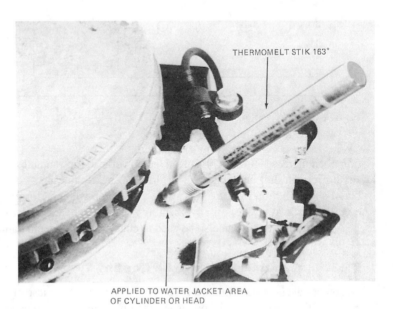

Figure 3–52 Thermomelt Stik applied to water jacket area of cylinder or head.

and synchronization to the carburetor and over-advanced timing should be checked. With the air-cooled system, there is generally little trouble with clogged cooling fins in the marine environment, but check and make sure that all shrouds that direct air flow have been installed.

Do not start an air-cooled outboard (with water pump) unless the lower leg is submerged in water. Running a completely assembled unit out of water, even for a short period of time for adjustments, can do serious damage to the water pump and lower leg (exhaust housing). If the outboard cannot be run in water, remove the powerhead and make necessary carburetor adjustments. Be careful not to overspeed the powerhead when the lower unit is removed, because this model does not have a governor to control top RPM. Propeller load determines RPM at wide-open throttle.

The water-cooled powerhead has a lot more problems to consider. Let's take the obvious ones first. The thermostat may be damaged or corroded, and the thermostat cover gasket may be leaking or damaged. Consider the pressure relief feature of the system. Will the spring allow the thermostat or poppet valve to move from its seat? One must be free to move.

The water pump is the heart of the cooling system, and requires a periodic inspection of the pump body, bottom plate, and impeller for scoring, which would prevent a seal. Check for grooves in the driveshaft where the seal rides. Any damage in these areas may cause air or exhaust gases to be drawn into the pump, putting bubbles into the water. In this case, air does not aid in cooling. Is the pump inlet clear and clean of foreign material or marine growth? Check that the inlet screen is totally open. How about the impeller? Try to separate the impeller hub from the rubber. If it shows signs of loosening or cracking away from the hub, replace the impeller. Have the impellers taken apart, and are the blade tips worn down or do they look burned? Are the side sealing rings on the impeller worn away? If so, replace the impeller. The life of the powerhead depends on this pump, so don't reuse any parts that look damaged. Are any parts of the impeller missing? If so, they must be found. Broken pieces will migrate up the water tube into the water jacket passages and cause a restriction that might block a water passage (Figure 3–53).

It can be expensive to locate the broken pieces in the water passages, but they must be found, or major damage could occur. The best insurance against breaking the impeller is to replace it at the beginning of each boating season, and don't run it out of water. When installing a metal-bodied pump housing, coat all screws with non-hardening sealing compound to retard galvanic corrosion. The water tube carries the water from the pump to the powerhead. Grommets seal the water tube to the water pump and exhaust housing at each end of the tube, and they can deteriorate. Also, water tube(s) should be checked for holes through the side of the tube, and for restrictions, dents, or kinks.

Overheating at high RPMs, but not under light load, may indicate a leaking head gasket. If a head gasket is leaking, water can go into the cylinder, or hot exhaust gases may go into the water jacket, creating exhaust bubbles and excessive heat. The aluminum heads have a tendency to warp, and need to be surfaced each time they are removed. Resurface them by using emery paper and a surface block moving in a figure-eight motion. Also, inspect the cylinders and piston for damage.

Other areas to consider are the exhaust cover gaskets and plate (Figures 3–54 and 3–55). Look for corrosion pin holes. This is rare, but if the outboard has been operated in

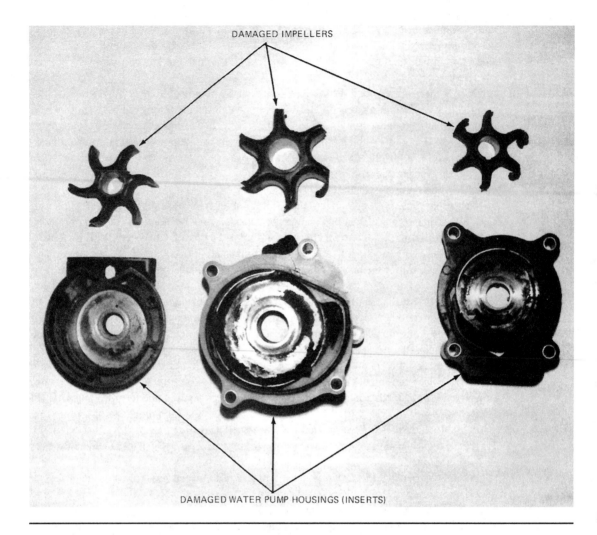

Figure 3–53   Damaged water pump housings and impellers.

saltwater over the years, there just may be a problem. If the outboard is mounted too high, air may be drawn into the water inlet or sufficient water may not be available at the water inlet. When underway the outboard anti-ventilation plate should be running at or near the bottom of the boat and parallel to the surface of the water. This will allow undisturbed water to come to the lower unit, and the water pick-up should be able to draw sufficient water for cooling.

When the outboard has been run in brackish or saltwater, the cooling system should be flushed. (Remove propeller for safety reasons.) The outboard should run at idle speed on the flushette for at least five minutes. This will wash the salt from the castings and reduce

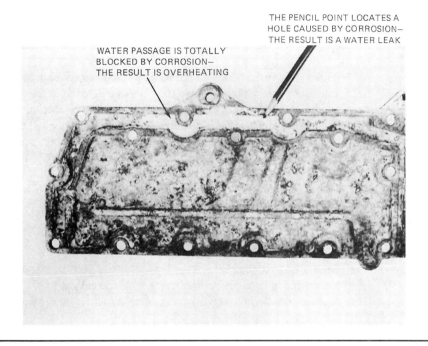

WATER PASSAGE IS TOTALLY
BLOCKED BY CORROSION—
THE RESULT IS OVERHEATING

THE PENCIL POINT LOCATES A
HOLE CAUSED BY CORROSION—
THE RESULT IS A WATER LEAK

Figure 3–54   Lack of flushing caused corrosion buildup and damage to the exhaust cover.

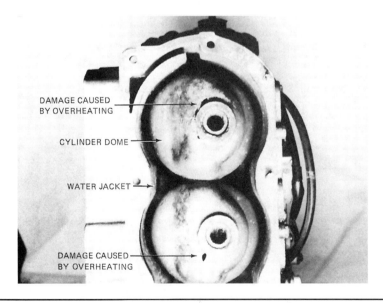

DAMAGE CAUSED
BY OVERHEATING

CYLINDER DOME

WATER JACKET

DAMAGE CAUSED
BY OVERHEATING

Figure 3–55   Overheating caused by water pump failure.

internal corrosion. If the outboard is small and there is no flushette that will fit, run the outboard in a tank, trash can, drum or bucket. There is no need to run in gear during the flushing operation. After the flushing job is done, rinse the external parts of the outboard off to remove the salt spray.

When service work is done on the water pump or lower unit, all the bolts that hold the lower unit to the exhaust housing and the bolts that hold the water pump housing should be coated with non-hardening gasket sealing compound from OMC or Mercury anti-corrosion grease, to guard against corrosion. If this is not done, the bolts will be difficult to remove or will be seized by galvanic corrosion the next time service work is performed.

Last, but not least, if the unit is equipped, check to be sure the overheat warning system is working properly. By grounding the wire at the sending unit, the horn should sound, and/or a light should turn on. To properly test this circuit for a specific year, make, and model, refer to the appropriate service manual.

KNOW THESE PRINCIPLES OF OPERATION

- Sealing of the crankcase.
- Measuring of cylinders, connecting rod, and crankshaft.
- Method of installing a crosshatch pattern into the cylinder wall.
- Identifying piston design.
- Recognizing piston conditions.
- Identifying the different types of piston rings used.
- Installation of loose and caged needle bearings.
- Connecting rod service.
- Crankshaft service.
- Oil injection system.
- Water pump service.
- Diagnosing cooling system problems.

Now let's look at the four-stroke powerhead.

Section II: Four-Stroke Powerhead Repair

Objectives

After studying the second section of this chapter, you will know:

- That valves help seal the combustion chamber.
- How to diagnose compression loss.
- How the valve guide enables the valve face to properly contact the seat.
- How to make valve adjustments.
- The sequence of torquing cylinder head bolts.
- How the valves and seats are refaced.
- How to install a crosshatch pattern on the cylinder wall.
- How to install piston rings.
- The purpose of crush height on insert main bearings.

3.14   Valve Job

Because the valves are located in the cylinder head overhead valve (**OHV**), a valve job can be done separately from a major overhaul. Often this is the case because a valve has burned and compression is low, and the rings, piston, and cylinder are still in good condition. That leaves you the option of doing just a valve job.

In the four-stroke motor the valves are opened and closed once in every two revolutions of the crankshaft (Figure 3–56)

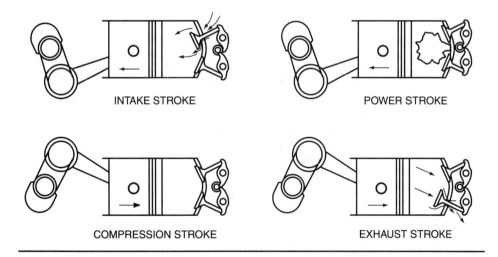

INTAKE STROKE

POWER STROKE

COMPRESSION STROKE

EXHAUST STROKE

Figure 3–56   The four-cycle engine requires two revolutions of the crankshaft, and therefore four strokes of the piston to complete the intake, compression, power and exhaust strokes.

During this cycle, the valves are subject to wear and combustion heat that must be quickly dissipated away from the valve. Heat is removed through exhaust flow, transferred to the valve seat, cylinder head, and into the cooling system water flow. The valves opening and closing every two revolutions of the engine eventually wear the valve face and seat, so the valve will no longer seal to the valve seat. Therefore, compression and combustion pressures begin to leak past the valve face and seat. As the powerhead runs, the leak at the valve seat gets worse and eventually compression is no longer satisfactory. This leakage lowers power output, resulting in poor engine performance.

3.15   Powerhead Diagnosis

First need is to diagnose the powerhead's poor performance by proving the mechanical condition of the motor using a compression test or a leakdown test. These tests will determine if compression is low and where it went. Did compression escape by the rings, past a head gasket, or is there a burned exhaust valve? Find out by doing one or both tests.

3.15.1   Compression Test

Remove the spark plugs and evaluate their condition. (See Chapter 10, "Tune Up".) Then install a compression gauge into the spark plug hole. Now pull hard and fast on the pull rope to rapidly turn the engine over while holding the throttle wide open, or crank the powerhead if it is an electric start model. (Battery should be in a good state of charge.) All cylinders should read within 15 PSI (103 kPa) of each other.

Analyze the results as follows:

- **Normal.** Compression builds quickly and evenly, using the same number of strokes on each cylinder.
- **Low reading.** Squirt a little motor oil into the cylinder and retest. If compression is noticeably improved, the piston rings are leaking and will have to be replaced. If compression does not increase, then there is leakage past a valve, indicating that a valve job is necessary.
- **Low reading on both cylinders.** This would indicate a leaking or blown head gasket between cylinders.
- **Excessively high reading.** This indicates carbon deposit buildup in the combustion chamber. This condition may be accompanied with preignition and detonation. A fuel additive applied through the carburetor opening may help to remove the carbon.

3.15.2   Leakdown Test

By using a leakdown test you can accurately determine which combustion chamber component is leaking. Is it a valve, head gasket, piston, or rings? If possible, warm up the powerhead by running in a water tank before making the test.

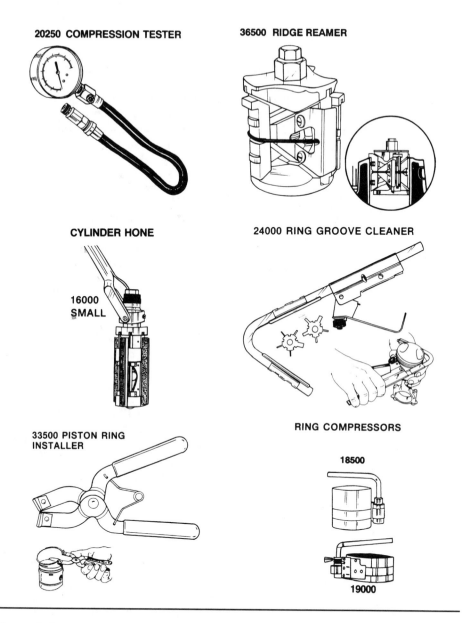

Figure 3–57 Specialty tools. *(Courtesy of Lisle Corporation)*

The leakdown test is a procedure that uses a regulator to put controlled low pressure compressed air into the combustion chamber (Figure 3–58). The piston is held at top dead center of the compression stroke by using a flywheel ring gear tool or a socket and flex handle on the crankshaft nut.

 **Warning**

The flywheel may turn in either direction when low pressurized air is applied.

You are simulating combustion by applying low air pressure to the combustion chamber. Position the piston at top dead center of the compression stroke and secure the holding tool so the crankshaft will not rotate. Adjust air regulator to a set point and read the outlet gauge. It should read in the "okay" zone. (See the manufacturer's instructions.) If not, with air flowing into the combustion chamber, listen for audible escaping air as follows:

- At the carburetor air horn: leaking intake valve.
- At the propeller and water flow indicator hole: exhaust valve leaking.
- At the motor oil fill opening: leaking piston rings and or damaged piston and cylinder wall.
- At the head gasket: leaking gasket, warped head. (You may want to spray soapy water around the edge of the head gasket and watch for bubbles.

Some minor leakage is normal in all powerheads new and used, but where you hear an audible escaping air flow, the compression component is not sealing properly. At this point an experienced technician knows exactly what type of repair is needed to correct the problem. Once the powerhead is proven to be in satisfactory mechanical condition you can do the necessary ignition procedures and service the fuel system by performing a major tune-up.

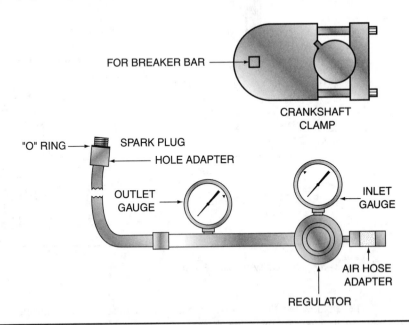

Figure 3–58   A leakdown tester, used to determine the location of combustion chamber leakage.

3.16   Valve Service

To start this major repair, clean up the powerhead by removing all external dirt. Dirt on the outside has a way of getting inside the powerhead while repairs are being made. Follow the service manual instructions for cylinder head removal and clean the combustion chamber and the piston head of carbon deposits. Remove the rocker arm shaft, camshaft, and with the use of a special valve spring compression tool, remove the valve springs. Mark valves for their location. Note that there may be valve spring seats under each valve spring. Grasp the head of each valve and move it back and forth, checking for excessive clearance between the valve guide and the valve stem. To determine stem-to-guide clearance use a small-hole gauge (Figure 3–59) inserted into the guide, making two or three measurements with a micrometer.

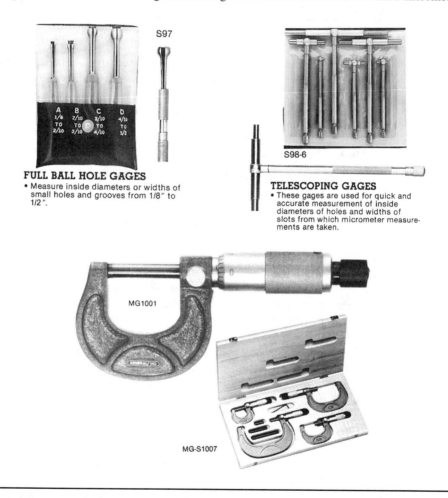

FULL BALL HOLE GAGES
• Measure inside diameters or widths of small holes and grooves from 1/8" to 1/2".

TELESCOPING GAGES
• These gages are used for quick and accurate measurement of inside diameters of holes and widths of slots from which micrometer measurements are taken.

Figure 3–59   Precision micrometers. An outside micrometer is for measuring external dimensions, diameters, or thicknesses *(Courtesy of General Tools Manufacturing, Co., Inc.)*

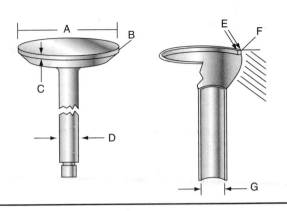

Figure 3–60 Diagram showing valve and seat details: valve head diameter (A); valve face angle (B); valve margin (C); valve stem diameter (D); seat width (E); 45-degree or 30-degree angle (F); and guide i.d. (G).

Check your measurements against the service manual specifications. Some guides can be replaced by pressing them out. If a valve guide is replaced the valve seat will have to be reground. When the valve guides are true, the valve will go straight in and out. The next step is to recondition the valve seats and reface the valves (see Figures 3–59, 3–60 and 3–61).

3.16.1   Refacing the Valves

With guides and valve seats reconditioned, the next step is to reface the valve. This is easily accomplished using a Gizmatic hand valve refacer, from Neway, which produces sharp, accurate, extremely clean valve surfaces with tungsten carbide blades (Figure 3–62). As the valve is refaced, the margin thickness must stay above minimum specifications (see Figure 3–60). If the margin is too thin, the valve will overheat and burn and may cause preignition to occur.

If the valves and seats were reconditioned using a valve refacing machine and a valve seat grinder, then the valves will have to be lapped in. This is done using a valve lapper and lapping compound on the valve face (Figure 3–63). The valve lapper is suction-cupped to the valve head and the valve is installed in the guide. With light pressure, rotate the valve back and forth until you see a grey narrow band all the way around the valve face. Do not let compound come in contact with the valve stem or get into the guide. Do a good job of cleaning up the compound from the seat and valve face when lapping is completed. The valve face and seat are now mated and capable of holding in the high pressures of combustion.

3.16.2   Valve Springs

Before valve springs are installed, they need to be checked for proper tension and squareness. A special tester can measure the tension and a small carpenter's square can check squareness of the spring. Place the square next to the spring standing on end and slowly rotate the spring, watching for the top coil to move away from the square more than 1/32-inch (Figure 3–64).

IT'S AS EASY AS 1, 2, 3.

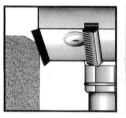 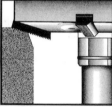 

FIRST CUT
The first cut cleans and
reconditions the area
below the seat.

SECOND CUT
The second cut cleans and
reconditions the area
above the seat.

THIRD CUT
A few revolutions of the
cutter produces a
precision seat.

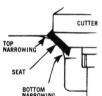

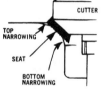

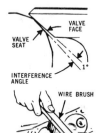

WHY GOOD VALVE SEATS ARE IMPORTANT

A. Good valve seats give more compression and a cooler running engine.

B. The better the valve seats and the valve faces mate, the better the valves perform their functions.

C. TWO WAYS of achieving this tight fit are:

1. SHARPLY DEFINED, CLEAN SEATS.
   These can best be achieved by first removing burned material. Above by top narrowing, below by bottom narrowing, and then cutting the seats.

2. INTERFERENCE ANGLE.
   We recommend an interference angle. This causes maximum pressure between the valve and the seat to occur at the outside diameter. Example: 45° valve - 46° Seat.

PREPARATION FOR CUTTING VALVE SEATS

A. CLEAN CYLINDER HEAD THOROUGHLY.

1. Remove all oil and grease with solvent.

2. Remove all carbon and combustion deposits with wire brush.

NOTE: **Follow recommended safety procedures when using a solvent.**

Figure 3–61   An introduction to cutting valve seats (continued on next page). *(Courtesy of Neway Manufacturing, Inc.)*

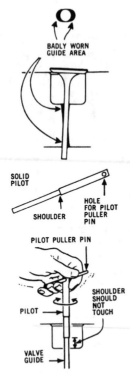

B. CLEAN VALVE GUIDES.

1. Remove all deposits with wire brush.

2. Check condition of valve guides. (See MAN-UFACTURER'S specifications and inspection procedures.)

BADLY WORN VALVE GUIDES

Always replace or resize if not within manufacturer's specifications.

SELECTION AND USE OF PROPER PILOT

A. SOLID PILOTS.

1. Select a pilot same diameter (fractional or metric) as valve stem.

2. Insert pilot in valve guide, twisting slightly, until very snug. Pilot shoulder should not touch valve guide.
   - If small, try next size larger.
   - If too large, try next smaller size.

B. EXPANDABLE PILOTS.

1. Select a pilot same diameter (fractional or metric) as valve stem.

2. Insert pilot in valve guide.
   TO AVOID PILOT DAMAGE . . .
   - Valve guide must be longer than expandable section of pilot (collet).
   - Expandable section of pilot (collet) must be inside valve guide.

3. Pilot shoulder should be about ⅛" (3mm) above valve guide.
   Insert pilot puller pin in pilot hole. Turn and expand pilot until snug, while holding nut.

NOTE: **Collet will expand maximum .020" (,5mm)**

Remember, the accuracy of the valve seat cutting depends upon a tight fitting pilot in a round straight guide.

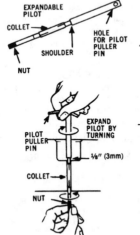

Figure 3–61   Continued on next page.

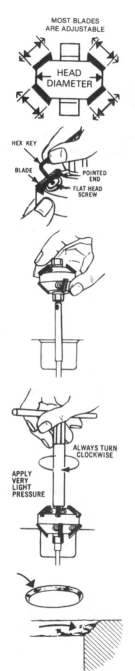

SELECTION AND USE OF PROPER VALVE SEAT CUTTERS

BEFORE USING CUTTERS, ADJUST BLADES TO REQUIRED DIAMETER THEN TIGHTEN ALL SCREWS THAT HOLD BLADES IN PLACE

Pointed ends must always point toward hub or center of head.

Note: Moving blades outward will increase cutting dia. to ¼" (6.5 mm) larger than cutter head diameter.

GENERAL INSTRUCTIONS

A. Select cutter approximately same size as valve head diameter and with correct angle.

B. Place cutter on pilot and slowly lower cutter to valve seat. DO NOT DROP CUTTER.

C. Place T-handle or power unit over hex of cutter.

D. Turn clockwise and apply very light pressure. Release the down pressure at end of each cut. Make one or two turns with no pressure.

CENTER THE CUTTING PRESSURE.

Maintain a downward pressure over centerline of pilot.

DETAILED CUTTING INSTRUCTIONS
A. INITIAL INSPECTION OF SEAT.

1. LOOK AT SEAT.

   The size of the PITS, BURN OUTS, and BLOW BYES will determine the amount of material that must be removed with the remaining cuts.

Figure 3–61   Continued on next page.

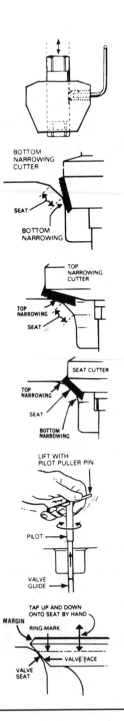

B. BOTTOM NARROWING CUT.

1. Cut lightly with narrowing cutter (usually 60° or 75°).

2. Cut until a fine continuous line is formed with valve seat. (This operation RAISES THE BOTTOM EDGE OF THE SEAT.)

   Note: Some narrowing cutters have moveable hubs so hubs will not rest on guide.

C. TOP NARROWING CUT.

1. Cut lightly with narrowing cutter.

2. Cut until seat width is slightly less than required. (This operation LOWERS THE TOP EDGE OF SEAT.)

D. FINAL SEAT CUT.

1. Cut lightly, with seat cutter.

2. Cut seat to proper width. This should take only 3 to 5 turns.

NOTE: Finished valve seat will have a machine textured finish (not highly polished or shiny). This provides a soft surface for final mating with valve in first seconds of engine operation.

E. INSPECT SEAT.

1. Remove pilot, using pilot puller pin.

2. Insert valve in valve guide.

3. Bang valve slightly up and down in the guide (holding it with fingers top and bottom — above and below the cylinder head). Do this until "Ring Mark" shows on the **valve face** surface.

4. Ring Mark should be positioned ⅓ of the way down valve face from margin.

5. If Ring Mark is too HIGH — cut top narrowing angle slightly to lower Ring Mark — if Ring Mark is too LOW cut seat angle slightly to raise mark.

Figure 3–61 Continued on next page.

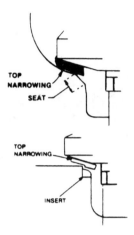

TOP
NARROWING
SEAT

TOP
NARROWING

INSERT

F. TOP NARROWING CUT.
(Hemisphere Chamfer & Recessed Seat).

1. Adjust radius blades to blend hemispheres to top narrowing cut.

NOTE: **If more than 1 radius blade is used, adjust to approximate same position.**

---

2. Adjust (3) special stepped blades to clear step in casting.

DIAL INDICATOR

CHECKS CONCENTRICITY
ALL AROUND SEAT

CHECKING VALVE SEAT CONCENTRICITY AND CLOSURE

A. CONCENTRICITY:
USE VALVE SEAT DIAL INDICATOR. Concentricity should be within .002″ (,05mm) (total indicator reading).

B. CLOSURE: PRUSSIAN BLUE METHOD:

1. Paint the valve face with Prussian Blue.

2. Remove pilot and insert valve into guide.

3. Turn the valve back and forth in the seat about ⅛″ (3mm) without pressure. A fine clean line will appear on the face of the valve.

4. If an "open" spot appears on the line, more than ½″ (12,7mm) long, return seat cutter onto pilot and blend in by turning the cutter 1 or 2 revolutions with the fingers.

5. If line has only shorter intermittent open spots, do not blend. It will peen itself in the first few seconds of engine operation.

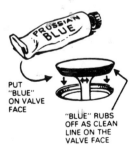

PUT
"BLUE"
ON VALVE
FACE

"BLUE" RUBS
OFF AS CLEAN
LINE ON THE
VALVE FACE

NOTE: **Use a properly refaced valve. Occasionally a newly refaced valve is a bit out of round. If so, try another valve.**

Figure 3–61   Continued.

Figure 3–62   The "GIZMATIC" efficient tool for refacing valves. *(Courtesy of Neway Manufacturing, Inc.)*

Figure 3–63   Valve lappers.
(Courtesy of Lisle Corporation)

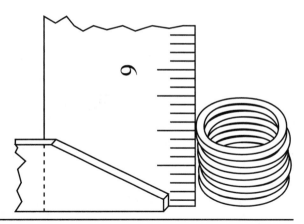

Figure 3–64 A small carpenter's square can be used to check the valve spring for being square (bent).

A valve spring that is out of square will cause premature wear of the valve guide. If a valve spring is out of square or has lost tension, discard it. Be aware that there may be a difference in intake and exhaust valve springs. Refer to the service manual for proper installation of spring seat and retainer installation.

3.16.3   Valve Adjustment

The end of the valve stem needs to be inspected for being flat. If it is worn, it should be resurfaced or the valve replaced permitting proper valve adjustment. This is the part of the valve that contacts the rocker arm and where valve clearance adjustment is made. To adjust the valves, the camshaft must be positioned (timed) to the crankshaft. Check the position of the timing marks of the camshaft sprocket and crankshaft pulley. When properly aligned the timing belt is properly installed and shafts are positioned for valve adjustment of cylinder 1 (Figure 3–65).

Using a feeler gauge, the rocker arm adjusting screw is adjusted to specified clearances. (OMC 15 HP, specifications: the smallest valve head is exhaust and is adjusted to .006 and the largest valve head is intake and is adjusted to .004.) When the valves for cylinder 1 are adjusted, then the crankshaft is rotated one revolution (360 degrees). This positions the camshaft at 180 degrees. Now the valves of cylinder 2 are in position to be adjusted. The OMC powerhead must be cold when adjustments are made. When each adjustment is completed, torque the tappet nut to specification. Valve clearance should be checked after initial 20-hour break-in period and annually thereafter.

3.16.4   Cylinder Head Flatness

The valves seal two openings in the combustion chamber. The piston, rings, head gasket, and cylinder head are also used to close the combustion chamber. While the cylinder head is off, check it for flatness (warpage). Use a feeler gauge and machinist straight edge. If a

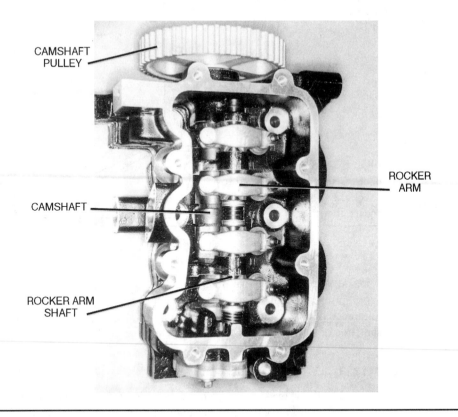

Figure 3–65 Cylinder head showing rocker arm shaft, rocker arms, and camshaft. *(Courtesy of Outboard Marine Corporation)*

specified feeler gauge (.004 inch) slips between the head and straight edge, which is placed end to end on the head, the head will have to be replaced or resurfaced (Figure 3–66).

When installing the head, use a new head gasket and the recommended gasket sealing compound as indicated in the service manual. Be aware that the head bolts may not all be the same length. Apply motor oil to the threads of the head bolts and torque in sequence and pattern. (OMC recommends SysteMatched Four-Stroke Outboard Break-In Lubricant be applied to head bolts).

3.17   Ring Job

With the cylinder head and crankshaft removed, the cylinder ridge can be removed using a ridge reamer. The cylinder ridge develops more in the four-stroke than the two-stroke, because there is no oil in the fuel to lubricate the upper cylinder wall. With the cylinder ridges removed, the pistons can be pushed out. Mark the pistons so they can be reinstalled into their original cylinders. Once this is done you can measure the cylinder for being over-

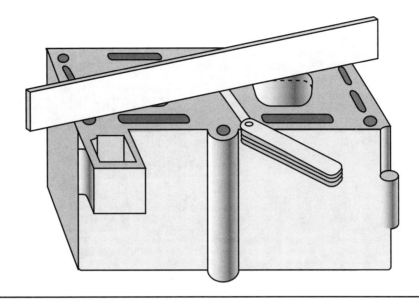

Figure 3–66   To check for cylinder head warpage, use a machinist's straight edge and a feeler gauge.

sized, tapered, and out of round. The procedure for measuring the cylinder is basically the same as for measuring two-stroke cylinders (see the first section, Figure 3–5).

3.17.1   Installing Crosshatch Pattern

When reconditioning the four-stroke cylinder wall, we are more concerned about the crosshatch pattern to control oil and lubricate the piston skirt. A medium grit flex-hone may be used to establish the pattern, so oil can be retained and lubricate between the piston skirt (oil also is held in piston-grinding marks) and cylinder wall crosshatch pattern. With a proper crosshatch pattern on the cylinder wall, the compression and oil control rings will quickly seat.

3.17.2   Installing Piston Rings

When installing new rings, you must determine if they will fit the cylinder. This is done after the crosshatch pattern is installed on the cylinder wall and the cylinder is cleaned up. Check the cylinder wall for honing grit by using the white-towel test (see the first section of this chapter). Place each compression ring in its respective cylinder bore and use a piston head to push and square the ring within the cylinder. Then measure the gap between the ring ends using a feeler gauge and compare it to specifications. Remember that one cylinder may be larger than the other, so keep rings separated per cylinder (Figure 3–67).

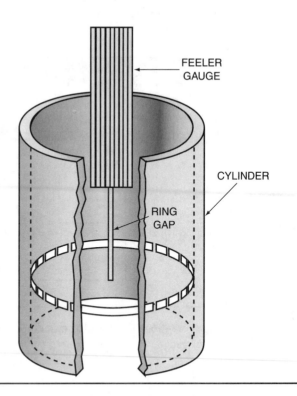

Figure 3–67   Ring gap measurement.

Each piston has three ring grooves and they may all be of different widths. Using a ring groove cleaning tool or the end of a broken ring, scrape out any carbon buildup from the grooves. Be careful not to damage the ring lands while doing this. The piston may also have "grinding" marks on the piston skirt that are used to retain oil between the cylinder wall and the piston skirt. If these marks are worn away (Figure 3–68), then the piston should be replaced. Also check the piston pin fit. Some pins may be loose on both sides and others may be a press fit on one side. Do not reuse the piston pin retainers if the pin is removed.

With carbon removed from all the ring grooves, the oil ring is installed first. It has three components—the segmented ring and two scraper rings. The segmented ring has horizontal slots which permit the oil being scraped from the cylinder wall to run back into the sump. Oil is scraped from the cylinder wall by the two scraper rings. They are installed below and above the segmented ring in the same lower ring groove.

The piston ring grooves need to be inspected for wear. With the carbon removed from the grooves, install each compression ring in its respective groove, as compression ring 1 and 2 may be different thicknesses. Using a feeler gauge, measure the side clearance between the ring and ring land (Figure 3–69). If clearance is to specifications, push rings all the way into their grooves to check for proper depth of the grooves. You should be able to bridge a machinist rule across the rings and it should contact the piston, above and below the ring grooves.

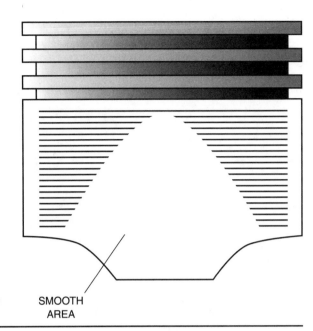

Figure 3–68 Replace the piston if the grinding marks are worn smooth as illustrated.

SMOOTH AREA

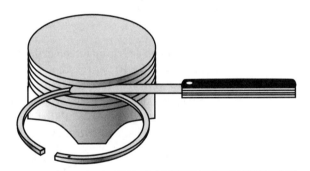

Figure 3–69 Checking top compression ring for clearance.

When the piston and rings are installed, placement of each ring gap is critical. As shown in Figure 3–70 the two scraper ring (oil ring) (1) gaps are placed opposite the segmented oil ring gap (3) and spaced apart, one-quarter of the cylinder diameter. The middle compression ring gap (2) is centered between the scraper ring gaps. The top compression ring gap (4) is placed 180 degrees across from the middle compression ring gap. The placement of these ring gaps is critical, so compression and combustion pressures will be controlled (Figure 3–70).

3.17.3   Oil Pump

The oil pump is a gerotor assembly, mounted to the bottom end of the cylinder head (OMC) and is driven by the lower end of the camshaft (see Figure 2–20). Motor oil is

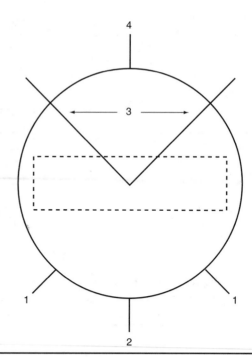

Figure 3–70   Placement of ring gaps for proper compression and combustion pressures.

pulled through a pick-up screen in the sump, then it is pressurized at the oil pump and is filtered (OMC 15 HP) before going to the engine bearings and overhead valve assembly. As engine RPM increases so does oil pressure.

By removing the oil pressure sending unit, a gauge can be installed and a reading of 20 PSI at 2000 RPM and 35–40 PSI at 4200–5200 RPM should be indicated. Problems which may occur and vary your test results are the following:

- Inadequate oil in sump. Check oil with outboard in vertical position.
- Defective pressure regulator.
- Inlet screen/filter and/or tube clogged or damaged.
- Restricted/clogged internal passages.
- Problems with the gerotor oil pump.

How often should you change the motor oil? This varies depending upon the type of oil used and conditions of operation. Johnson and Evinrude recommend changing their oil every season or at 200 hours, or every 100 hours when using an after-market brand SAE 10W 30 SG, SH oil. It is best to change the oil after the motor has been warmed up, as more contaminates will come out with the oil.

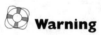 **Warning**

Always supply water to the engine when running it!

3.17.4   Main Bearings

Remember that insert bearings are used for main bearings on the crankshaft (similar to automotive engine bearings). These bearings are slipped into the crankcase cover and the cylinder block. They are aligned by lugs (tangs) protruding from the bearing halves that are slipped into machined recesses in the block and crankcase cover. The lower main bearing is also the thrust bearing and controls crankshaft end play. Bearing crush height is used to press and secure the bearing halves into the bearing saddles (see Figure 2–15). The bearing lugs and crush height prevent the bearing inserts from moving within the bearing saddle.

Main bearing clearance can be checked using a product known as *plastigage*. A small strip of plastigage is placed on the main bearing insert and the main bearing cap is installed and torqued to specifications. (Be careful not to rotate the crankshaft.) This flattens the plastigage strip. The main bearing cap is now removed and the flattened plastigage is measured using a paper indicator, revealing the bearing clearance.

3.17.5   Connecting Rod

The rod bearing is an integral part of the aluminum connecting rod (OMC). Therefore, the connecting rod must be reinstalled back on the same crankpin. Mark the cylinder number on the rod when you remove it. Measure the rod bearing inner diameter for wear and check for scoring. If the rod bearing is not to specifications, the rod will have to be replaced. Note that the rod cap and rod have alignment marks that are matched upon reassembly. Do not over torque the rod bolts, as distortion to the connecting rod will occur. The connecting rod can be checked for a "twist," by placing it on a flat surface and checking it using a feeler gauge. The crankpin on the crankshaft should be inspected for scoring and being out of round. Oil all parts with SAE 30 motor oil upon reassembly.

The mating surfaces of the crankcase and cylinder block must be carefully cleaned. They can be easily damaged by scraping tools. Some manufacturers want you to use a special sealant (gasket) remover that will soften the old material. Check the service manual for the type of remover and sealant to be used on these surfaces during clean up and reassembly.

3.17.6   Seals

Neoprene seals are used in the block/crankcase at the upper and lower crankshaft bearings. These seals are used to retain the motor oil and seal out water and exhaust at the lower crankshaft bearing. Use a seal installer, point the lips of the seals in, and lubricate them with motor oil.

 Note

Check the service manual about this—it may vary with different makes and models.

Use the recommended gasket sealing compound on the seal cases when installing the seals. Some models may use an O-ring internally in the splined recess, where the driveshaft/

crankshaft splines mate. The O-ring should be replaced and proper grease applied to the splines whenever the powerhead is removed. If only the lower unit is removed for service work, apply the recommended grease to the driveshaft splines.

KNOW THESE PRINCIPLES OF OPERATION

- How to make compression and leakdown tests.
- Procedures for grinding valves and seats.
- Checking cylinder head flatness.
- Measuring cylinder for being tapered and out of round, or oversized.
- Installing the cylinder crosshatch pattern.
- Piston ring installation.
- Inspection and measuring the connecting rod.
- Inspection and measuring the crankshaft journals.
- Inspection and replacement of main bearings.
- Installation of oil seals.

REVIEW QUESTIONS

Chapter 3, Section 1: Two-Stroke Powerhead

1. The crosshatch pattern on the cylinder wall
   a. provides a path for lubrication under the ring bearing area.
   b. aids in seating the rings.
   c. Both a and b are correct.
2. After honing of the cylinder
   a. the finish is measured and compared to specifications.
   b. the cylinder is wiped with an oily cloth as a final and only clean up.
   c. the cylinder is cleaned up with hot water and soap, and checked for cleanliness with a white cloth.
3. The piston rings
   a. transfer heat, control blow-by gases, and seal the gap between the piston and cylinder wall.
   b. control oil consumption in two-cycle powerheads.
   c. maintain clearance between the cylinder wall and piston.
4. Ring gap measurement is taken to
   a. determine ring fit to the cylinder.
   b. determine if there is enough gap for ring expansion.
   c. Both a and b are correct.
5. Ring side clearance is taken
   a. with a worn ring.
   b. before the ring is installed on the piston.
   c. after the ring groove is cleaned and a new ring is installed in the ring groove.

6. Piston diameter is measured at the
    a. piston skirt.
    b. piston bosses.
    c. head of the piston.
7. Piston pin retainers are
    a. used to help retain the rings.
    b. never used as the pins are pressed into the rod.
    c. installed at the end of the piston pin.
8. Piston pins are
    a. a bearing race.
    b. installed in piston bosses.
    c. Both a and b are correct.
9. The two-stroke connecting rod
    a. connects to the main bearing journal.
    b. is a bearing race.
    c. controls crankshaft throw.
10. The oil holes in a two-stroke connecting rod are
    a. positioned down to drain the oil.
    b. installed toward the crankshaft.
    c. positioned toward the tapered end of the crankshaft.
11. The two-stroke connecting rod with fractured parting lines is
    a. aligned by the shoulder on the connecting rod.
    b. torqued before alignment is made.
    c. aligned by using a scribe on the chamfered corners.
12. The connecting rod bearing race is the
    a. crankshaft rod journal (crankpin).
    b. connecting rod big end inner bore.
    c. crankshaft main bearing journal.
    d. Both a and b are correct.
13. Two-stroke crankshaft labyrinth sealing rings are used to
    a. prevent fuel transfer from one crankcase to another.
    b. retain positive and negative pulses within a given crankcase.
    c. Both a and b are correct.
14. The crankshaft converts horizontal motion into rotary motion.
    a. true
    b. false
15. The two-stroke crankcase must be sealed against both pressure and vacuum.
    a. true
    b. false
16. Two-stroke crankshaft end play is controlled by
    a. shims under the bearing caps.
    b. a thrust washer installed on the crankshaft.
    c. Both a and b are correct.

17. The two-stroke crankshaft may be
    a. turned and polished.
    b. used with heavy chatter marks on the bearing journal.
    c. cleaned up with 320 emery paper where loose needle bearings are used.
18. The exhaust and the cylinder must be sealed against
    a. water.
    b. pressure.
    c. Both a and b are correct.
19. When a white powder is noted around the lower unit attaching bolts, what is a good tool to use to loosen the bolts? (The bolts are seized.)
    a. a breaker bar
    b. localized heat applied to the casting from a torch
    c. drill and use an easy out
20. When removing the powerhead from the lower motor cover (adapter) you find that the gasket has broken free, but the powerhead still cannot be lifted form the lower motor cover. What is the problem?
    a. The water tube is stuck to the powerhead.
    b. The crankshaft and driveshaft splines have rusted together.
    c. The shift shaft is stuck in the lower unit.
21. A pattern is installed on the cylinder wall to
    a. remove score marks.
    b. control engine oil burning.
    c. aid in ring seating and oil retention on the cylinder wall.
22. After honing or reboring, how is the cylinder cleaned up?
    a. A solvent wet rag is pushed through the cylinder several times.
    b. It is scrubbed with a brush using hot soap and water.
    c. Use a rag with 50/1 oil and push it through the cylinder several times.
23. A two-stroke cylinder that has a chrome finish is inspected for
    a. wearing through of the chrome.
    b. scoring and flaking away of the chrome.
    c. meeting factory specifications, plus both a and b.
24. A two-stroke piston that directs incoming fuel to the top of the cylinder is of the _____ design.
    a. crown
    b. flat
    c. deflector
25. On a piston with the deflector head design, which side is placed toward the exhaust port?
    a. the sloping side of the head
    b. the deflector side of the head
    c. the side with the power port

26. In the OMC piston that has the word *loose* stamped into the inside of the piston skirt, the piston pin is pressed out with the loose boss positioned_____.
    a. down
    b. up
    c. either up or down

27. Piston damage may be caused by
    a. maladjusted timing/synchronization.
    b. detonation/preignition.
    c. Both a and b, plus lean mixtures and a damaged water pump.

28. The major thrust side of the piston is
    a. in the direction of rotation.
    b. opposite the direction of rotation.
    c. controlled by the piston oil retainer.

29. Heat is transferred from the piston through the
    a. piston boss.
    b. fuel mixture.
    c. rings and piston skirt.

30. Piston pins are secured to the piston by
    a. a press fit.
    b. a piston pin retainer (snap ring).
    c. Both a and b are correct.

31. What are the functions of the piston ring?
    a. They are a seal and maintain an oil film on the cylinder wall.
    b. They seal against the piston head deflector to control blow-by gases.
    c. They control oil consumption.

32. The type of ring used in a two-cycle power head is
    a. an oil control ring.
    b. a compression ring.
    c. neither type of the above.

33. What two measurements are taken to determine if the piston ring fits the cylinder bore and piston?
    a. ring side clearance
    b. ring gap
    c. Both a and b are correct.

34. The use of an automotive-type ring compressor on a two-cycle piston and rings will generally
    a. allow clearance for ring installation.
    b. break the new rings.
    c. damage the piston ring area.

35. The connecting rod transfers combustion pressure from the piston pin and bearings to the
    a. cage of the connecting rod bearing and to the crankshaft rod journal.
    b. connecting rod bearings and to the crankshaft rod journal.
    c. labyrinth seal and the crankshaft main bearing journal.
36. In the fractured design of connecting rod, the cap is aligned by
    a. running a #2 pencil over the bearing race of the connecting rod.
    b. torquing the cap to specifications.
    c. running a #2 pencil over the chamfered corners until they are smooth on all corners and then torquing to specifications.
37. Caps on connecting rods can be interchanged with another connecting rod.
    a. true
    b. false
38. Oil holes placed in the side ends of the connecting rods are placed
    a. down towards the bottom of the powerhead.
    b. up toward the flywheel end.
    c. It doesn't make any difference which way they face.
39. Two-stroke connecting rods are inspected for
    a. straightness plus b and c.
    b. chatter marks and scoring.
    c. pit marks and color change.
40. In larger two-stroke outboards, the connecting rod and main bearings are of the _____ design.
    a. sleeve
    b. insert
    c. caged and loose needle
41. Main bearings are used to control
    a. axial movement.
    b. crankcase pressure.
    c. piston major thrust.
42. The two-stroke crankshaft is case-hardened at the crankshaft journal; therefore, it should not be
    a. welded and turned to standard size.
    b. turned undersize.
    c. Both a and b are correct.
43. What type of grease is used to hold the needle bearings during assembly?
    a. chassis grease
    b. needle bearing grease (a bee's wax preparation)
    c. moly lube
44. When sealing rings are mounted on the two-stroke crankshaft, they are to
    a. support the crankshaft.
    b. control pressures between crankcases.
    c. control blow-by gases.

45. The tapered end of the crankshaft is used to mount the
    a. upper main bearing.
    b. flywheel.
    c. timer base.
46. TC-W3, two-stroke water cooled lubricant is designed
    a. as an ashless oil.
    b. to control rusting of the bearings and metal surfaces.
    c. to minimize carbon build-up, plus a and b.
47. A powerhead designed to run on 50/1 oil mix, will have cleaner exhaust ports if
    a little more oil is added (a rich mix of oil).
    a. true
    b. false
48. An OMC economixer supplies oil to the powerhead *directly* for immediate
    bearing lubrication.
    a. true
    b. false
49. The economixer receives electrical signals from which source?
    a. battery, ignition
    b. tachometer, ignition battery
    c. battery, ignition, alternator (tachometer), and carburetor blade position
50. What determines the oil ratio in a Mercury 135-200 HP oil injection?
    a. powerhead temperature
    b. sensor movement by throttle linkage
    c. oil reservoir level
51. What moves the oil from the remote tank to the under cowl oil tank? (Mercury
    oil injection)
    a. electric oil pump
    b. fuel pump vacuum
    c. crankcase pressure
52. What will happen if the remote oil tank cap is not sealed? (Mercury oil injec-
    tion)
    a. oil will drain from the tank
    b. oil will not be moved to the under cowl oil tank
    c. under cowl oil tank will overflow
53. What will cause the horn in the Mercury oil injection system to "beep" intermit-
    tently?
    a. low oil level at the remote tank
    b. no oil in the remote tank
    c. low oil level in the under cowl tank
54. The thermostat, pressure relief valve, and water pump RPM control powerhead
    temperature.
    a. true
    b. false

55. What is the purpose of the exhaust holes in the exhaust housing (mid-section)?
    a. to gain a higher RPM in the operating range
    b. to relieve the exhaust for easier starts
    c. to reduce exhaust sound
56. Water pumps utilizing an impeller with longer blades act like a _____ pump at high speed.
    a. centrifugal
    b. full displacement
    c. positive displacement
57. Water pump housings are made of
    a. aluminum.
    b. plastic (nylon).
    c. Both a and b are correct.
58. When the powerhead overheats on larger outboards, a horn may sound at the helm. The temperature switch _____ to sound the alarm for a warning.
    a. opens
    b. neutralizes
    c. completes the circuit to ground
59. To check for proper thermostat operation and overheating a
    a. Thermomelt Stiks rated at 125 degrees and 175 degrees are applied to the cylinder area.
    b. Thermomelt Stik rated at 145 degrees is applied to the cylinder head area.
    c. Thermomelt Stiks rated at 125 degrees and 163 degrees are applied to the top side of cylinder 1 or at the thermostat pockets.
60. When disassembled for service, if there are pieces missing from the water pump impeller
    a. a chrome water pump kit should be installed.
    b. an impeller drive shaft key should be modified.
    c. powerhead water passages should be inspected to locate and remove the missing parts.
61. Overheating at wide-open throttle but not at low speed, may indicate that
    a. the tell-tale hole is plugged up.
    b. a head gasket is leaking.
    c. the water pump impeller key is rusted.
62. Bolts used to attach the water pump and lower unit are coated with gasket sealing compound from the bolt tip to the underside of the bolt head.
    a. true
    b. false
63. The water pump impeller is inspected for
    a. breaking away from the hub.
    b. cracked and damaged impeller blades.
    c. Both a and b are correct, plus checking the pump housing for scoring.

64. Heat is carried away from the operating air-cooled powerhead by means of
    a. radiation.
    b. conduction and convection.
    c. conduction.

REVIEW QUESTIONS

Chapter 3, Section II Four-Stroke Repair

1. The intake valve is opened once in every two crankshaft revolutions.
   a. true
   b. false
2. What position should the throttle valve be in when making a compression test?
   a. wide open
   b. halfway open
   c. closed
3. During a compression test, oil is squirted into the cylinder and compression does not improve. What is your diagnosis?
   a. Rings are sealing.
   b. Carbon has built up on the cylinder head.
   c. A valve is not sealing.
4. During a leakdown test, air is heard leaking from the carburetor air horn. What part has failed?
   a. rings
   b. head gasket
   c. valve
   d. piston
5. If the valve guide is replaced
   a. the valve seat will have to be reground.
   b. the valve face will have to be refaced.
   c. Both a and b are correct.
6. As the valve is refaced, the margin will decrease.
   a. true
   b. false
7. When reconditioning the valve seat, the first cut is
   a. made in the area above the seat.
   b. made on the seat face.
   c. made in the area below the seat.
8. The accuracy of the valve seat cutting depends upon
   a. using an undersize pilot in a straight guide.
   b. using an adjustable pilot in a worn guide.
   c. using a tight-fitting pilot in a straight guide.

9. What determines the size of seat cutter to be used?
   a. the size of the valve stem
   b. the same approximate size as the valve head
   c. the angle of the valve seat
10. What determines the amount of material that must be removed to clean up the valve seat?
    a. the angle of the cut
    b. the size of the pits, blow-bys, and burnouts
    c. the size of the valve seat
11. What is the angle of the bottom-narrowing cut on the seat?
    a. 45–46 degrees
    b. 60–75 degrees
    c. 30–31 degrees
12. The top-narrowing cut on the valve seat is continued until
    a. seat width is slightly less than required.
    b. there is slightly more than required.
    c. used seat contact area is removed.
13. The finished valve seat will have a
    a. machined textured finish.
    b. soft surface for final mating.
    c. highly polished look.
    d. Both a and b are correct.
14. Grasping the valve head and stem with your fingers and banging the valve up and down in the guide will establish a seat "ring mark" on the valve face surface.
    a. true
    b. false
15. Valve seat concentricity can be checked using
    a. a micrometer
    b. a dial indicator
    c. Prussian Blue
    d. Both b and c are correct.
16. The valve seat ring mark should be one-third of the way down the valve face from the margin.
    a. true
    b. false
17. The exhaust valve can be adjusted when the crankshaft is positioned at TDC of the intake stroke.
    a. true
    b. false
18. Valve springs can be checked for squareness by using a small carpenter's square.
    a. true
    b. false

19. A crooked valve spring will cause wear at the
    a.   valve head.
    b.   valve guide.
    c.   valve retainer.
20. The cylinder head surface is checked for flatness by
    a.   placing a machinist's straight edge end-to-end and trying to push a feeler gauge under it.
    b.   using a feeler gauge and then a micrometer.
    c.   placing the head on a flat surface and sanding with emery paper.
21. If the torque sequence is not followed, distortion of the cylinder head will occur.
    a.   true
    b.   false
22. Why does the cylinder ridge develop more in a four-cycle engine?
    a.   The motor runs hotter.
    b.   There is less oil on the cylinder wall.
    c.   It runs at higher RPMs than a two-stroke.
23. What is the purpose of the cylinder wall crosshatch pattern?
    a.   to increase piston skirt to cylinder wall clearance
    b.   so oil can be retained and lubricate between piston skirt and cylinder wall
    c.   so the oil ring can remove more oil from the cylinder wall
24. Grinding marks on the piston skirt are used to retain oil.
    a.   true
    b.   false
25. The segmented piston ring gap is placed in the ring groove 2.
    a.   true
    b.   false
26. Piston ring side clearance is for ring expansion and movement.
    a.   true
    b.   false
27. What piston ring gap is centered between the oil scraper ring gaps?
    a.   top compression ring
    b.   segmented ring
    c.   middle compression ring
28. OMC 15 HP oil pump is of the gerotor type and mounted
    a.   in the oil sump.
    b.   inside the crankcase.
    c.   at the lower end of the camshaft.
29. How often is the OMC motor oil changed in an OMC 15 HP?
    a.   100 hours
    b.   200 hours
    c.   once a season regardless of hours

30. What type of main bearing is used in an OMC 15 HP?
    a. insert bearing
    b. ball bearing
    c. needle bearing

31. Match marks on the connecting rod are used to
    a. determine what cylinder the rod goes into.
    b. align the rod cap to the connecting rod.
    c. align to the major thrust side of the piston.

32. Mating surfaces of the block/crankcase are easily damaged when cleaning.
    a. true
    b. false

33. The crankshaft upper main bearing seal is installed with the seal lip pointing
    a. out.
    b. in.

34. When installing the powerhead onto the exhaust housing, motor oil is applied to the crankshaft/driveshaft splines.
    a. true
    b. false

CHAPTER 4

Electricity/Electronics for the Technician

Objectives

After studying this chapter, you will know:

- The meaning of *voltage, current*, and *resistance*.
- The movement of electrons.
- Ohm's Law.
- How AC voltage is converted to DC voltage.
- How to measure voltage, amperes, and resistance.
- Electron theory.
- Current theory.

4.1   Basic Electricity

Electricity is an efficient energy work horse. It is a challenge to the technician because it is not visible, but the effects of its work are quite evident. The effects seen every day are heat; magnetism; and chemical reaction, such as when charging a battery. Electricity is the free movement of electrons which are present in atoms and make up the material of the stator, switches, ignition coil, and the wire conductors. Free electrons are like a line of dominoes standing close together (Figure 4–1)—when the first one is pushed over, the others continue to fall. The resulting action resembles electrical flow through the conductor.

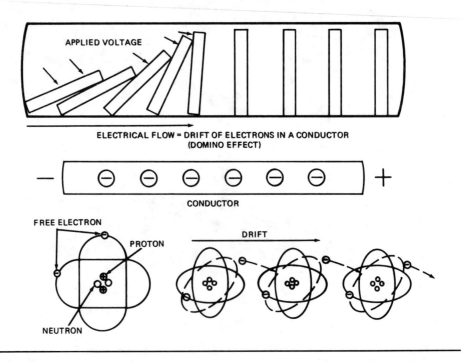

Figure 4–1   Atom structure showing electron drift.

Electrical flow is from negative to positive when considering electron theory. However, current flow is another theory, which has the current flowing positive to negative. This is most commonly used in outboard and automotive repair. Conductors can be made up of aluminum, steel, silver, carbon, copper, and may become stranded wires, switches, contacts, silicon controlled rectifiers, and diodes that are used to control circuits. All these offer little resistance to current flow. Stranded wires are used in the outboard to carry electrical current because they withstand the movement and vibration. They are sealed in plastic for insulation. Insulators (insulation) are nonconductors, such as rubber, glass, paper, or plastic and offer high resistance to current flow.

The following terms are associated with electricity as it is used in the outboard.

4.1.1   Voltage

Voltage is the pressure (push) that causes the electron to flow through the conductor. A *volt* is a measurement of that pressure. Voltage is measured with a voltmeter when installed across the battery or circuit (positive to negative). Without voltage there is no electricity (see Figures 4–4 and 4–6).

4.1.2   Current

Current is the amount of free electrons which are forced by voltage to drift from one atom to another in a conductor. The amount of electrons that can flow (drift) in a conductor are limited by available voltage, the conductor size, and/or resistance. The larger the conductor, the less resistance to current flow, and the greater the current flow will be. An *ampere* is a measure of the rate of electrical flow. Current is measured by placing an ammeter in series in the circuit or by using an induction ammeter. Current is produced mechanically or chemically (see Figure 4–2 and refer to Figures 4–4 and 4–6).

There are two types of current—direct current (DC) and alternating current (AC). Direct current is the flow of electrons in one direction at a constant rate, such as the current that comes from the battery. It is used to operate starter motors, solenoids, lights, etc. It can also be stored in a capacitor. Alternating current is the flow of electrons, first in one direction and then in the other direction. It is used to run AC motors, household lights, etc. It can be rectified to direct current, as in the case of an alternator stator.

4.1.3   Resistance

Resistance is any restriction to the flow of free electrons in an electrical circuit. All electrical circuits have a designed resistance. Increasing or decreasing the resistance from the designed resistance will cause a change in current flow. A loss of resistance (load) will cause the current to go wild and overheat the conductor. An increase of resistance caused by corroded connections or frayed wires will decrease current flow, and will upset the circuit (Figures 4–3 and 4–6).

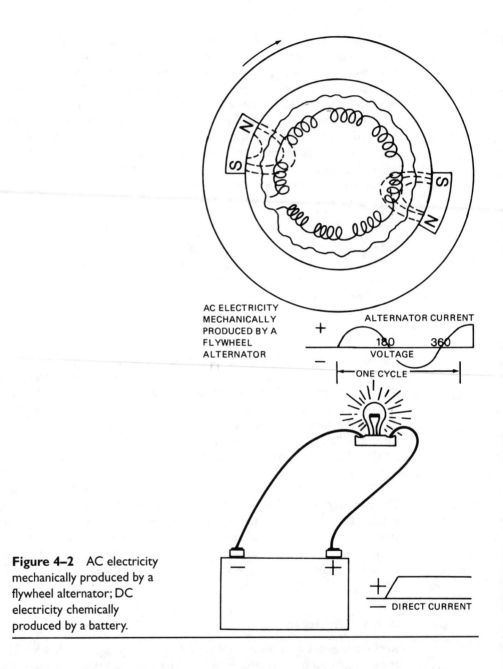

AC ELECTRICITY
MECHANICALLY
PRODUCED BY A
FLYWHEEL
ALTERNATOR

Figure 4–2 AC electricity mechanically produced by a flywheel alternator; DC electricity chemically produced by a battery.

The lower the temperature, the lower the resistance to electron flow in a conductor. Conversely, heat will increase the resistance to electron flow in a conductor. Resistance is measured in **ohms**, with an **ohmmeter**. Its presence can also be detected by making voltage drop tests on the circuit, using a voltmeter.

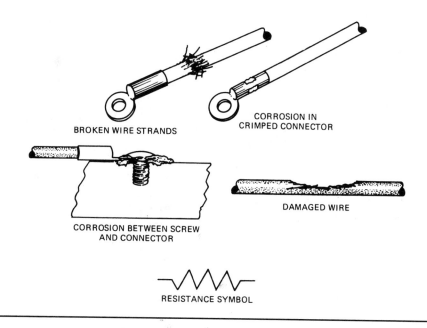

Figure 4–3   Examples of resistance.

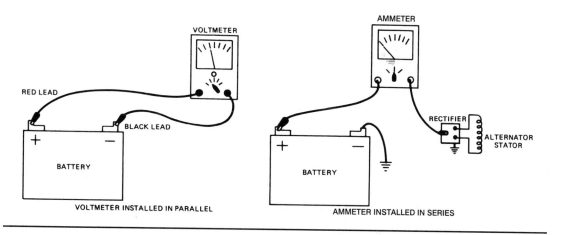

Figure 4–4   Voltmeter installed in parallel; ammeter installed in series.

4.1.4   Circuits

A *circuit* is an electrical path for current to flow through to a motor, ignition coil, or starter (hereafter called *load*). Electricity will not flow satisfactorily unless the circuit is com-

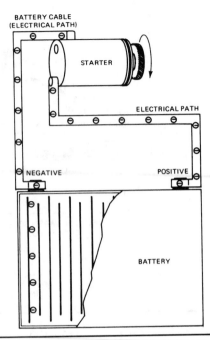

Figure 4–5   Battery voltage is applied, then current movement begins.

pleted by having a return path and having the necessary designed-in resistance (Figures 4–5 and 4–6). There are three types of circuits: parallel, series, and series parallel. The first two are common to the outboard and are defined here.

A *series circuit* is a circuit that provides only a single path for current to flow through, from electrical source through all the circuit's components and back to the electrical source. If just one component fails, the entire circuit will shut down. An example of circuit resistance is: (R) = switch, 1 ohm + light, 4 ohms + resistor, 2 ohms = 7 ohms.

Another example of a series circuit is the battery cables, solenoid, and the starter motor as shown in Figure 4–7. Component parts in a circuit add resistance. To measure resistance in this heavy amperage draw circuit, a voltage drop test is performed, using a voltmeter on the positive and negative sides of the circuit.

Series circuits follow laws that include the following:

1.   The total resistance in a series circuit is equal to the sum of the individual resistors (components).
2.   Amperage flow through a series circuit is the same throughout the circuit.
3.   Voltage across a series circuit is the sum of voltage across each separate unit.

In a *parallel circuit*, current has two or more electrical paths. Thus, if one circuit is open, the current still flows through the other circuits. Total current flow through the circuit is the sum of the current flow in each circuit. Voltage in a parallel circuit is the same throughout. Because of the many paths, total resistance to current flow is much less.

Volts, or voltage describes the amount of pressure needed to push electrical energy through a circuit. The higher the voltage, the greater the pressure. Electrical pressure is comparable to water pressure. When we move water through a pipe we must apply pressure to it. The water pressure is measured in pounds per square inch. The pressure that pushes electrical energy through a circuit is called volts or voltage. Water pressure or electrical voltage can exist with no flow as shown.

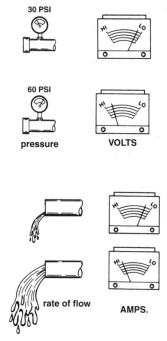

Amperage can be described as the rate of flow of electrons. The flow of water through a pipe is expressed in gallons per minute while the rate of electrons is expressed in amps or amperes. This is why when a high rate of electron flow is required such as to start a motor, the size of the conductor or (usually a No. 6 gauge wire) mass is increased.

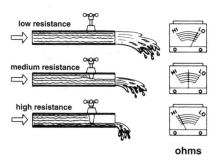

A water pipe and faucet can be used to describe resistance. The top view shows the faucet wide open, showing low resistance. If the faucet is turned half way off, we have a medium degree of resistance and 3/4 of the way closed, a high resistance of flow. This is also true electrically.

Figure 4–6   Volts, amps, and ohms, in a water comparison. *(Courtesy of Tecumseh Products Co.)*

The key switch in the starter control circuit is installed in a parallel circuit. The current that flows through it is not the same current that flows through the solenoid contacts and starter (see Figure 4–7). Parallel circuits commonly found on the outboard are ignition, power trim, and the lights on the boat. To check this type of circuit, a voltmeter is used to determine if voltage is present in the wire.

 Note

A continuity light may also be used, but it is not accurate.

How do we get the voltage to push the current through the designed resistance? Electricity, or the movement of free electrons, is initiated by either mechanical or chemical means. Through mechanical means, an alternator provides the voltage and thereby produces electron flow. This is done by passing a magnetic field through the conductor (flywheel magnets are passing the stator or coil). If the powerhead is starting, the battery (through chemical means) supplies the voltage and current flows through the starter solenoid and starter motor. The current flow will be normal if the resistance in the circuit is normal and there is a *return path* to the battery (Figure 4–7).

4.1.5   Load

Every electrical circuit must have a load (resistance). A *load* can be a light bulb, motor, ignition coil, or an ignition module. All of these parts are resistance units that perform a

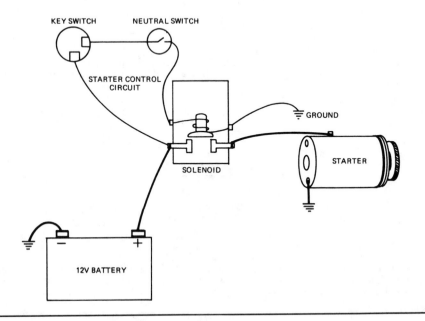

Figure 4–7   Starter motor control circuit.

function and put a designed resistance into the electrical circuit. The engineer that designed the circuit determined the needed resistance for the circuit, thus, the installation of a certain size wire and certain part-numbered solenoid, ignition coil, etc., for a given circuit. Without this designed load, the circuit would flow too much current, heat to the point of burning the insulation, and burn the wire into two pieces.

4.2   Ohm's Law

A formula known as Ohm's law allows the engineer to put the electrical circuit into mathematical terms. Ohm's law states: For any circuit or part of a circuit under consideration, the current in amperes is equal to the electromotive force in volts divided by resistance in ohms. Put another way, it takes one volt to push one ampere through one ohm of resistance. Formulas from Ohm's law are below:

$$\text{Amperes} = \frac{\text{Volts}}{\text{Resistance}} \quad \text{or} \quad I = \frac{E}{R}$$

Cover with your finger the one factor you need to find, then divide or multiply.

$$\text{Resistance} = \frac{\text{Volts}}{\text{Amperes}} \quad \text{or} \quad R = \frac{E}{I}$$

$$\frac{E}{I \mid R}$$

Volts = Amperes X Resistance
$$\text{or } E = I \times R$$

Power (watts) = Volts X Amperes
$$\text{or } W = V \times A$$

Electromotive force = volts = E

Current = Amperes = I

Resistance = Ohms = R

Power (watts) = W

A simple way to remember the relationship between E, I, and R is to set them in the following form: $\frac{E}{I \times R}$. When you need to find an unknown factor, cover the unknown symbol and read the relationship between the other two. Probably the only time that you would need to use this formula would be in building a circuit from scratch. However, it should be evident that the basic relationship between voltage, current, and resistance is most important in any systematic approach to troubleshooting the electrical circuits on the outboard.

  Note

Note the circuits given in Figure 4–8. Apply Ohm's law to them, working through the mathematical equations.

(a) $3 + 3 + 7 + 2 = 15$ ohms of total resistance.

(b) $9 + 7 + 2 = \dfrac{12V}{18 \text{ ohms}} = .6$ amp of current flow

(c) $\dfrac{6 \times 12}{6 + 12} = \dfrac{72}{18} = 4, \quad \dfrac{4 \times 4}{4 + 4} = \dfrac{16}{8} = 2$ ohms

Calculator Formula for c, d, e, and f.
Step 1: $1 \div 6\ \text{M} + 1 \div 12\ \text{M} + 1 \div \text{RM} = 4$
Step 2: $1 \div 4\ \text{M} + 1 \div 4\ \text{M} + 1 \div \text{RM} = 2$ ohms total resistance

(*Note*: This example is taken from diagram C below)

(d) $\dfrac{6 \times 3}{6 + 3} = \dfrac{18}{9} = 2, \quad \dfrac{2 \times 2}{2 + 2} = \dfrac{4}{4} = 1,$

$12V$ divided by $1 = 12$ amps

$(.75 + 1.25 = 2)$

(e) $\dfrac{2 \times 2}{2 + 2} = \dfrac{4}{4} = 1, \quad \dfrac{3 \times 6}{3 + 6} = \dfrac{18}{9} = 2,$

$1 + 2 = 3 =$ total resis tan ce

(f) $\dfrac{1 + 1 + 1 = 3,}{1 + 2 + 3 = 6} \quad \dfrac{4 \times 4}{4 + 4} = \dfrac{16}{8} = 2,$

$\dfrac{5 \times 6}{5 + 6} = \dfrac{30}{11} = 2.72$ total resis tan ce

$\dfrac{12V}{2.72 \text{ ohms}} = 4.4$ amps

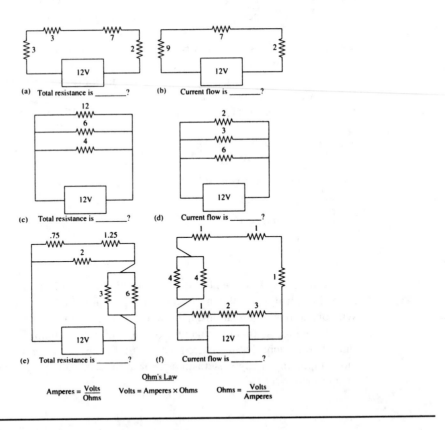

(a) Total resistance is _____?

(b) Current flow is _____?

(c) Total resistance is _____?

(d) Current flow is _____?

(e) Total resistance is _____?

(f) Current flow is _____?

Ohm's Law

$\text{Amperes} = \dfrac{\text{Volts}}{\text{Ohms}} \qquad \text{Volts} = \text{Amperes} \times \text{Ohms} \qquad \text{Ohms} = \dfrac{\text{Volts}}{\text{Amperes}}$

Figure 4–8   Circuit mathematical equations. Apply Ohm's law and figure the total resistance of (a), (c), and (e), and figure the current flow of (b), (d), and (f).

One of the main reasons for failure of a given electrical circuit is unwanted resistance. This resistance to the flow of free electrons may be caused by corrosion, loose terminals, or the use of undersized wire for the current to be carried in (Figure 4–3). Resistance can also develop beyond design specifications in ignition coils, points, spark plug cables, etc. Most of the circuit repairs are to bring the resistance back to the design specification by cleaning the connection and replacing coils, switches, and other parts. The marine environment is responsible for a lot of the problems because moisture accelerates corrosion of connections, while heat, age, and vibration also destroy component parts.

Other than resistance, there are three basic causes for the failure of electrical components:

1. **Shorts:** Shorts occur when two adjacent bare conductors (wires or hot wire to ground) make a contact and therefore the electrical contact allows a bypass of the load. The short reduces circuit resistance, and excessive current will flow beyond design limit of the circuit. The conductor warms up because of the excessive current flow. Depending on how and where the short occurs, it is possible to blow a fuse or open a heat-sensitive circuit breaker. One type used in 12V marine applications will automatically reset as the circuit breaker cools down. An *ohmmeter* is used to check resistance. Components within a starter motor or power trim motor, such as an armature, may short. These armatures can be checked with an ohmmeter, but for accuracy, an armature tester should be used.

2. **Ground.** In a single wire electrical system, the insulated wire (conductor) in the circuit carries the positive (hot) current to the load (current theory). After passing through the load, electricity is routed to ground, through the powerhead assembly and negative battery cable, then to the negative battery terminal (a common return path). An unintentional ground exists when the positive lead touches the powerhead or negative battery cable before reaching the load designed into the circuit. When this happens excessive current will immediately flow. A short in an auxiliary circuit would blow a fuse or open a heat-sensitive circuit breaker, which is installed in the circuit to protect against accidental heavy current flow. Grounds are checked by using a continuity light or with an ohmmeter when electricity is off.

3. **Open.** An open is like a bridge over a river that has washed out. There is a gap, and you cannot cross over it. An open is a break in the wire, a wire that is disconnected, one that accidentally came loose from a connector, or one that has been purposely cut. With a break, there is no continuity in the circuit, so there is no current flow. The electricity is unable to get back to the source, which is the battery or the alternator. Opens are always checked with a continuity light, voltmeter, or when the electricity is off, use an ohmmeter.

4.3   Insulation

A plastic coating around a wire is an insulator. The insulation protects the wire and keeps the wire from contacting other wires or the powerhead. This prevents an accidental short to other circuits or to ground.

4.4   **Semiconductors**

A *diode* is an electrical device that will permit current to flow through itself in one direction. The "arrowhead" symbol shows the direction of current flow, from positive to negative (Figures 4–9 and 4–10). Diodes can be made of silicon or germanium, which in pure form will not pass electrons. Therefore, "doping" or adding an impure element to the pure crystal, is accomplished in its molten state. The kind of impurity added to the silicon determines whether it will have an excess or absence of electrons. For example, doping the pure silicon with phosphorus will result in an N-type material, or material with an excess of electrons. Doping with boron will create a P-type material with the absence of electrons. When voltage is applied to this doped crystal of silicon, there is an orderly flow of electrons from one terminal of the voltage source to the other. Voltage moves an electron into one end of the crystal, and an electron exits at the opposite end, thus maintaining a constant number of electrons in the material (diode).

When a diode is connected in the opposite manner (reverse bias) it will not conduct current. It cannot do so since the N-material side of the diode is connected to the positive battery terminal and the P-material side to the negative battery terminal. The electrons in the N-material are attracted to the positive battery terminal side away from the diode junction. At the same time, the holes in the positive diode material are attracted to the negative battery terminal of the diode away from the junction area. This creates an open circuit, which blocks the flow of electrical current.

Diodes are used in the circuitry of alternators and electronic ignition systems.

A *transistor* is an electrical device that acts as an electrical switch. It can be used to turn circuits on or off and has no moving parts. Therefore, it is a completely static device and is entirely electrical. Consequently, there are no contact points or rubbing blocks to wear out in modern-day electronic ignitions. While the diode has one P-type material, in the transistor there are two areas of P material and one area of N material to dope the silicon crystal (transistor).

In Figure 4–11, a bipolar transistor is shown symbolically, indicating the manner in which it operates. Only when there is current flow in the emitter-base circuit (gate) is there current flow in the emitter-collector circuit. The emitter-base circuit (gate), which requires only a minimal current, can act as the trigger to turn this "switch" on or off, controlling a heavier current flow in the emitter-collector circuit.

The *silicon-controlled rectifier* (SCR) is similar to a bipolar (PNP) transistor but has a fourth layer and therefore three PN junctions. It is sometimes called a four-layer PNPN diode since it passes a current in only one direction. When the anode (Positive side) of an SCR is made more positive than the cathode (Negative side), the two outermost PN junctions are forward biased. The middle PN junction, however, is reverse biased and current cannot flow. By its nature, the SCR is a semiconductor that does not allow current flow in either direction. However, when the SCR is "triggered" by directing a positive voltage to its gate, it allows current to flow from anode to cathode. Unlike the transistor "switch," the benefit is that only a momentary trigger is needed at the gate to "start" the SCR conducting. As long as voltage across the anode and cathode remains high enough, the SCR will continue to conduct (Figure 4–12). A bipolar transistor requires voltage at the base at

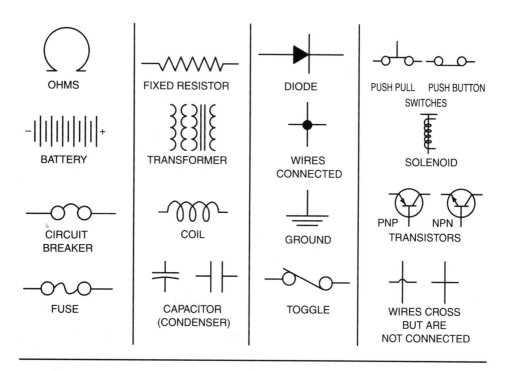

Figure 4–9   Electrical symbols used on schematic diagrams.

all times to remain "on." The SCR is incorporated into the outboard ignition module and triggers a release of voltage from the capacitor through the ignition coil (transformer) for high voltage to the spark plugs.

Resistors come in various sizes and shapes, but they all do the same thing: they resist or limit current flow. They are striped with color bands to express the resistance value of the resistor. You can use the color chart to determine a given value of a resistor, or you can measure it using an ohmmeter (Figure 4–13).

Capacitance is the ability or capacity to store voltage on two conducting plates separated by a dielectric (insulator). The plates can be made of aluminum foil, zinc, steel, or copper. The insulator can be paper, ceramic, glass, or plastic. (This definition describes a capacitor used in ignition systems.) Capacitors are rated in microfarads. The capacitor can also be called a *condenser*, because electric charges collect on the plates, like moisture fogging a glass window pane.

As magnets in the outboard flywheel rotate past the ignition charge coil, electrical energy is produced (see Figure 4–18). This energy is transferred to the capacitor, where it is stored until it is released through the ignition coil to fire the spark plug (Figure 4–13).

Since the diode, transistor, and resistor may come in a sealed ignition module, you are unable to make repairs on them. (Repair involves troubleshooting and replacing the

CONVERTING ALTERNATING CURRENT TO DIRECT CURRENT

"Because alternating current is rapidly changing we place a diode in the circuit to cause it to flow in only one direction."

A single diode produces half wave rectification which may be used in charging a battery.

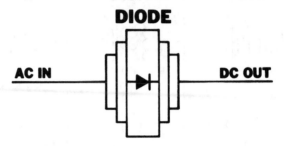

The figure below shows how the diode blocks out the negative half of the alternating current.

In other systems, multiple diodes are used in the full wave rectifier as shown below.

Figure 4–10   Converting alternating current to direct current. *(Courtesy of Tecumseh Products Co.)*

module.) However, it is helpful to know what they do and how they operate. This will help you understand the operation of alternators and electronic ignition in later chapters.

Magnetism is a property possessed by certain materials that allows these materials to exert mechanical force on neighboring masses of magnetic materials. As a child you probably played with a magnet, picking up things like screws, tacks and nails. This same magnetic attraction applies to a wire conductor. Magnetism and electricity are closely related. This can be demonstrated by wrapping a coil of wire around a piece of iron bent into a horseshoe shape, and supplying low voltage electricity to the coil of wire. The result is a magnetic field created in and around the iron bar. Therefore, it becomes an electromagnet.

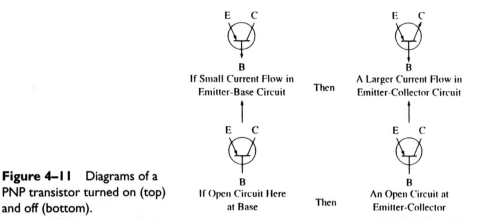

Figure 4–11 Diagrams of a PNP transistor turned on (top) and off (bottom).

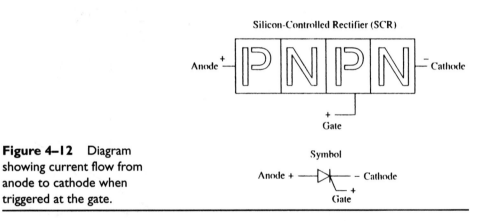

Figure 4–12 Diagram showing current flow from anode to cathode when triggered at the gate.

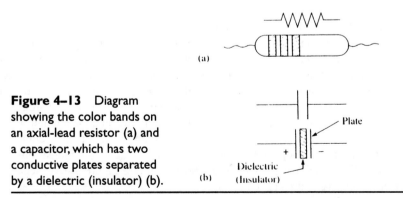

Figure 4–13 Diagram showing the color bands on an axial-lead resistor (a) and a capacitor, which has two conductive plates separated by a dielectric (insulator) (b).

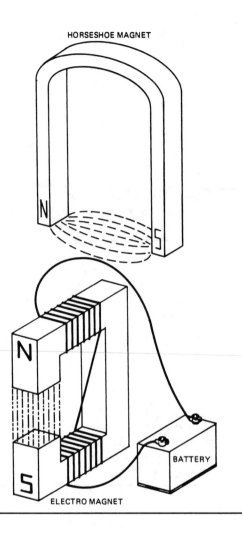

Figure 4–14   Magnetism.

The magnetic circuit is south to north internally and north to south externally (Figure 4–14). By placing a moving wire (coil of wire) **_conductor_** within a magnetic field, or by moving the magnetic field over a conductor, electricity will be generated in the wire conductor. There must be three things present for this to happen: motion, magnetic field, and a conductor.

Electricity flowing in a conductor will create its own magnetic field around that conductor, as shown in Figure 4–15. The strength of the magnetic field around a conductor is dependent upon wire size, the number of turns of wire, and the amount of current flowing in the wire. Reverse the current flow through the wire, and the direction of the magnetic field will also be reversed. Whether it is a regular magnet or an electromagnet, like poles will repel and unlike poles will attract each other (see Figure 4–16). Negative and positive electrical charges attract each other.

Electricity for the Technician

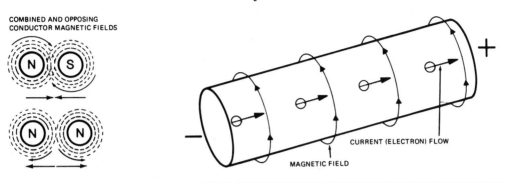

COMBINED AND OPPOSING
CONDUCTOR MAGNETIC FIELDS

CURRENT (ELECTRON) FLOW

MAGNETIC FIELD

Figure 4–15   Current flow creates a magnetic field around the conductor (left-hand rule applies—electron theory).

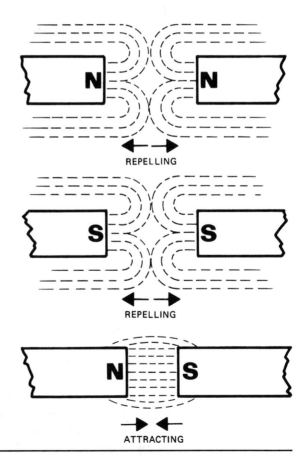

REPELLING

REPELLING

ATTRACTING

Figure 4–16   Magnetism.

Mutual magnetic induction occurs if a wire is present within the variable magnetic field of the current-carrying wire. Voltage is induced into that secondary wire by the action of the moving magnetic field. Induction only takes place when the current in the primary wire is turned on or off or when it reverses direction, as in AC current. The ballooning and collapsing effect of the magnetic field is necessary to have induction. In this case, that is the motion (Figure 4–17). Induced alternating current can be generated by mechanical means. This is done by passing the flywheel magnets past the stator coils, thus the alternator (Figure 4–18).

How does this apply to the outboard? The starter motor turns because of magnetic fields repelling and attracting, which forces the rotation of the starter motor armature. The flywheel carries magnets that are moving past the stator charging coils inducing alternating current into the stator winding; thus, we have alternator output. The ignition coil, with its primary and secondary windings, is a step-up transformer. This steps up the secondary voltage because of a moving magnetic field, producing a spark at the spark plug for ignition. These are some of the practical applications that are applied to make the outboard motor what it is today.

4.5 Solutions from the Wiring Diagram

Consider a road map that will guide you from Anaheim, California, to Chicago, Illinois. How many roads will take you to Chicago? There are many choices but you will choose only one route and follow it; in this way, you will not get lost.

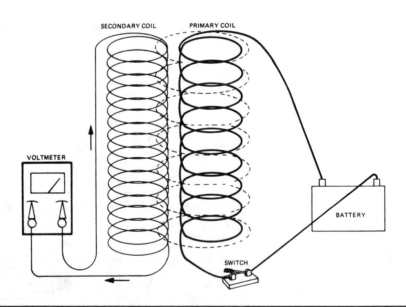

Figure 4–17 Electromagnetic/mutual induction. As the field shown here collapsed, it cut the secondary coil as well, inducing a voltage and current flow within that coil. This principle is used in ignition coils.

Using principles of electricity and magnetism, if a magnetic field is produced around a current - carrying conductor, The reverse is also true. When a magnetic field cuts a conductor there will be a current flow. Using this fact we can see how electricity is induced into a coil whenever a magnetic field passes across the coil. When a magnet passes the end of a coil, an electrical flow is generated within the coil. The amount of this current (or flow) is determined by the magnet, wire size, size and shape of steel laminations, number and direction of the turns in the coil, and speed of the engine.

Here we show how two coils are acted upon by the North Pole magnet and the South Pole magnet. Note that the coils are series connected.

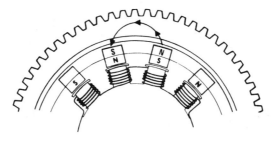

Figure 4–18   Electrical induction. *(Courtesy of Tecumseh Products Co.)*

Wiring diagrams are similar in design. They are illustrated in block schematic form, which illustrates the electrical components and connections, but generally not their physical location on the outboard or boat. This is done to simplify the diagram, and is the same as a road map. We search out one route, one circuit on the map, and follow it to the destination. We do not try to view the total maze on the map. In the same way, we follow a wiring diagram, by searching out one circuit and following it from the component that is not working to the source of power. We follow the schematic from the non-operating part through the circuit (line on schematic) to the battery. We have, therefore, an understanding of all the connections and components that are in the circuit. With this understanding we are able to schematically check the circuit. Remember, in order to operate, an electrical device must have a completed electrical path to it from the electrical source and a return path to the source. This return path is the ground wire in the boat or the powerhead itself, which takes the place of a return wire. Locate the non-operating component on the schematic and follow the wires (lines) to the current source. There will be switches, connections, and possibly a fuse along the way. Don't be misled and branch out of the circuit into another one.

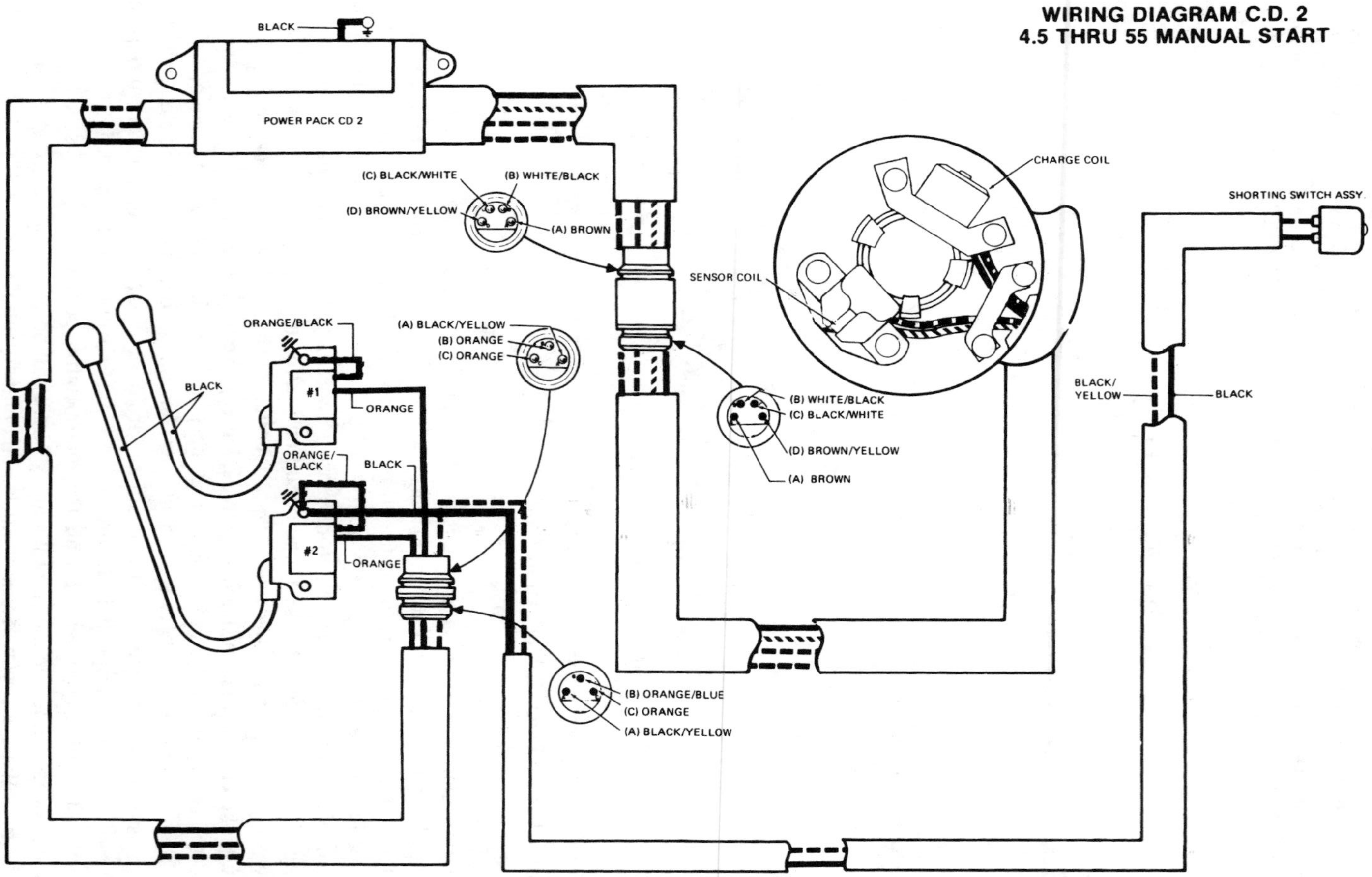

Figure 4-19 Wiring diagram C.D. 2., 4.5 through 55 manual start. *(Courtesy of Outboard Marine Corporation)*

196

Many circuits are in parallel and share the same electrical source at a switch or a circuit breaker (see Figure 4–19).

4.5.1   Systematic Elimination

If a motor is not working we generally know it. This is the starting point for service checks. In a step-by-step manner check for the presence of electricity at the motor and at each convenient location along the circuit. This is similar to stopping for gas on your trip. If we get a voltage reading at the motor lead we know that the motor is bad or the return path (the ground) is in question. Jump the motor with a jumper lead to ground and see if it will operate. If it does, repair the ground. If no voltage is present, then backtrack on the circuit until voltage is found, and make the necessary repair. Go back to the motor and test for voltage and then run the motor.

There are times when voltage may be picked up with a test light. However, it is hard to tell sometimes if the light is bright or dim. If it is a little dim, then there may be just enough voltage to light the bulb but insufficient current flow to run the motor, because of resistance in the circuit. Working with a voltmeter and ammeter is the best, so the amount of voltage and current is known.

The process of elimination is a rapid means of determining the cause of circuit failure. The idea is to plan a sequence of tests and follow it. Jumping around is haphazard testing and will lead to frustration.

 Safety

Always test with electrical test instruments and do not spark a wire to see if current is there—this can damage electronic components.

4.5.2   Electrical Measurements

Instruments that can measure the electrical circuits are the voltmeter, ammeter, and ohmmeter.

The *voltmeter* is a high-resistance instrument used to measure the electrical pressure in a circuit. It is hooked up in the same way you would install a light into a circuit. It is placed in parallel with the battery, which is red lead on positive (hot) and black lead on negative ground, or from a hot wire to a good ground (Figure 4–20).

The *ammeter* is a low-resistance instrument used to measure the current flow in an electrical circuit. An ammeter is installed in series; this means that the circuit is disconnected, and the ammeter is spliced into the circuit. There are induction-type ammeters where this does not have to be done. This type is placed along the side or the induction pick-up is clipped around the wire or cable to obtain the flow of amperes. All of the current flows through the ammeter (when installed in series) from the battery or alternator to the rest of the circuit. The ammeter is designed so that the internal resistance in the meter is very low, and so its presence in the circuit will have minimal effect on the total circuit resistance.

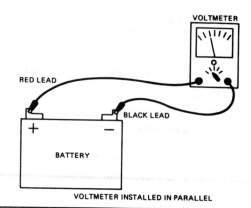

Figure 4–20   A voltmeter
installed in parallel.

VOLTMETER INSTALLED IN PARALLEL

 Warning

The ammeter should never be installed in parallel, as you would a voltmeter. Because of the low resistance of the ammeter, high current would flow through and "fry" the instrument (see Figures 4–21 and 4–22).

The **ohmmeter** is not used in a circuit when electricity is present, so you must *disconnect the circuit* from the battery. The ohmmeter uses an internal battery. The instrument will measure the resistance in an electrical circuit. There are generally several resistance ranges incorporated in one meter.

The ohmmeter is equipped with batteries to supply a given current to the circuit through the test leads and to the indicating meter. Any circuit or resistance placed between the test leads will reduce the current flow back to the meter, and will be indicated at less than full scale meter movement, indicating the resistance in ohms (see Figure 4–23).

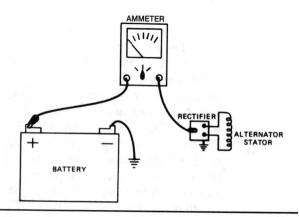

Figure 4–21   Ammeter
installed in series.

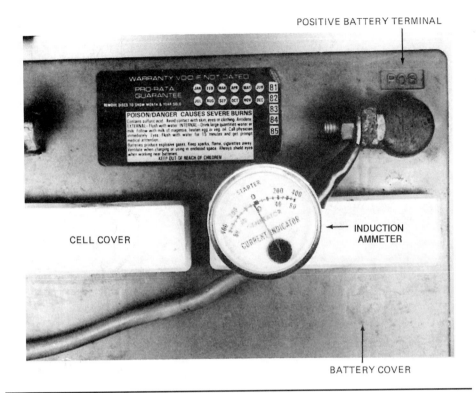

Figure 4–22   Use of an induction-type ammeter.

4.6   **Electrical Principles**

The following principles of electricity are helpful to understand when working with electricity as it is used in the outboard.

1. Atoms consist of positive charges (protons) and negative charges (electrons).
2. Applied voltage to a conductor causes movement of the electrons through a conductor, and is called an *electric current.*
3. The higher the voltage, the greater the electron flow (which is measured in amperes with an ammeter).
4. Conductors are made of materials that contain free electrons. Free electrons move from one atom to another when electrical voltage (pressure) is applied and the circuit is complete.
5. Electron flow is from negative to positive (electron theory).
6. Current flow is from positive to negative (current theory).

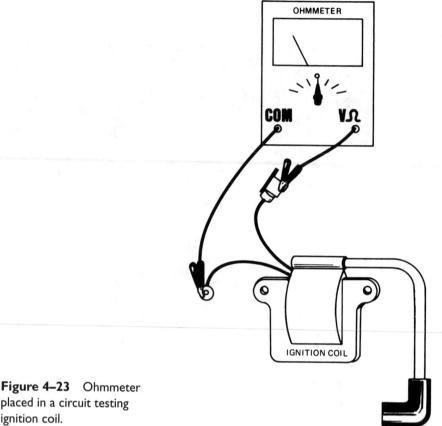

Figure 4–23   Ohmmeter placed in a circuit testing ignition coil.

7. Current flows from the positive battery terminal into a circuit through a switch to the load, and returns through ground back to the negative battery terminal. This is a completed electrical path (current theory).

8. Resistance is measured in ohms. All circuits (conductors) resist the flow of current to some extent. This depends upon:
    - Conductor temperature—resistance usually increases with temperature.
    - Conductor size—the smaller the cross section, the higher the resistance.
    - Conductor length—the longer the conductor, the higher the resistance.
    Conductors are made of silver, copper, aluminum, steel, and iron because it is easy to create an electrical imbalance with applied voltage to move the free electron in a completed electrical circuit.

9. Insulators are made of plastic, nylon or glass.

10. The voltage drop across any circuit element is directly proportional to its resistance and the amperage flowing through it. Ohm's law explains the interaction of voltage (E), current (I), and resistance (R) in any closed electrical circuit:

Volts = Amperes × Resistance $E = I \times R$

$$Ohms = \frac{Volts}{Amps} \qquad\qquad R = \frac{E}{I}$$

$$Amperes = \frac{Volts}{Ohms} \qquad\qquad I = \frac{E}{R}$$

$$\frac{E}{I \mid R}$$

Knowing any two values it is possible to figure the third value.

11. Magnetic fields have a south and north pole. Like poles (north-north) repel each other. Unlike poles (north-south) attract other.

12. An electromagnetic field is created when electricity flows through a coil of wire. Reverse the electrical flow and the poles will reverse.

13. The magnetic field is from south to north internally within the magnet, and from north to south external to the magnet.

14. Left-hand rule: You can determine the direction of lines of force around a conductor by taking it in your left hand, with thumb pointing the direction in which the current is flowing. Wrap your fingers around the conductor. This is the same direction that the lines of force are revolving.

15. Left-hand rule for polarity of an electromagnet: This rule states that when the fingers of the left hand are wrapped around a coil in the direction of current flow, the thumb will point to the north pole of the electromagnet. Of course, if the direction of current flow were reversed, polarity too would be reversed.

16. An electromagnetic field will attract iron toward the magnet (coil).

17. Strength of an electromagnetic field is dependent upon:
   • The amount of current flowing through the coil.
   • The type of coil core (air or iron).
   • The number of turns of wire on the coil and the gauge of the wire.

18. ***Electromagnetic induction:*** A voltage can be induced by physically passing a conductor through a permanent magnetic field, or by passing a magnetic field past a conductor. The amount of voltage depends upon:
   • The speed of cutting the magnetic field by the conductor (or field). The greater the speed, the higher the voltage.
   • The strength of the magnetic field.
   • The number of conductors. The more wraps of wire the higher the voltage.

19. If a coil's alternating magnetic field (AC) cuts the turns of a second coil, the voltage induced in the second coil will be more or less than the voltage in the first coil, depending upon whether there are more turns or less turns on the second coil.

20. Conductors located within a magnetic field try to move.

21. A capacitor or condenser can temporarily store electricity, releasing it when triggered by a sensing circuit (trigger circuit).

22. Outboards have AC current in the stator and it is rectified to DC current at the ignition module (power pack) and at the rectifier.

23. The battery stores electricity in chemical form. Cell voltage for a fully charged battery is 2.1V. The cells are usually connected in series so the total battery voltage for six cells is 12.6V.

24. Two 12-volt batteries connected in series, negative to positive, will produce 24V. Series multiplies voltage.

25. Two equal 12-volt batteries connected in parallel will double capacity of one battery.

26. Battery rating "Marine Cranking Amperes": This is the number of amperes which a battery can deliver at 32 degrees Fahrenheit for 30 seconds and maintain at least a voltage of 1.2V per cell or higher. This differs from cold cranking amperes which are measured at 0 degrees Fahrenheit.

27. Battery rating "Reserve Capacity": The number of minutes for which a new fully charged battery will deliver 25 amperes at 80 degrees Fahrenheit. This represents the time which the battery will continue to operate essential accessories in the event of alternator failure.

KNOW THESE PRINCIPLES OF OPERATION

- Basics of electricity.
- Ohm's law.
- Parallel, series, and series parallel circuits.
- Electromagnetic induction.
- Magnetic fields.
- Semiconductors.

REVIEW QUESTIONS

1. Current theory indicates that electricity flows
   a. negative to positive
   b. positive to negative
2. Conductors are made of plastic and nylon.
   a. true
   b. false
3. Define the term *current*. (Write your answer on the back of the answer sheet.)
4. A larger conductor offers more resistance than a smaller conductor to equal flow of electricity.
   a. true
   b. false
5. List two types of current/voltage.
6. Define the term *resistance*. (Write your answer on the back of the answer sheet.)

7. The _____ the temperature, the lower the resistance to electrical flow in a conductor.
   a. higher
   b. lower
8. Resistance is measured in _____ .
   a. volts
   b. amperes
   c. ohms
9. Current is measured with a _____ .
   a. voltmeter
   b. ammeter
   c. volt/ohmmeter
10. Voltage is the _____ that causes the electrons to flow through a conductor.
   a. pressure
   b. current
   c. resistance
11. Electricity will not flow satisfactorily unless the circuit is_____.
   a. opened
   b. completed
   c. magnetized
12. A circuit is an electrical
   a. path.
   b. magnetic path.
   c. flow of protons.
13. Circuits that split the current are called
   a. series circuit.
   b. parallel circuit.
14. A designed electrical load for an outboard electrical circuit may be an
   a. ignition module, ignition coil.
   b. trim motor, starter motor.
   c. stator winding and/or any of the above.
15. Express Ohm's law in writing. (Write your answer on the back of the answer sheet.)
16. Failure of electrical circuits (components) may be classified under what three broad terms? (Write your answer on the back of the answer sheet.)
17. A short _____ the circuit and excessive current will flow beyond the designed limit of the circuit.
18. List two devices that are installed in electrical circuits to prevent excessive current flow.
19. An open is a _____ in a wire.
   a. resistance
   b. break
   c. connector

20. A diode is an electrical _____ .
     a.   switch designed to open with a positive pulse to the gate
     b.   check valve
     c.   relay coil
21. Electricity flowing through a wire wrapped into a coil, with a soft iron bar in the center of the coil, becomes a _____ .
     a.   magnet
     b.   silicon-controlled rectifier
     c.   electromagnet
22. Strength of a magnetic field around a conductor is dependent upon
     a.   wire size and the amount of current flowing.
     b.   the amount of turns of wire.
     c.   Both a and b are correct.
23. Electricity can be induced through mechanical means by
     a.   quickly passing a flywheel magnet within a few thousandths of the ignition coil.
     b.   quickly passing an electromagnet near the flywheel of an outboard.
     c.   Both a and b are correct.
24. The electrical wiring diagram is similar to a road map.
     a.   true
     b.   false
25. The voltmeter is a high-resistance meter.
     a.   true
     b.   false
26. What turns on a silicon-controlled rectifier?
     a.   the ignition switch
     b.   a positive pulse to the gate
     c.   the tilt switch
27. A capacitor can be used to
     a.   store electricity
     b.   open a circuit
     c.   pulse the gate of the SCR
28. An ohmmeter is used to check resistance in a circuit when the electricity is_____ .
     a.   on
     b.   off
     c.   intermittent
29. Current flow is measured with an induction-type ammeter, or an ammeter is placed in _____ with the circuit.
     a.   parallel
     b.   series-parallel
     c.   series

30. Mutual electrical induction is the transfer of energy from one object to another
    a. when objects are in physical contact.
    b. without the objects being in physical contact.
    c. when magnetic fields oppose each other.
31. Current flowing from a 12V battery into a circuit that has 9, 7, and 2 resistance units is
    a. 18 amps.
    b. .6 amps.
    c. .9 amps.
32. Write the formula to determine the flow of amperage in a circuit.
    Amperes = _____
33. Write the formula to determine resistance.
    Resistance = _____
34. For a silicon-controlled rectifier to remain on, it must continue to pass current from the gate to the cathode.
    a. true
    b. false

CHAPTER 5

The Marine Battery

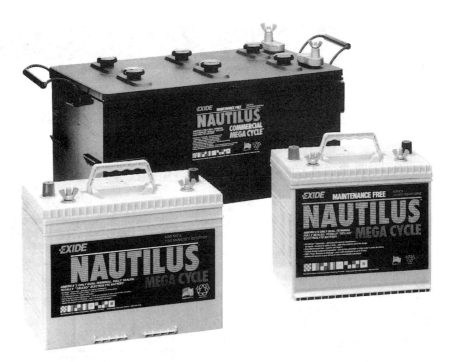

(Courtesy of Exide Corporation)

Objectives

After studying this chapter, you will know:

- Battery construction.
- Safe handling of the battery.
- The dangers of electrolyte (sulfuric acid).
- How the battery produces electricity.
- How to test the battery's state of charge.
- How to charge the battery.
- How to load-test the battery.

5.1   Function of the Battery

The battery is a reliable source of portable direct current, and is available in several sizes and reserve capacity. Care must always be used around the battery (see Figure 5–1). The marine battery must perform the following three main functions:

SAFETY WARNING

Batteries contain sulfuric acid electrolyte. This is a highly **CORROSIVE POISON**. They also produce a mixture of hydrogen and oxygen gases which will **EXPLODE** if ignited.

WHEN WORKING ON OR NEAR BATTERIES, MIXING OR POURING ACID SOLUTIONS, ALWAYS WEAR PROTECTIVE CLOTHING AND PROTECT EYES WITH SAFETY GOGGLES. KEEP SPARKS, FLAMES AND CIGARETTES AWAY.

KEEP BATTERIES AND ACID OUT OF REACH OF CHILDREN.

If acid contacts skin or eyes, flush affected parts with clean water immediately and repeat for 15 minutes. Then seek prompt medical attention.

If acid is taken internally, call medical help immediately. Drink large quantities of water, milk or milk of magnesia, beaten eggs or vegetable oil.

Acid spilled on clothing, workbench or floor may be neutralized with baking soda or ammonia solutions. Do not store acid, or mix acid solutions in metallic containers.

Use only glass, ceramic or acid resisting plastic vessels. Never discard used containers before they have been rinsed clean, then puncture them to prevent further use.

When charging batteries, keep area well ventilated and bar general access. Connect/disconnect batteries only when charge is switched off. Make sure tools cannot short-circuit battery terminals. Keep vent caps on battery during charging.

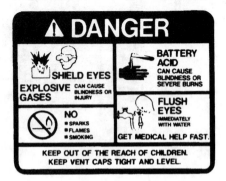

Figure 5–1   Battery safety information. *(Courtesy of Exide Corporation)*

BATTERY CONSTRUCTION

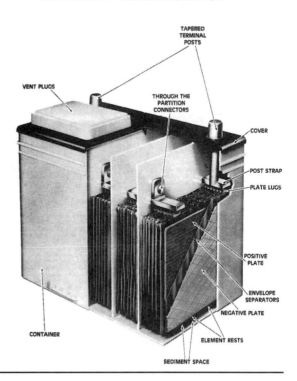

Figure 5–2   Battery construction. *(Reproduced with permission
from the Battery Council International)*

1.   Supply power to the starter for cranking the powerhead.
2.   Supply extra power when the alternator system is unable to supply the demand.
3.   Act as a voltage stabilizer in the electrical system.

If properly selected and installed, the battery will supply 12 volts of electricity for many
boating seasons.

5.2   Battery Ratings

There are two minimum SAE-rated (Society of Automotive Engineers) specifications that
apply to the outboard starting battery that will keep the cranking RPM within specifications. Outboards should use a battery with 360 Cold Cranking Amperes (amps) (CCA)
with a minimum of 90 minutes reserve capacity at 80°F. A higher rated battery is an Exide
Nautilus marine battery which has a reserve capacity of up to 180 minutes at a 25-amp
draw. This assures Loran pilot, lights and radio operation for emergency operation, if the
alternator should fail. Deep-cycle batteries are suitable *if* they meet or exceed the minimum
Marine Cranking Amps (MCA) requirements. These ratings with the battery charged will
supply adequate amperage for the starter motor and maintain a voltage above 9½ volts (V).

This is important so that the starter motor will not overheat because of low voltage, and will crank the powerhead above the minimum cranking RPM specification. Slow cranking RPM means a hard start because of low ignition output to the spark plugs, lower compression, and that the starter motor will run hotter.

Remember that the reserve capacity rating also means safety, radio communications, and Loran pilot operation, which is very important to guide the boat operator and passengers back to the harbor when there are alternator problems.

5.3   The Marine Battery

Automotive batteries are *not* recommended for marine use. Some marine batteries have specially designed glass mat separators to reduce shedding of the active material and damage from vibrations when operating the boat in choppy water. Some Exide marine batteries have "gelled" electrolyte and the "Gel Cell" is totally maintenance free—install it and forget it. These batteries contain no liquids to spill or leak. They are factory-sealed to prevent water leaking into the battery if submerged to 40 feet. They also resist the effects of vibration. However, they require the use of a special battery charger.

Marine Cranking Amps (MCA) specifications are higher than automotive Cold Cranking Amps (CCA). Here is an example using 550 MCA versus 410 CCA:

550 MCA: The number of amps a new battery can deliver for 30 seconds at 32°F without the terminal voltage falling below 7.2V for a 12-volt battery.

410 CCA: The number of amps a new battery can deliver for 30 seconds at 0°F without the terminal voltage falling below 7.2V for a 12-volt battery.

The MCA rated battery will better meet the demands of the boat and outboard electrical needs. It has 140 additional cranking amps at the same voltage.

5.3.1   Construction

A storage battery element is made up of alternating positive and negative plates made of two dissimilar leads (see Figures 5–2 and 5–3). They are kept apart by separators, which are porous, and made up of resin-impregnated cellulose fiber, microporus rubber, polyethylene or plastic. The separators permit the passage of charged ions of electrolyte between negative and positive plates. To add capacity to the battery, a desired number of *negative* plates and a post strap are welded together. The desired number of *positive* plates and post strap are also welded together. The separators are then inserted between the plates with the ribs of the separator placed next to the positive plate. When assembled by placing one positive plate group and one negative plate group together with separators, it is known as an *element*. There is one element per cell (Figure 5–4).

Cell voltage becomes 2.1V. To make a 12-volt battery, six cells are installed in series using through-the-partition connectors (Figure 5–5). This brings the fully charged battery voltage up to 12.6V. The more plates (square inch area) a battery has, the more capacity is built into the battery.

The plate grid is a framework to hold the active material, and is made of lead alloy and some antimony to strengthen and stiffen the soft lead (see Figure 5–3). The positive plate

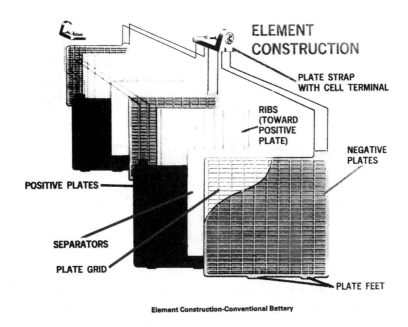

Element Construction-Conventional Battery

Figure 5–3   Element construction. *(Reproduced with permission from the Battery Council International)*

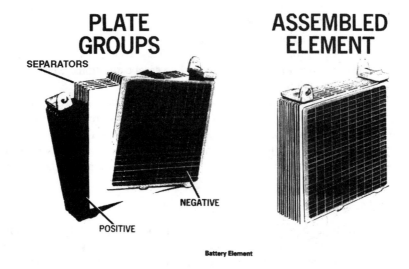

Battery Element

Figure 5–4   Battery element. *(Reproduced with permission from the Battery Council International)*

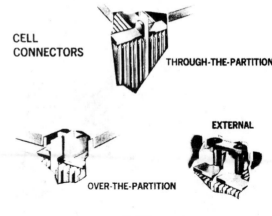

Figure 5–5 Cell connectors.
(Reproduced with permission from the Battery Council International)

active material is lead oxide, which is pressed into the grid in "mud" form and is chocolate brown in color. It changes to lead dioxide (PbO2) as the battery is charged and to lead sulfate as the battery discharges. The negative plate active material is sponge lead (pb), which is pressed into a different grid, and is grey in color. During the discharge cycle the negative plate turns to lead sulfate and back to sponge lead as the battery is charged. The grids conduct the current to and from the active materials. The grids have feet on them which position over the element rests (bridges). The space below the plates contains the element rests which run the full length of the battery bottom. The space below the top of the rests acts as a sediment chamber for collecting active material which sheds from the plates. Loss of the active material through shedding is part of the normal wearing process of battery life. The battery case may be made of polypropylene which is very light and durable. Cell covers provide a vent to allow hydrogen gas to escape during the charging process, and provide a means of adding water as needed.

Electrolyte is a mixture of water and sulfuric acid and is approximately 36 percent sulfuric acid by weight or 25 percent by volume. It is premixed to 1.265, as it comes with the dry-charge battery. As the electrolyte is poured into the battery cells, it becomes an electrical carrier and the battery becomes active. The battery is charged immediately after filling to force trapped air from the plates. This prevents oxidation in the plates.

5.3.2 Specific Gravity

Specific gravity can tell us something about each battery cell. Simply defined, *specific gravity* (SG) is the thickness of a liquid. Water at 65°F (20°C) is assigned the number of 1. Anything that is thicker than water will be assigned a higher number. Therefore, a liquid with an SG of 2 is twice as thick as water, while a liquid with an SG of 0.5 is half as thick as water. Consider the battery's electrolyte which is sulfuric acid (H_2SO_4). It is composed of sulfur and water and is obviously thicker than water alone. Actually, it has a specific gravity of 1.265 (usually read as "twelve sixty-five"). Later in the chapter you will learn how to measure the battery's specific gravity using a hydrometer.

5.3.3   The Working Battery

Chemical reaction between the plates and the electrolyte supplies electrical energy to the circuits. During this discharge process the active material on the positive plate (made of lead dioxide) and the negative plate (made of sponge lead) are changing to lead sulfate as acid is going into the plate material, and the plates are becoming more alike. While this conversion is taking place, part of the sulfuric acid is changed into water. This action then reduces the specific gravity of the electrolyte. Therefore, voltage becomes lower, since it depends on the difference between the two plates and acid concentration. The state of charge of the battery can be determined by measuring the specific gravity of the electrolyte using a hydrometer (Figures 5–6 and 5–7). A battery left in a low state of charge allows lead sulfate crystals to form on the plates, which will prevent the battery from coming back up to full charge. This will reduce the battery's capacity and shorten its life.

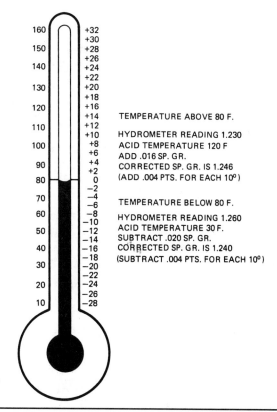

Figure 5–6 Hydrometer temperature correction.

TEMPERATURE ABOVE 80 F.

HYDROMETER READING 1.230
ACID TEMPERATURE 120 F
ADD .016 SP. GR.
CORRECTED SP. GR. IS 1.246
(ADD .004 PTS. FOR EACH 10°)

TEMPERATURE BELOW 80 F.

HYDROMETER READING 1.260
ACID TEMPERATURE 30 F.
SUBTRACT .020 SP. GR.
CORRECTED SP. GR. IS 1.240
(SUBTRACT .004 PTS. FOR EACH 10°)

5.3.4 Charging

To bring the battery state of charge back to full charge at 1.265, an electric DC current is sent through the battery in the reverse direction. (See the "Battery Charging Guide" in Figure 5–7.) The voltage for charging must be slightly higher than the voltage of the battery, about .5V per cell. Each cell is 2.1V and with six cells in the battery, this brings the battery voltage to 12.6V when fully charged. Thus, the battery could be fully charged using 15V. As the battery is charged, "shop" type chargers begin to taper back in current flow. This happens because of internal battery resistance. The battery is increasing its state of charge. After the battery state of charge stops increasing for three consecutive readings (as measured with a hydrometer at one-hour intervals), the battery is fully charged.

Battery State of Charge

State of Charge	Specific Gravity as Used in Cold and Temperate Climates	Specific Gravity as Used in Tropical Climates
Fully Charged	1.265	1.225
75% Charged	1.225	1.185
50% Charged	1.190	1.150
25% Charged	1.155	1.115
Discharged	1.120	1.080

12-Volt Battery Charging Guide
for
Fully Dicharged Batteries

Rated Capacity	Slow Charge*	Fast Charge†
80 Minutes or Less	14 Hours @ 5 Amperes	1 ³/₄ Hours @ 40 Amperes
	7 Hours @ 10 Amperes	1 Hour @ 60 Amperes
Above 80 to 125 Minutes	20 Hours @ 5 · Amperes	2 ¹/₂ Hours @ 40 Amperes
	10 Hours @ 10 Amperes	1 ³/₄ Hours @ 60 Amperes
Above 125 to 170 Minutes	28 Hours @ 5 Amperes	3 ¹/₂ Hours @ 40 Amperes
	14 Hours @ 10 Amperes	2 ¹/₂ Hours @ 60 Amperes

*If time is available, the lower charging rates in amperes are recommended.
†Initial rate for standard taper charger

Electrolyte Freezing Temperatures

Electrolyte Specific Gravity	Freezing Temperature	
	F.	C.
1.280	−92	−69
1.265	−71.3	−57.4
1.250	−62	−52.2
1.200	−16	−26.7
1.150	+5	−15
1.100	+19	−7.2

A 3/4 charged battery is in no danger of freezing. Therefore, batteries should be kept at least 3/4 charged and checked every 30 days.

Figure 5–7 Battery charging guide.

If the battery condition is poor, the battery may not come up to a full charge. The battery may be sulfated, meaning the reverse current flow cannot reverse the chemical action. If the battery condition is good, you could expect a reading of 1.265. As the battery is charged, hydrogen gas is liberated from the cells and needs to be ventilated.

 **Warning**

These fumes are very *explosive. Never create a spark or smoke a cigarette over the battery.* The fumes may ignite and blow up the battery in your face!

5.4   **Usage of the Battery**

5.4.1   **The Starting Battery**

Two types of batteries can be considered for pleasure boating: (1) the starting battery (rated in MCA), which is similar to the automotive battery and (2) the deep-cycle battery. The starting battery is designed with porous active material in the plates (see Figure 5–8). The

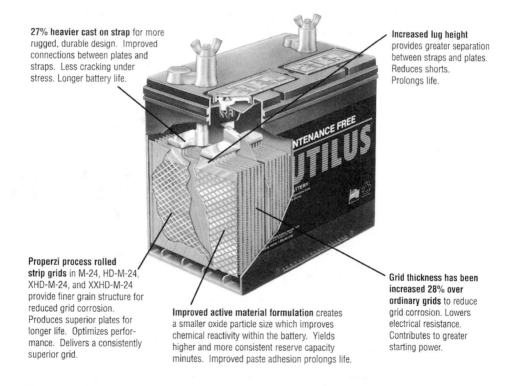

27% heavier cast on strap for more rugged, durable design. Improved connections between plates and straps. Less cracking under stress. Longer battery life.

Increased lug height provides greater separation between straps and plates. Reduces shorts. Prolongs life.

Properzi process rolled strip grids in M-24, HD-M-24, XHD-M-24, and XXHD-M-24 provide finer grain structure for reduced grid corrosion. Produces superior plates for longer life. Optimizes performance. Delivers a consistently superior grid.

Improved active material formulation creates a smaller oxide particle size which improves chemical reactivity within the battery. Yields higher and more consistent reserve capacity minutes. Improved paste adhesion prolongs life.

Grid thickness has been increased 28% over ordinary grids to reduce grid corrosion. Lowers electrical resistance. Contributes to greater starting power.

Figure 5–8   Exide Nautilus Marine Starting Battery. *(Courtesy of Exide Corporation)*

plates are thin and designed for quick acid penetration and high amp discharge for starting. The positive plate is weakened by repeated lower amp drains. An example of this kind of drain is the electric thruster used for bass fishing. The active material tends to fall off the grid and thus reduces battery capacity and shortens battery life. Many outboards have an unregulated charging system which is compatible with the marine starting battery. This is not a maintenance-free battery, and the electrolyte level in the cells needs to be checked at 30-day intervals. Some features of this battery are thin plates, a polypropylene case, a carrying handle, specially designed glass mat separators and rust/corrosion-resistant quick-disconnect terminals (Figures 5–8 and 5–9).

5.4.2 Deep-Cycle Battery

The deep-cycle battery is specially constructed to overcome the shortfalls of the marine starting battery. This battery has denser active material and thicker plates, which help keep the active material in the grid during repeated deep discharge and recharge cycles (see Figures 5–9 and 5–10). Glass mat separators may be used to reinforce the plates and reduce vibration damage and shedding of the active material from the grid. The deep-cycle battery should be used for bait tank pumps, powering lights, fish finders, and Loran and electric trolling motors (Figure 5–10). Some deep-cycle batteries can be used for a starting battery, and also be deep-cycled. It is not compatible with the unregulated

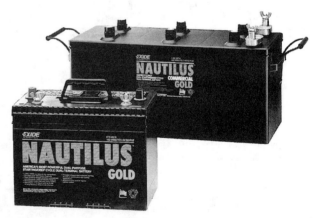

DUAL PURPOSE

- Fibrous mat, microporous, encapsulated separators (3 times more costly than standard separators) provide maximum starting power and unsurpassed deep cycling performance.

- The Nautilus Gold battery is built with "Pack-Tite™" construction in a shock resistant polypropylene case, and also has a built-in, fold down handle.

- The Nautilus Gold has 2 sets of terminals permitting uncluttered harnessing for starting cables and deep-cycle wiring connection.

Figure 5–9 Dual purpose Exide Nautilus Gold Marine Starting and Deep-Cycle Batteries. *(Courtesy of Exide Corporation)*

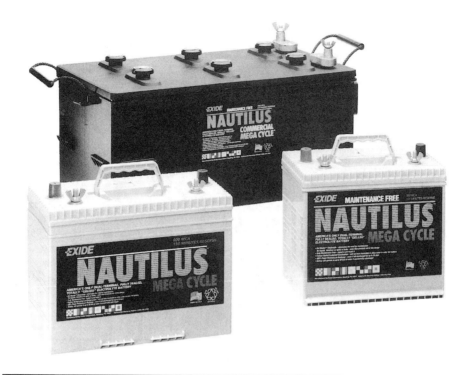

Figure 5–10   Exide Nautilus Mega Cycle Dual Terminal "Gell Cell" Starting and Deep-Cycle Batteries featuring "gelled" electrolyte; dual terminal; faster, safer recharging; low discharge/100% recovery; operation under water up to 40 feet. *(Courtesy of Exide Corporation)*

charging system. Immediately after extended use, the battery needs to be placed on slow-charge with a standard charger, until 1.265 specific gravity reading is obtained. Don't overcharge the battery, as this will cause grid corrosion and reduce battery life (Figure 5–7). Features of this battery are the thick plates with high capacity, reinforced glass mat separators, a polypropylene case, a carrying handle, and rust/corrosion-resistant quick connectors.

Two deep-cycle batteries can be placed in parallel and thereby double the reserve capacity and yet remain at 12 volts. This will allow for prolonged fishing with an electric outboard (thruster). They may also be placed in series, which will deliver 24V and last for the peak reserve capacity of just one battery (Figures 5–11 and 5–12). This would supply voltage for those thrusters that require 24V.

Distilled water is, without a doubt, the best for routine water additions to the battery. However, any water that is safe for drinking (except mineral water) is safe to use in the battery. Water of known high mineral content should not be used.

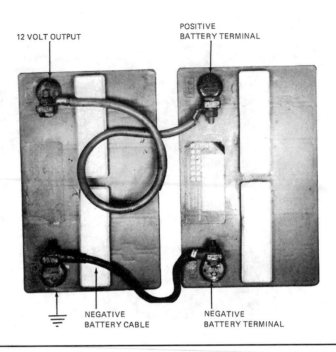

Figure 5–11   Batteries connected in parallel.

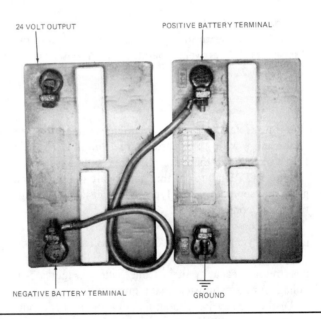

Figure 5–12   Twelve-volt batteries connected in series.

5.5   **Load-Testing the Battery**

With the battery state of charge above 1.225 a load test can be run (see Figure 5–13). This test loads the battery with a variable carbon pile to one half the cold cranking amps rating of the battery for 15 seconds, and the voltage should not fall below 9.6 volts at 70°F. (See Figure 5–14 for estimated electrolyte temperature and minimum voltage.) If it does, the battery capacity is down and it should be charged and retested. If it stays above 9.6V at 70°F for 15 seconds, the battery is good but may need an additional charge (Figure 5–13). Check it with a hydrometer.

Another way to load-test the battery without using a carbon pile is to use the starter motor as a load.

 Danger

Always wear a safety face shield and hook up a voltmeter for this test.

Crank the powerhead and watch for a boiling action in any one of the cells. If boiling occurs in a cell, that cell is shorted and the battery will have to be replaced. Voltage will also drop rapidly and fall below minimum specifications. If no boiling occurs in any cell and voltage remains above 9.6V, the battery is good. If voltage drops below 9.6V without boiling in any cell, check the state of charge and the starter motor. *This test has a short-coming*—it is possible that the starter motor is drawing too much current or the powerhead is "dragging." Of course, you can quickly check the powerhead. Remove the spark plug wires from the spark plugs, place them in a spark checker, and turn the powerhead over by hand.

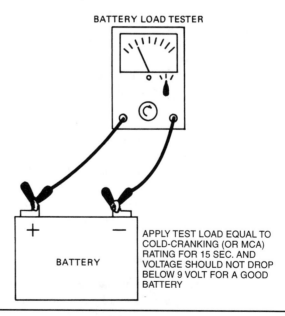

Figure 5–13   Battery load test.

5.6   Surface Discharge Test

The battery case and posts should always be kept clean. As the battery charges and discharges, electrolyte accumulates on top of the battery. This electrolyte will conduct electricity across the battery top causing a battery drain. A surface discharge test will prove this. With the red voltmeter lead attached to the positive post, drag the black lead across the battery top. A low voltage reading will be registered if the battery is dirty. If a reading is indicated, clean the battery top with a mixture of bicarbonate of soda (baking soda) and water, using an old paint brush. Use 1 tablespoon to each pint of water. *Do not* let this solution get inside the battery! When the solution stops bubbling, wash the battery off with water.

5.7   Battery Drain Test

On occasion batteries that are installed and in good condition go dead for no apparent reason. The alternator system checks out fine and the battery charges up to full charge, but the battery can be dead again in a day or so. This will generally happen in a larger boat where the battery is left hooked up. To find the reason for the dead battery, a drain test should be made. The drain test will determine if there is a drain on the battery. Disconnect the negative battery cable and hook the black voltmeter lead to the battery negative post. Hook the red voltmeter lead to the disconnected negative battery cable. If there is a reading of 12V, there is a drain on the battery. To find the circuit causing the drain, separate the boat accessory lead from the outboard starting battery. If the voltage drops back below 12V, the problem circuit is in the accessories. Hook the accessory lead back up and start pulling the fuses out one at a time. When the voltage drops as a particular fuse is pulled, that is the problem circuit. Make a close inspection of that circuit and make the necessary repair. If the voltage did not drop when the accessory wire was removed, the problem is in the outboard harness, key switch, accessories hooked to the control, or the alternator rectifier is shorting. Start by disconnecting the easiest components first. When voltage drops below 12V you have found the problem circuit (Figures 5–14 and 5–15).

The modern outboard has many electronic circuits, including switch boxes, an alternator rectifier, and add-on circuits. These circuits may show a drain on the battery for a short period of time (five minutes) during the test and then the voltage drops. If the boat is not going to be used for a few weeks, disconnect the battery. Battery quick disconnects are available at the parts store.

On larger outboards and boats, the use of a battery isolator will allow the alternator charging system to charge both the starting and an auxiliary battery at the same time. This will prevent accessories connected to the auxiliary battery from drawing power from the starting battery (see Figure 5–16).

5.8   Installation

Exide recommends that the battery always be installed in a well-ventilated area. Batteries release gases during the charging phase that if exposed to spark or flame, can explode.

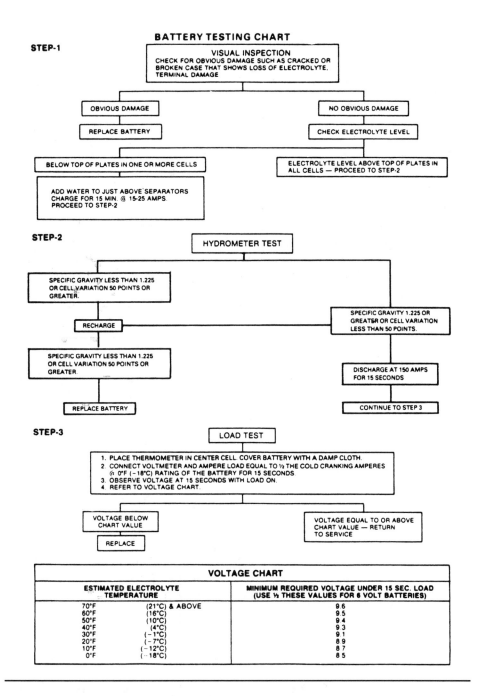

Figure 5–14   Battery testing chart. *(Reproduced with permission from the Battery Council International)*

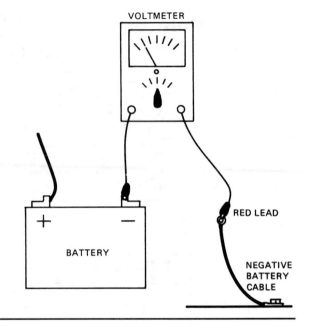

Figure 5–15 Battery drain
test. (Twelve volt reading indicates a short. If you suspect
electronics in the system wait
five minutes for them to bleed
off. Any reading below 12V is
 acceptable.)

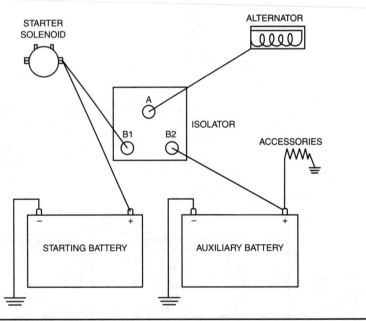

Figure 5–16   A battery isolator will allow the charging system to charge both the
starting battery and an auxiliary battery at the same time while preventing accessories
connected to the auxiliary battery from drawing power from the starting battery.

When installing a battery in your boat, it is important to use either a box or tie-down system to keep the battery stationary once underway. This will reduce unnecessary vibration. Make sure all connections to the battery terminals are tight. Loose wires will cause sparking and arcing which can damage the boat's electronics, melt wires, or cause an explosion. Additionally, it is important to coat the terminals and connections with a corrosion inhibitor. The corrosion inhibitor should be reapplied every several months. Failure to do this will result in poor connections and wire corrosion, especially in saltwater environments. Corrosion increases resistance in the wires, reducing the current drawn to run electrical equipment. When installing a new battery, be sure to remove any plastic battery terminal protectors before attaching wires.

5.9   Maintenance

Exide recommends that all marine batteries, except gel types, have removable vent caps so that electrolyte levels can be checked regularly. You should check the electrolyte every month. When storing a battery for the winter, check and fill it with distilled water as needed, recharge the battery fully, and store it in a cool place. When preparing the battery after winter storage, recharge the battery to its full charge state.

KNOW THESE PRINCIPLES OF OPERATION

- How to safely handle a battery.
- Dangers of electrolyte and hydrogen gas.
- Battery construction.
- How the battery produces electricity.
- What happens chemically as the battery discharges and charges.
- Procedures for testing the battery.
- Charging the battery.

REVIEW QUESTIONS

1. A storage battery element is made up of alternating positive and negative plates made of dissimilar leads.
   a. true
   b. false
2. Battery cell separators are made of
   a. hard rubber.
   b. microporous rubber, polyethylene, and plastic.
   c. nylon.
3. The plate grid is a framework to hold
   a. filtered sediment.
   b. active material.
   c. acid.

4. A fully charged battery cell voltage is_____.
    a.   2V
    b.   2.1V
    c.   2.2V

5. The positive plate is brown in color and is made up of
    a.   lead sulfate.
    b.   lead acid.
    c.   lead dioxide.

6. The active materials of both battery plates are converted into a different chemical compound during the discharge cycle. This compound is called
    a.   lead oxide.
    b.   lead sulfate.
    c.   lead dioxide.

7. Electrolyte is a mixture of sulfuric acid and
    a.   lead oxide.
    b.   water.
    c.   lead dioxide.

8. When is the battery fully charged?
    a.   when a hydrometer reading taken at each cell is 1.265
    b.   after the battery state of charge stops increasing for three consecutive readings, as measured with a hydrometer at one hour intervals
    c.   Both a and b are correct

9. When a battery is sulfated it means the reverse current flow from the battery charger cannot reverse the chemical action.
    a.   true
    b.   false

10. Hydrogen gas is the gas liberated from the cells of a charging battery.
    a.   true
    b.   false

11. The freezing point of electrolyte with a specific gravity of 1.200 would be corrected to_____.
    a.   −62°F
    b.   +5°F
    c.   −16°F

12. If a hydrometer reads 1.200 specific gravity and the thermometer registers 110°F, the corrected specific gravity reading would be_____.
    a.   1.212
    b.   1.242
    c.   1.232

13. The cells of a battery are connected in series.
    a.   true
    b.   false

14. A flame or lighted cigarette over a charging battery may cause an explosion.
     a.   true
     b.   false
15. What are two major differences between a starting battery and a deep-cycle battery?
     a.   the plates are thicker
     b.   denser active material
     c.   Both a and b are correct
16. A starting battery is charged to 1.255 SG reading. The battery load test registers 8V after 15 seconds. What would you recommend to the customer?
     a.   recharge the battery
     b.   install the battery in a 115 HP outboard
     c.   replace the battery
17. A battery is load-tested at one half the cold cranking amps at 50°F. What is the minimum required voltage under the 15 second load?
     a.   9.6V
     b.   9.4V
     c.   9.1V
18. Two deep-cycle batteries may be used in parallel to operate a 24V electric thruster motor?
     a.   true
     b.   false

CHAPTER 6

Starter System

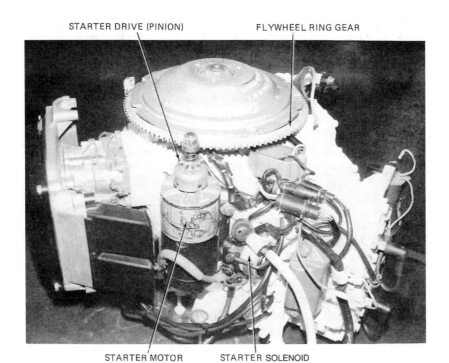

STARTER DRIVE (PINION)   FLYWHEEL RING GEAR

STARTER MOTOR   STARTER SOLENOID

Objectives

After studying this chapter, you will know

- What the starter circuit consists of.
- The function of the starter solenoid.
- Starter motor principles of operation.
- Troubleshooting using the starter draw and system voltage tests.
- Manual starter principles.
- Troubleshooting the manual starter.

6.1   Starter Motor Circuit

The function of the starter motor is to convert electrical energy into mechanical energy. The battery supplies electrical energy causing the starter motor armature to rotate. This forces the drive assembly out and into mesh with the flywheel ring gear, thus cranking the powerhead. This action is controlled by a key switch at the helm control, which passes the electricity to a neutral switch controlled by the shift lever or armature plate position. With the shift lever in neutral, electricity is passed on to a solenoid which is activated, completing the function of the control circuit. With the solenoid closed, the starter circuit between the battery and the starter motor is completed, allowing high amperage to flow into the starter motor to crank the powerhead (Figure 6–1).

The key switch is multipurpose, completing several functions at the same time. These functions include opening the ignition circuit, closing the starter control circuit, and sometimes closing the circuit to the primer. Of course, accessories may be switched on and off, as well.

The neutral switch is a switch that opens and closes the starter control circuit. Being in neutral, the shift lever allows the switch to close, completing the circuit. In forward or in reverse the neutral switch is in the open position, and the starter control circuit cannot be activated. This prevents starting in gear and the possibility of personal injury or damage to the boat (Figure 6–1).

6.2   Wiring Harness

Consider the wiring harness used on large outboards. The battery cables are routed through this harness, as well as five other wires. The additional wires are for the choke (primer), temperature, starter control, a fused battery lead, and ignition.

The wiring in this harness is sufficient, however, the battery lead to the control is not large enough to handle all the modern accessories you may want. If you choose to install some accessories, then install another battery lead of sufficient size, complete with fuse or circuit breaker. If you care to, you can fuse individual circuits at the helm. Remember that electricity has to have a return path, so install the same size wire for ground back to the negative post of the battery. By installing these two wires, you will stay away from low voltage problems associated with the battery harness lead.

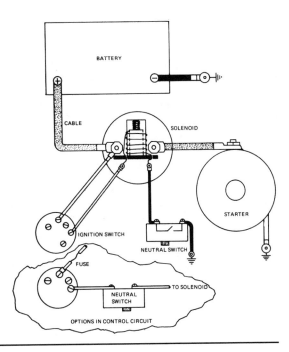

Figure 6–1   Starter circuit.

   Note

Be careful when you plug in the harness to the outboard connector. The prongs can be bent or pushed aside so there is no contact. If you have trouble slipping the connector together, you might try lubricating it with multipurpose silicone lubricant or a small amount of dielectric grease to ease the installation.

Battery cables are the means of carrying the electrical current to the starter motor. For large horsepower outboards, these cables must be at least 6-gauge in size to adequately carry the current required for the starter motor. However, the size depends on placement of the battery in the boat. If the battery is near the transom, then 6-gauge is fine. If the battery is located some distance from the outboard, then 4-gauge should be used. The longer the cable run, the more resistance there is in the cable. There can be only a maximum of a .7V drop in the total positive starter circuit. This is from the positive battery post to the starter motor terminal, and .3V on the negative side (see Figure 6–9).

6.3   Solenoid

The job of the solenoid is to complete the circuit between the battery and the starter motor (Figure 6–2). It does this by closing the starter circuit electromagnetically, when activated by current from the key switch. This is a completely sealed switch, meeting the SAE 1171 standards for marine application.

Safety

Do not substitute an automotive-type solenoid for this application (as it is not sealed and fumes can be ignited upon starting the powerhead).

The starter solenoid consists of a coil winding, plunger, return spring, contact disc, and four externally mounted terminals. The solenoid is installed in series with the positive battery cable and the starter motor positive cable with both cables separately mounted to the two larger solenoid terminals. The smaller terminals connect to the neutral switch and ground (see Figure 6–1).

To activate the solenoid, the shift lever is placed in neutral, closing the neutral switch. Electricity coming through the ignition switch (start terminal) goes into the solenoid coil winding creating a magnetic field. The electricity then goes on to ground in the powerhead. The magnetic field surrounds the plunger in the solenoid, which draws the contact disc into contact with the two larger terminals. Upon contact of the terminals, the heavy amperage flows through the positive battery cables to the starter motor which is then activated. When the key switch is released, the solenoid magnetic field is no longer supported and the magnetic field collapses. The return spring working on the plunger lifts the contact disc, opening the circuit to the starter.

When the armature plate (under the flywheel) is out of position or the shift lever is moved into forward or reverse gear, the neutral switch is placed in the open position and the starter control circuit cannot be activated. This prevents accidental starting in gear.

STARTER PINION GEAR   RING GEAR

STARTER MOTOR   SOLENOID

Figure 6–2 Starter motor relay.

6.3.1   Troubleshooting the Solenoid

The solenoid is practically trouble free. If you suspect problems, first load-test the battery. Then look up the starter circuit in the wiring diagram. Note that there are two circuits, the high amperage starter circuit and the low amperage control circuit. When the key switch is turned, listen for a click at the solenoid. If you hear the click, the control circuit is good. Then test for voltage at the starter motor. If there is voltage at the starter, a reading below 9V with no starter action indicates a bad starter or high resistance in the circuit. A reading at or near battery voltage indicates that the starter has an open circuit inside. To test for resistance in the solenoid and cables, follow the voltage drop tests shown in Figure 6–9, or as given in the service manual. If no click is heard at the relay, use a jumper wire to jump between the battery terminal of the solenoid and the "S" terminal. If it now works, the problem is in the control circuit. Using a voltmeter, test for voltage at the neutral switch in the control (at both terminals). If there is no voltage, test at the key switch, with the key switch in the start position. Depending upon the year, make, and model, there may be a fuse between the battery and the ignition switch (Figure 6–3).

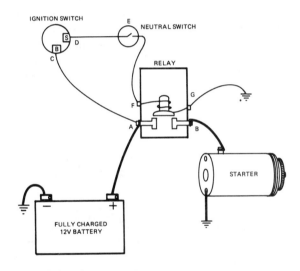

SAFETY: TO PREVENT UNEXPECTED CRANKING REMOVE STARTER TO SOLENOID CABLE!*

All tests are made with black voltmeter lead grounded.
Connect to point A, reading should be 12V.
Connect to point B, turn key to start, relay should click with a reading of 12V.
No click or 12V. reading move to test point C, should read 12V.
Connect to point D, turn key to start, reading should be 12V.
Connect to E, turn key to start, reading should be 12V, at both terminals.
Connect to F, turn key to start, reading should be 12V.
Install a jumper wire to ground at terminal G, turn key, relay should click and 12V at B.

If at any point there is not a 12V reading, repair the wire or replace the switch between the last test point.

*Neutral switch may be located between terminal G and ground.

Figure 6–3   Starter circuit testing.

6.4   Starter Motor Principles

The starter consists of two major parts: the housing assembly and the armature. The housing holds the ceramic magnets and allows for the commutator end cap and drive end cap to fit against it. Within the commutator end cap there are brushes and a bushing. The drive end cap has a bushing and bosses for mounting the starter assembly. The bushings hold the armature in a central position within the housing (Figures 6–4 and 6–5). The starter drive is mounted to the end of the armature shaft.

In order to run the starter motor, there must be magnetic fields that attract and repel each quarter of a turn of the armature. The ceramic magnets in the housing are epoxied to the housing, and create a permanent magnetic field across the housing. There are two magnets, one on each side of the housing (see Figure 6–5). The armature is inserted between these two magnets. Its many loops of heavy wire are wrapped on a laminated soft iron core, which in turn is mounted on a shaft that is centered in bushings for rotational support. The heavy wire loops are clamped and soldered to the commutator bars, and the carbon brushes ride on the bars making a sliding electrical contact. This allows the *current* to flow through the armature, creating a magnetic field around each conductor that interacts with the housing magnetic field, forcing a rotation (Figure 6–6a). Each loop is attracting on one side, while the other side of the loop is repelling: an attraction for a quarter of a turn, a repelling for a quarter of a turn, an attraction for a quarter of a turn, and a repelling for a quarter of a turn. Thus, the armature makes one revolution (see Figures 6–6 and 6–7).

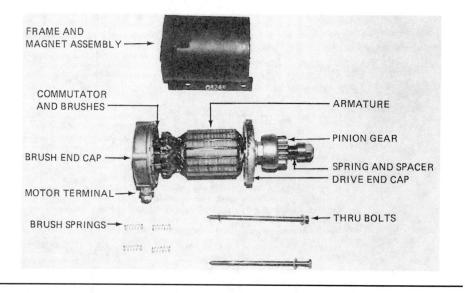

Figure 6–4   Exploded view of OMC starter motor.

PINION

FRAME AND
FIELD ASSEMBLY

ARMATURE

MAGNETS (2)

COMMUTATOR

Figure 6–5   Cutaway starter mounted on OMC V-6.

Just as each loop comes up to the neutral point, the following set of commutator bars move into contact with the brushes. This produces the same electromagnetic force on each following conductor loop. As the armature continues to turn, the brushes are changing commutator bars, which continue the magnetic force turning the armature. The magnetic strength of the housing magnets and the magnetic field of the armature (applied voltage) with its heavy conductors determine the maximum torque of the starter motor. Generally speaking, the more current flow, the stronger the armature magnetic fields and the torque.

Remember that there are three things that must be present for the induction of electricity: motion, a magnetic field, and a conductor. All three are present in the starter motor. As the armature spins within the magnetic field, a **counter electromotive force (CEMF)** (voltage) is generated within the armature loops. This counter voltage opposes the voltage from the battery (Figure 6–8). The counter voltage is strongest at high RPM, and this limits

(a)

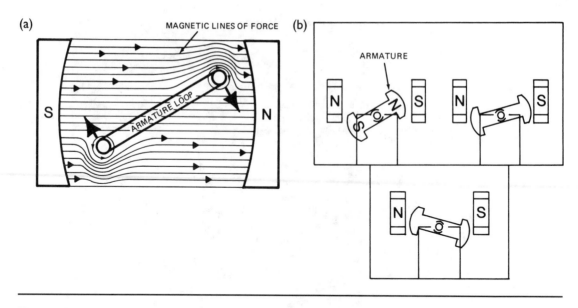

(b)

Figure 6–6 Motor principles (a); the commutator disconnects the armature (b).

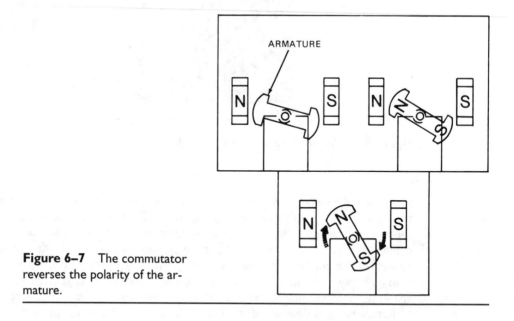

Figure 6–7 The commutator reverses the polarity of the armature.

the amount of current from the battery which can move through the armature loops. This counter voltage also limits the maximum speed and torque under normal conditions. This is one good reason to have a battery that is in good condition and charged up. When the

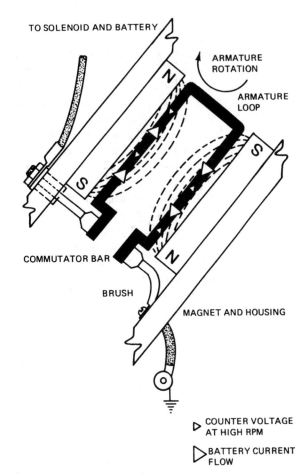

TO SOLENOID AND BATTERY

ARMATURE ROTATION

ARMATURE LOOP

COMMUTATOR BAR

BRUSH

MAGNET AND HOUSING

▷ COUNTER VOLTAGE AT HIGH RPM

▷ BATTERY CURRENT FLOW

Figure 6–8   Counter electro-motive force (voltage) opposing battery voltage.

starter is operating at the designed RPM for short periods of time, it will not get hot. If you use a battery that is in a low state of charge, the voltage will be down, and the starter RPM and counter voltage will be low. This allows for more current flow from the battery, which causes a heat buildup in the starter. The cranking speed will also be slow, the secondary ignition output will be low, and starting will be difficult.

If there are 200A of current at 10V going into the starter motor during cranking, the horsepower of the motor would be approximately 2.8. The simplified circuit illustrations that support the theory of operation are intended for instructional purposes only.

6.4.1   Troubleshooting the Starter Motor

Instrumentation is needed to effectively diagnose the starter circuit. During routine tune-up, a tachometer may be connected to the powerhead to test the cranking RPM. The cranking RPM should come up to or exceed the specifications (450 RPM). If it does, there is nothing wrong with the starter circuit. If it does not meet specifications, load-test the battery to prove the battery's ability to do the cranking job (Figure 5–15). A starter draw test is next. This test uses an induction-type ammeter placed on the positive cable, which determines the amount of current going into the starter (see Figure 7–6). If current flow is high, crank the powerhead by hand to see if it is dragging. If the powerhead is okay, the starter armature is dragging or shorted. Test out the starter armature on a growler with careful inspection of the bushings. If current flow seems low and cranking RPM is slow with a good battery, then there is too much resistance in the cables, solenoid, and/or their connections. Conduct voltage drop test to determine if the cables are bad or if connections are corroded (Figure 6–9).

If there is no action in the starter circuit at all and the battery is proven good, then possibly the control circuit is not working. This circuit can be tested with a voltmeter, which will determine if electricity is getting to the solenoid, neutral switch, and to the ignition switch (see Figure 6–3). For further information on these tests refer to the specific service manual for your outboard. Also refer to the wiring schematic to determine the switches or fuses in the circuit, and what the wiring color code is (see Figures 6–10 and 6–11).

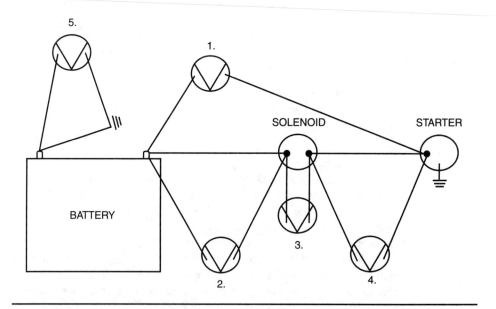

Figure 6–9 Starter circuit voltage drop test. Disable ignition and hook up voltmeter as in V-1. Crank engine: If reading is under .7V, circuit OK. If higher, run tests 2, 3, 4, and 5. Readings should be under 0.3V for #2, 0.2V for #3, 0.2V for #4 and #5, 0.3V. If readings are too high, clean connections, replace solenoid or cables.

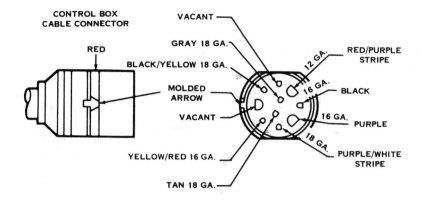

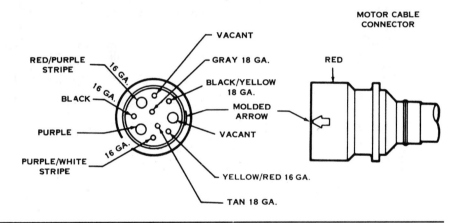

Figure 6–10   Connector for 70-75 HP. *(Courtesy of Outboard Marine Corporation)*

6.5   **Starter Drive**

The starter armatures are designed to turn at relatively high speeds to minimize amperage draw. The drive unit located on the end of the armature shaft and the ring gear on the flywheel provide the means to gear down the crankshaft. The starter drive ratio is approximately 10:1 to 14:1. The starter drive must be able to engage the gears when the powerhead flywheel is not turning, and must be able to disengage from the flywheel as the powerhead starts.

The starter drive unit operates on the spinning bolt and nut principle (inertia). As the armature shaft accelerates, the mounted pinion gear moves up the helix and engages the flywheel ring gear teeth, cranking the powerhead. As the powerhead starts, the acceleration of the larger diameter ring gear sends the pinion gear back down the helix, clearing the ring

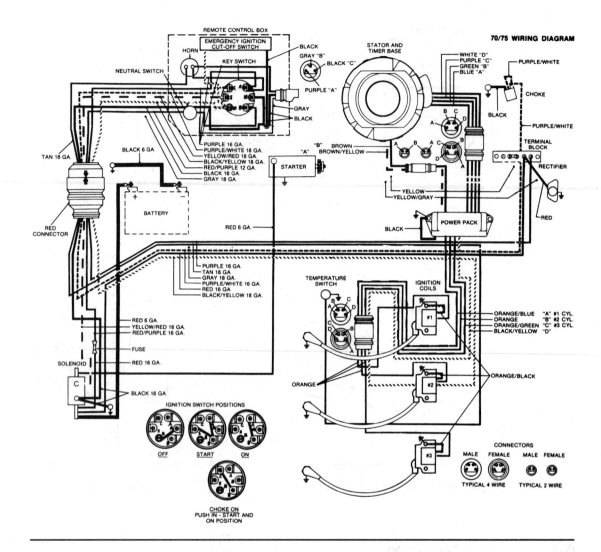

Figure 6–11   70/75 wiring diagram. *(Courtesy of Outboard Marine Corporation)*

gear. The pinion gear is held in the disengaged position during powerhead operation by a spring. Periodic lubrication is required on the helix of the drive unit in the marine environment (Figures 6–12 and 6–13).

6.6   Manual Starter

The main parts of the manual starter are the cover, rewind spring, and starter pawl arrangement. Pulling the rope rotates the pulley, winds the spring, and cams the starter pawl into

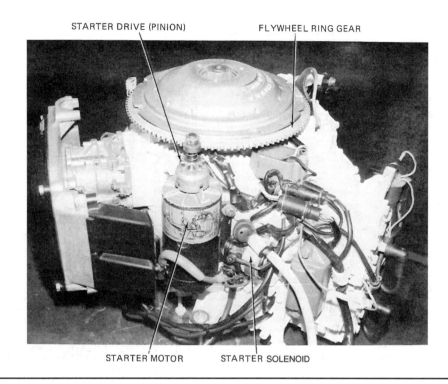

STARTER DRIVE (PINION)   FLYWHEEL RING GEAR

STARTER MOTOR   STARTER SOLENOID

Figure 6–12   Starter drive (gear ratio).

engagement with the starter hub, which is attached to the flywheel. Once the pawl contacts the hub, the powerhead is spun as the rope unwinds from the pulley. Releasing the rope on rewind starters moves the starter pawls out of mesh with the hub. The powerful clock-type spring recoils the pulley in the reverse direction to rewind the rope to the original position (Figure 6–14).

There is a newer design manual starter that employs the principles of inertia, as found in electric starter motors. The nylon pinion gear slides upward and engages the flywheel ring gear as the starter rope is pulled. At the same time, the rewind spring is tightened so there can be a recoil of the starter rope. The ratio between the starter pinion gear and the flywheel provides a maximum cranking speed for easy starts and easy pulls. Some models have a cam follower that prevents activation of the manual starter until the shift lever is in neutral and/or the throttle is placed in the start position (Figure 6–15).

6.6.1   Troubleshooting the Manual Starter

Repair of this unit is generally confined to the rope, nylon pinion gear, and occasionally spring replacement.

ELECTRIC STARTER & SOLENOID
AMERICAN BOSCH 08142-23-M030SM

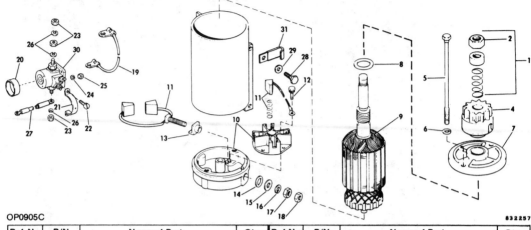

OP0905C

832257

Ref. No.	P/N	Name of Part	Qty.	Ref.No.	P/N	Name of Part	Qty.
*	387094	ELECTRIC STARTER ASSY.	1	17	387678	. . NUT AND WASHER	1
1	393285	. NUT, Spring & spacer	1	18	321642	. . LOCKNUT, Input stud	1
2	316740	. . NUT	1	19	580733	LEAD ASSY., Ground, solenoid to	
4	387683	. DRIVE ASSY.	1			bracket	1
5	387938	. THRU BOLT PKG.	1	20	311217	SLEEVE, Solenoid	1
6	321568	. LOCKWASHER	2	21	311218	CLAMP, Solenoid mounting	1
7	386639	. CAP ASSY.	1	22	308762	SCREW, Solenoid mounting	2
8	316750	. WASHER, Thrust armature	1	23	311349	NUT, Terminal	3
9	387682	. ARMATURE	1	24	120052	LOCKWASHER, Terminal	2
10	386645	. CAP ASSY., Commutator	1	25	306556	NUT, Terminal	2
11	385952	. . BRUSH AND SPRING SET	1	26	306325	LOCKWASHER, Terminal	3
12	387680	. . BOLT AND LOCKWASHER	2	27	581986	LEAD, Starter to solenoid	1
13	316743	. . BUSHING, Insulation	1	28	302479	SCREW, Str. mounting	3
14	316748	. . WASHER, Insulation	1	29	306325	LOCKWASHER	3
15	316749	. . WASHER, Plain	1	30	383622	STARTER SOLENOID ASSY.	1
16	316744	. . LOCKWASHER	1	31	310440	CLAMP, Lead to motor	1

* Not Shown

Figure 6–13   Electric starter and solenoid, American Bosch 08142-23-M030SM. *(Courtesy of Outboard Marine Corporation)*

 **Warning**

Follow the service manual instructions when disassembling the rope pull starter. The *recoil spring is under tension and can be dangerous* if not released properly.

Some older units may have small springs to aid engagement of the pawls into the hub. If these become weak or broken, the pawls will not engage to rotate the flywheel.

Other manual starters may have a friction spring and links connected to a pawl. If these become bent, the friction will not be correct, and the pawl will not be moved into engagement.

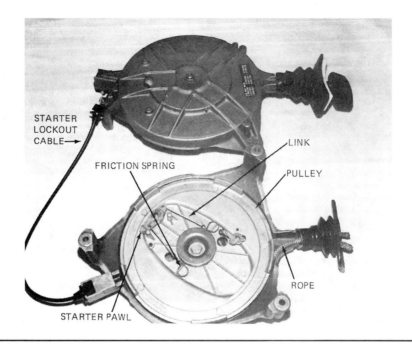

Figure 6–14   Manual starter. *(Courtesy of Outboard Marine Corporation)*

Figure 6–15   Inertia design manual starter.

Figure 6–16 Starting neutral switch.

Some models are equipped with a starter interlock. This system is designed to limit the RPM in neutral. The manual starter must not function when the throttle is advanced beyond the start position. If this happens, powerhead RPM could be excessive and damage the powerhead (see Figure 6–16 with a pencil pointing to the neutral switch).

The manual starter must not function when the shift handle is *not* in the neutral detent. This prevents starting in gear and possibly throwing the occupants to the deck or overboard. This is a safety system and should be checked after repairs are made (see Figure 6–17).

KNOW THESE PRINCIPLES OF OPERATION

- What makes the starter motor armature rotate.
- What causes the starter drive to engage the ring gear.

Figure 6–17 Starter interlock prevents starter function when throttle is advanced beyond start position.

- How the solenoid completes the starter motor circuit.
- How to make the starter draw and voltage drop test.
- The dangers involved when removing the manual starter recoil spring.
- Function of the starter interlock.

REVIEW QUESTIONS

1. There are two switches in the starter control circuit. They are
   a. ignition switch and solenoid.
   b. ignition start switch and neutral switch.
   c. key switch and trim switch.

2. When plugging the wiring harness connector into the lower motor cover electrical connector, you should
    a. be careful to align all electrical prongs.
    b. route the wiring harness properly so it will not have a strain placed on the connector.
    c. Both a and b are correct

3. The starter solenoid completes the circuit between the battery and the
    a. neutral switch.
    b. ignition switch.
    c. starter motor.

4. When is the neutral switch in the open position?
    a. in forward gear
    b. in neutral
    c. Both a and b are correct.

5. The starter brushes ride on the
    a. armature.
    b. drive end plate.
    c. commutator.

6. What causes the starter armature to rotate?
    a. repelling and attracting commutator bars
    b. repelling and attracting armature loops
    c. the force of inertia on the pinion gear

7. What three things are necessary for the induction of electricity?
    a. magnetic field, conductor, and a magnet
    b. conductor, motion, and a rectifier
    c. magnetic field, conductor, and motion

8. The starter draw test determines the amount of
    a. resistance in the starter motor.
    b. RPM of the starter motor.
    c. current flow into the starter motor.

9. The starter pinion gear is forced into the flywheel ring gear by
    a. inertia.
    b. the spinning nut and bolt principle.
    c. Both a and b are correct

10. On a newer type manual starter, the cam interlock may prevent starter operation when the shift lever is in the forward position.
    a. true
    b. false

11. What holds the starter pinion gear in position when the outboard is running?
    a. weight
    b. spring
    c. cam

12. Within the armature loop, the voltage is highest at
    a. highest RPM.
    b. lowest RPM.
    c. the neutral position.
13. The tachometer may be used to check the starter circuit for normal operation.
    a. true
    b. false
14. Starter heat builds up fastest when the battery is
    a. fully charged.
    b. in a low state of charge.
    c. shorted.
15. When performing the starter draw test, the induction ammeter is placed
    a. alongside the battery cable (or around the cable).
    b. on the starter motor.
    c. alongside the battery lead running to the rectifier.

CHAPTER **7**

Alternator Charging System

RECTIFIER

STATOR ASSEMBLY

Objectives

After studying this chapter, you will know:

- What components make up the charging circuit.
- Principles of operation.
- How AC voltage is rectified.
- Basics in testing the alternator.

7.1   Components of the Alternator Charging System

The alternator takes mechanical energy and converts it into electrical energy. Four major parts make up the alternator charging system: the flywheel, which holds the magnets; the stator coils mounted under the flywheel; the rectifier, which is mounted to the side of the powerhead; and the battery (Figure 7–1).

7.2   Principles of Operation

These four basic parts provide the basics for induction of electricity, by using a magnetic field in motion and a conductor. Motion comes from the spinning flywheel, which carries the magnetic field through the stator windings. This forces the free electrons to flow in the stator winding. The rectifier converts the AC electricity into usable DC electricity, which

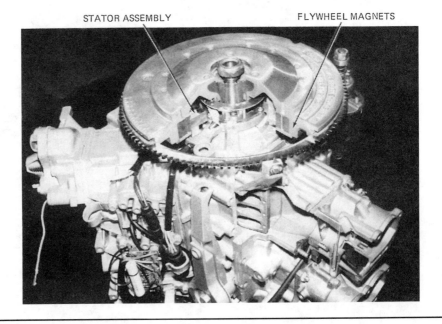

Figure 7–1   Alternator components.

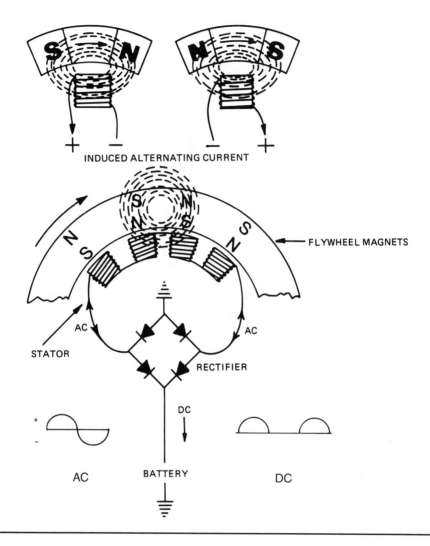

INDUCED ALTERNATING CURRENT

FLYWHEEL MAGNETS

STATOR

AC   AC

RECTIFIER

DC

AC   BATTERY   DC

Figure 7–2   Alternator charging circuit.

recharges the battery. Movement of free electrons (voltage) ceases when the flywheel stops its motion. As the magnetic field moves from north to south across the stator, the induced voltage also changes direction. The sign wave shows this positive-negative wave of alternator voltage (Figure 7–2).

AC voltage is routed to the rectifier, which routes the voltage through a one-way electrical check valve called a diode. As electricity can only pass in one direction, the positive wave (pulse) passes through and the electricity becomes direct current. This

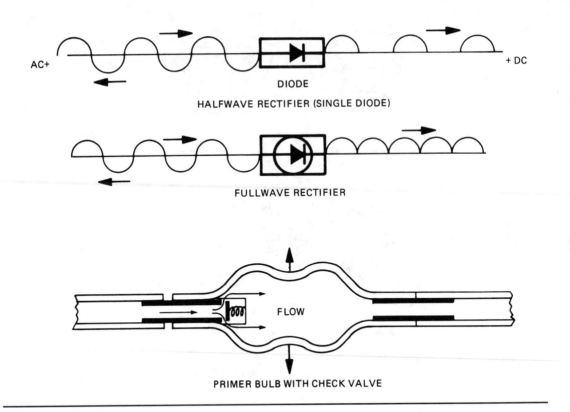

Figure 7–3   AC rectified.

current is pushed by the alternator voltage to the battery, and forces a reverse current flow through the battery, charging it. The negative portion on the stator voltage is stopped and rerouted, so that the whole wave is in one direction. Battery voltage is applied against the diode upon shutdown of the alternator (powerhead). Battery voltage is repelled and not allowed to pass through the diode. The diode can be compared to the fuel line check valve in the primer bulb. The check valve allows gasoline to flow forward, but not back to the tank (Figure 7–3). A voltage regulator is used to limit the amount of current delivered to the battery. The rectifier accomplishes this by shorting one stator lead to ground.

The battery voltage cannot drain back through the diode into the stator and to ground. Reversal of the battery leads will cause a direct short, and immediate damage to the rectifier diodes will occur. This circuit is simple to repair. If any one of the three basics fail, there will be no output. If there is no motion from the flywheel, there will be no moving magnetic field, and there will be no output. It is very rare for a magnet to fail, so the only item left is the conductor, which includes the stator, rectifier, and the wires. They are very simple to check using an ohmmeter and service manual illustrations. The simplified circuit illustrations which support the theory of operation are intended for instructional purposes only.

7.2.1   Troubleshooting the Alternator System

You just learned that if any one of the three basics fail, there will be no output from the system. To check out the stator, an ohmmeter or peak voltmeter is used. Peak voltmeters are usually found only in dealerships, so that procedure is not covered here. By disconnecting the stator wires (see wiring diagram for your outboard or Figure 6–11), you can test just the stator. Check for resistance, an open, and a ground. The total series of coils that make up the stator windings must not be shorted across one to the other or output will be lost or low. Using an ohmmeter, hook to the end of each stator wire and read the low scale. You are measuring resistance to see if it is up to specifications. This reading tells us that the stator coils are whole and complete, that there is continuity, and the resistance value is up to specifications. The stator wires are laid one on top of the other, as they are wound, and have nothing but varnish between them. Vibration can move a wire just a little bit, which can rub the varnish off, and cause a short within the stator coils. Next, one test lead is moved to ground where the stator is mounted. Using the high scale, there should be no reading. This check is to see if the stator wires are touching ground (Figures 7–4 and 7–5).

The rectifier is a one-way check valve; it can be checked using an ohmmeter. Hook up the ohmmeter leads, one to a common ground and one to a lead (terminal) coming from the rectifier. Using the appropriate scale, look for a reading, then reverse the ohmmeter leads,

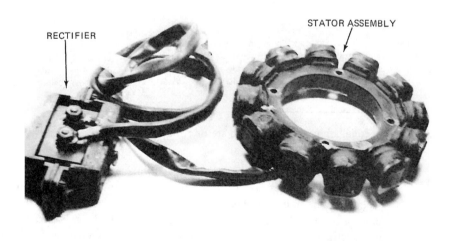

Figure 7–4   Alternator rectifier and stator assembly (as used by Mercury).

RECTIFIER OHMMETER TESTS

RECTIFIER

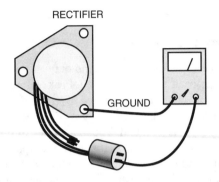

Connect one lead to rectifier ground. Connect the other lead to rectifier yellow/gray lead (OMC 15 HP). Note reading. Reverse the ohmmeter connections and note reading. A high reading in one direction and a low reading in the other direction—diode is okay.

RECTIFIER

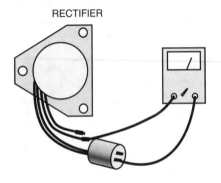

Move ohmmeter ground lead to rectifier red lead (OMC 15 HP). A high reading in one direction and a low reading in the other direction—diode is okay. Two high or two low readings—diode damaged. Next test between red and yellow leads. Results should be the same. Next test between red and gray leads. Results should be the same.

STATOR RESISTANCE TESTS

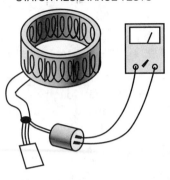

Disconnect battery. Disconnect yellow/gray and yellow connectors (OMC 15 HP). Calibrate ohmmeter and test between leads. 12 A stator resistance should read 0.45–0.54 Ω.

STATOR

To check if stator is shorted to ground, calibrate ohmmeter and connect one lead to stator ground. Connect the other ohmmeter lead alternately between both stator leads. Ohmmeter should register high readings. A low reading indicates the stator or stator leads are grounded. Repair or replace stator.

Figure 7–5   Rectifier ohmmeter tests and stator resistance tests.

and there should be no reading (infinity). This means that the diode is good. Repeat the test between the red lead and the other two leads (terminals). A normal diode will show a reading in one direction only. An infinite (very high) reading in both checks indicates the diode is open, and when both readings are zero, the diode is shorted. If there is a failure, the total rectifier is replaced as an assembly and there is no repair as the diodes are potted in. If there is a regulator problem, check with the owner on how the battery was hooked up. Reversing the battery leads can damage the regulator.

Note

A good place for corrosion to develop is where the battery and stator leads are hooked to the rectifier terminals. They should be coated with liquid neoprene to seal the connections against moisture.

To check alternator output and therefore the magnetic strength of the magnets, it is necessary to place an induction ammeter on the battery cable or install an ammeter in series (Figure 7–6).

7.3   Alternator Output

We will assume at this point that the rectifier, wiring, battery, and stator are good. With a 40-amp ammeter installed in series with the rectifier/regulator lead, at the battery side of the starter relay, and with the outboard in a test tank or in the water, the alternator

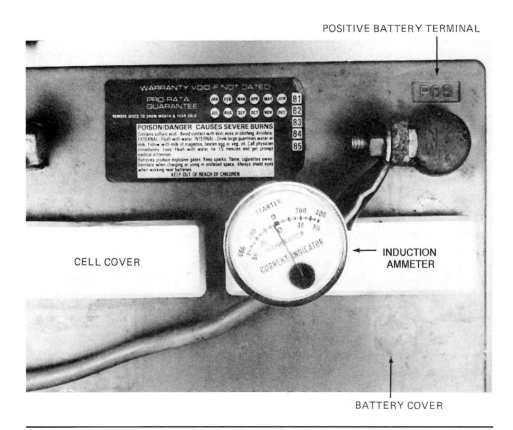

Figure 7–6   Induction ammeter used during alternator output test.

output test can be made. The battery should be load-tested and partially discharged or you may not get full alternator output, because of the battery's full state of charge. If possible, use a Stevens Load Bank tester. With a tachometer and Load Bank installed, run the powerhead in the 4500 RPM operating range, drawing the battery down the equivalent to the stator's full output. The ammeter should read at specification. If the reading is below specification then you know that the magnets are weak, because all other parts in the circuit are fine. They can be strong enough to give some induction but possibly not strong enough to give full induction. Therefore, the full amperage output will not be produced.

 Note

Testing the magnets with a large screwdriver is not an accurate way of making a test of the magnets.

Always give the alternator circuit a practical running output test after repairs are made. When the system comes up to total amp output called for in the service manual, this will prove the repair.

The boat tachometer is hooked up to the stator windings. To give an RPM reading, the tachometer senses the pulse coming from the stator. When the tachometer is working, it is an indication the stator is working.

 Safety

Problems with the alternator circuit can damage the boat tachometer, and problems with the tachometer can damage the alternator circuit.

Always check out the boat tachometer when there are charging circuit problems (Figures 7–7 and 7–8). If a battery isolator is installed in the alternator circuit, check for charging voltage going to both batteries (see Figure 5–16).

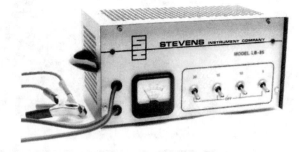

Figure 7–7 LB-85 Load bank. The load bank provides a simple, effective method of loading a charging system to determine if it is functioning at its rated capacity. *(Courtesy of Stevens Instrument Company)*

Charging Circuit Checks

Where To Look	Cause	What To Do	Result
		Undercharged Battery	
Battery	1. Defective battery or worn out battery	Check state of charge	
	2. Low electrolyte level	Add water and recharge	
	3. Corroded terminal connections	Clean with wire brush and coat with petroleum jelly	
	4. Loose terminal connections	Tighten securely	
	5. Excessive electrical load	Check total current draw in circuit	
Wiring	1. Corroded or loose connections	Clean and tighten connections	
	2. Short or ground in leads to stator	Check leads for defective lead insulation or connectors	
	3. Ground in other parts of circuit	Check for worn through insulation	
Rectifier	1. Defective rectifier diodes	Disconnect rectifier leads and check HI ohms scale	High reading on one check and low reading on other - diode OK Zero reading on both checks - diode shorted High reading on both checks - diode open
Alternator	1. Shorted or open windings	Run motor at full throttle operating range and note ammeter reading Note: Disconnect all accessories	If ammeter shows no charge, check stator windings
	a. Stator winding	Connect ohmmeter leads to either yellow or yellow/gray lead and ground	Infinite reading - winding OK. Any reading, stator shorted to ground
		Connect ohmmeter leads to yellow and yellow/gray leads	Reading to specification - windings OK Infinite reading - windings open circuit
Regulator (V4 and V6)	1. Defective regulator	Disconnect voltage regulator at terminal block Run motor at 4300-4600 rpm and note ammeter reading	If ammeter now shows charge, replace regulator If ammeter shows no charge, check alternator and rectifier
		Overcharged Battery	
	1. Defective regulator	Motor and regulator must be warm (run at least 20 minutes) and battery fully charged	Voltmeter reading 14.4 to 15.0 volts - regulator OK
	2. High resistance between battery and terminal in connecting leads	Run motor at 4300-4600 rpm and measure battery voltage Visually inspect for corroded connections	Voltmeter reading in excess of 15.0 volts replace regulator

Figure 7–8   Charging circuit checks. *(Courtesy of Outboard Marine Corporation)*

● If electrical parts are replaced or even removed from the motor, check the following:

    Wire and High Voltage Lead Routing
        ● as shown in service manual
        ● away from moving parts which could cut wires or wire insulation
        ● away from motor cover latches which can catch and cut insulation from High Voltage spark plug leads.

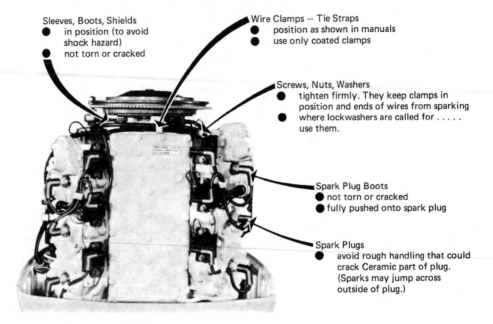

Sleeves, Boots, Shields
● in position (to avoid shock hazard)
● not torn or cracked

Wire Clamps — Tie Straps
● position as shown in manuals
● use only coated clamps

Screws, Nuts, Washers
● tighten firmly. They keep clamps in position and ends of wires from sparking
● where lockwashers are called for . . . . . use them.

Spark Plug Boots
● not torn or cracked
● fully pushed onto spark plug

Spark Plugs
● avoid rough handling that could crack Ceramic part of plug. (Sparks may jump across outside of plug.)

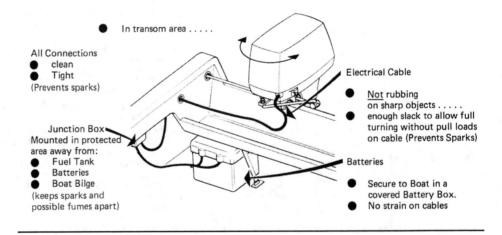

● In transom area . . . . .

All Connections
● clean
● Tight
(Prevents sparks)

Junction Box
Mounted in protected
area away from:
● Fuel Tank
● Batteries
● Boat Bilge
(keeps sparks and
possible fumes apart)

Electrical Cable
● Not rubbing on sharp objects . . . . . enough slack to allow full turning without pull loads on cable (Prevents Sparks)

Batteries
● Secure to Boat in a covered Battery Box.
● No strain on cables

Figure 7–9   Removing and replacing electrical parts. *(Courtesy of Outboard Marine Corporation)*

YOU SHOULD KNOW THESE PRINCIPLES

- Circuitry of the alternator charging circuit.
- Fullwave and halfwave voltage rectification.
- How to make resistance tests on the alternator circuit.
- The alternator output test.

REVIEW QUESTIONS

1. The following components make up the alternator charging system: rectifier, stator, switch box, battery, and flywheel.
   a. true
   b. false

2. Induction of electricity is a product of
   a. motion, a magnetic field, and a rectifier.
   b. motion, a conductor, and a magnetic field.
   c. motion, a stator, and a rectifier.

3. The rectifier changes the stator's current (voltage) to
   a. A/C.
   b. milliamps.
   c. D/C.

4. A diode is a one-way electrical check valve, and allows current to flow
   a. in both directions.
   b. to ground only.
   c. in one direction.

5. The rectifier can be tested with (powerhead not operating)
   a. an ohmmeter.
   b. an ammeter.
   c. a voltmeter.

6. The stator can be tested using (powerhead not operating)
   a. an ammeter.
   b. a voltmeter.
   c. an ohmmeter.

7. When the outboard is not running and the battery is connected to the harness, battery voltage is applied against the
   a. stator.
   b. rectifier.
   c. time base sensor.

8. To test the stator for ground, you would place the ohmmeter leads on
   a. each end of the stator leads.
   b. the rectifier "B" terminal and a stator lead.
   c. one stator lead and ground at the stator mounting.

9. Liquid neoprene should be applied to the rectifier terminals after service work is completed.
   a. true
   b. false

10. To check for alternator maximum output, you should
    a. place an induction-type ammeter against the stator lead.
    b. install a voltmeter on the battery, and run the outboard at high speed.
    c. install an ammeter in series at the positive solenoid terminal, and run the outboard at high RPM (operating range).

11. A tachometer that is not working may indicate that the charging system has a fault in it.
    a. true
    b. false

12. The magnets in the flywheel are tested by performing the
    a. spark test.
    b. starter draw test.
    c. alternator output test.

CHAPTER **8**

Ignition Systems

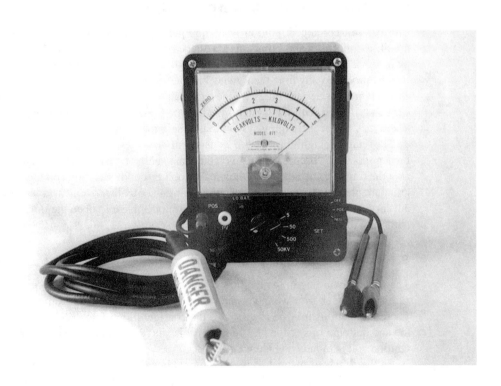

Objectives

After studying this chapter, you will know

- Standard magneto ignition systems.
- Low-tension magneto ignition systems (OMC).
- Magneto capacitor ignition systems (OMC).
- Capacitor discharge-3 ignition systems (OMC).
- Capacitor discharge-4 ignition systems (OMC).
- Capacitor discharge-6 ignition systems (OMC).
- Capacitor discharge troubleshooting principles.
- Alternator driven ignition systems (Mercury).
- Two-Cylinder Capacitor Discharge Module CDM ignition systems (Mercury).

Because of the many older model outboards still in operation, this chapter starts with older ignition systems and works through the years to the modern capacitor discharge ignition system.

8.1   Standard Magneto Ignition System

This is a self-contained ignition system, requiring no assistance from the battery. There is a coil, a condenser and points, and a spark plug for each cylinder. With more than two cylinders, a distributor system is used. On powerheads that do not use a key switch, spark may be produced any time the flywheel is rotated. For this reason the *spark plug wires are removed* from the spark plugs whenever repairs are made, so there will be no inadvertent start-ups. If used with this system, the battery's only function is to crank the powerhead. There may not be a charging system to maintain the battery.

The magneto ignition system consists of the flywheel magnets, coil, spark plugs, breaker points, condenser, the cam on the crankshaft, and a kill switch (Figure 8–1). To enable the timing to be advanced, the breaker points, condenser, and coil are mounted on the armature plate. At a predetermined point of armature plate movement, and therefore timing advance, a cam on the plate moves the carburetor linkage to open the throttle. This is synchronized to the timing advance.

8.1.1   Primary Ignition Circuit

The primary winding in the ignition coil consists of several hundred turns of fairly heavy, varnished copper wire wound around the coil laminations made from special alloy steel. These laminations may be shaped into the configuration of an "E" lying on its back. The top and bottom arms of the "E" become heels, and the center arm is the core on which the ignition coil is mounted (Figure 8–2).

As the flywheel magnets pass the laminations they carry the magnetic circuit around and through the ignition coil. One end of the primary winding is grounded to the laminations. The rest of the wire is wound around the center core and the remaining lead is attached to the ignition points. This completes the circuit to ground. A turning crankshaft carrying the cam opens the points, and the point spring closes the points to make and break

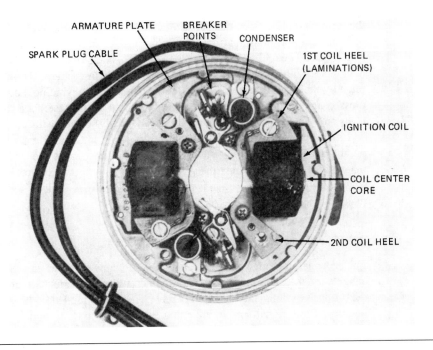

Figure 8–1   Standard magneto armature plate assembly.

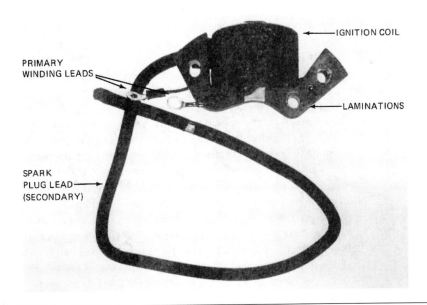

Figure 8–2   Standard magneto ignition coil.

the primary circuit. A condenser is attached to the circuit in parallel at the points. It provides a place for the primary current to go into, upon point opening. The electrical charge is immediately expelled in the reverse direction, and quickens the collapse of the coil magnetic field. Within the condenser, two alternate windings are made of aluminum foil, separated by dielectric paper. A lead is pressed against one winding, and the other winding is pressed against the wave spring and is grounded to the case. After the windings and the leads are installed in the case, they are hermetically sealed against moisture.

Either the kill switch or the key switch is wired into the primary circuit at the ignition points, and grounds this circuit when pushed (turned off), preventing a collapse of the magnetic field, which stops the powerhead. When the kill switch is released or the key switch is turned on, the ground is removed. Once again this allows for timed collapse of the magnetic field.

8.1.2 Secondary Circuit

The secondary circuit shares the same coil and consists of several thousand turns of fine-gauge varnished wire. One end is grounded at the coil lamination and then wrapped over the primary winding and insulated from it. The remaining lead is heavily insulated and terminates at a terminal spring, attaching to the spark plug terminal. The spark plug completes the secondary circuit.

8.1.3 Ignition Operation

The primary and secondary ignitions come alive as the flywheel magnets carry the magnetic field through the first heel, center core, and through the coil windings. This induces a current flow through the primary winding, closed points, and to ground (see Figure 8–3). As the flywheel continues to turn, the second heel of the laminations comes under magnetic influence and the magnetic circuit is then through the first and second heel. It tries to move up the center core of the coil but cannot. At this time, the magnetic field is temporarily opposed in the core (Figure 8–4).

The attendant magnetic field in the primary winding resists this polarity change and primary voltage peaks. At the time of this peak voltage, the ignition points are timed to open, interrupting the primary current flow. The ignition coil's attendant magnetic field instantly collapses across the secondary and primary windings, and the heavy magnetic field surges through the center core. This collapse induces a very high voltage surge through the secondary winding, through the spark plug wire(s) and spark plug(s) and it arcs across the electrode gap, returning through ground to the secondary winding (Figure 8–5). At the same time, there is action in the condenser. As the points open, the condenser is charged (250V), preventing a bridging of the point gap. It immediately expels this charge, starting a voltage flow back through the ignition coil primary winding. This speeds the collapse of the attendant magnetic field, and improves the spark at the spark plug. Voltage created by the condenser discharge continues to surge back and forth in the primary circuit, continually diminishing to zero voltage, unloading the coil primary winding. The circuits are now dead, waiting for the magnetic field to pass again.

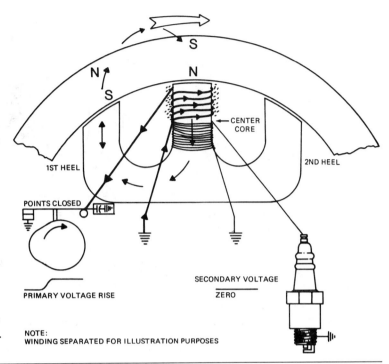

Figure 8–3 Induced primary current with an attendant magnetic field.

NOTE:
WINDING SEPARATED FOR ILLUSTRATION PURPOSES

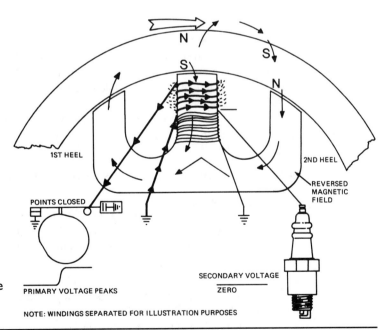

Figure 8–4 Core has become an electromagnet and opposes change in magnetic flow.

NOTE: WINDINGS SEPARATED FOR ILLUSTRATION PURPOSES

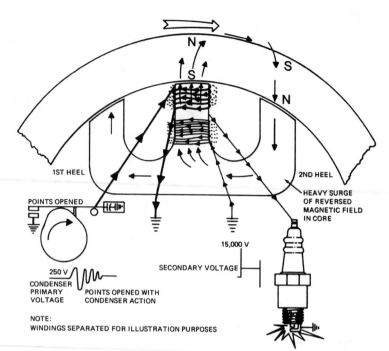

Figure 8–5 Collapse of attendant magnetic field with induced high secondary voltage.

This timed ignition spark must be synchronized with the carburetor throttle opening. There can be an overadvanced timing for the fuel delivered by the carburetor circuits, which will cause preignition and/or detonation that will damage pistons. During a tune-up, the synchronization adjustment should be checked to make sure that the timing and carburetion are working as a team.

The simplified circuit illustrations that support the theory of operation are intended for instructional purposes only.

8.1.4 Troubleshooting the Standard Magneto Ignition System

If your powerhead won't start, check it first for spark. The spark must jump a 1/4-inch gap. Use a spark tester set to this gap. If a spark tester is not available, open the spark plug electrode gap of a good plug to 1/4-inch, and ground (external) the spark plug to the powerhead. This wide gap requires maximum output from the coil, and will stress the coil. Pull the rope *hard* and *fast*, or crank the powerhead with a good battery. Remember, slow cranking gives a weak spark or no spark at all. If there is a good spark, you have proven the total ignition system, except for the spark plugs and timing. If there was no spark, check the kill switch circuit with an ohmmeter. If that is okay, pull the flywheel using a flywheel puller which attaches to the bolt pattern in the flywheel hub.

A good spark in the secondary circuit is totally dependent on a good primary circuit. There must be good centralized contact of the ignition points. Contact surfaces should be free of *excessive* pitting, corrosion, and carbon. Set points to a .020 inch specification. Some pitting and discoloration will occur during normal operation. Oxidation will occur while the outboard is in storage. Pitting is generally associated with condenser action. As the points open and close, metal is vaporized and transferred to the opposite point. This transfer will be controlled by the condenser capacity. Little transfer takes place with the proper capacity condenser installed. Excessive metal transfer at the points is caused by loose leads or a condenser with improper capacity. Without a good clean break at the points, the collapse of the coil magnetic field will be lazy, affecting the spark.

The point rubbing block is lubricated by the lubricating wick. This wick should be replaced each time the points are replaced. It supplies just enough lubrication to reduce friction between the rubbing block and the cam. If the lubrication is not present on the cam (applied by the wick), friction will develop heat causing a very rapid wear of the rubbing block. Also, rust may form, producing an abrasive surface on the cam, and accelerating the wear of the rubbing block. This changes point gap and timing. In a two-cylinder powerhead, the points must be set to open at exactly 180 degrees apart. This sets the timing of each cylinder. The point opening can be set using an ohmmeter or a battery-powered continuity light. A timing fixture is required to establish the point opening at the mark embossed on the armature plate (on some OMC models) (see Figure 8–6).

It is important that each point set opens at the prescribed time, so there will be no advanced or retarded timing. As the piston advances in the compression stroke, the points must open at a precise piston position. If a point(s) is misadjusted and advances timing, preignition and/or detonation will occur. This will cause damage to the piston head, ring

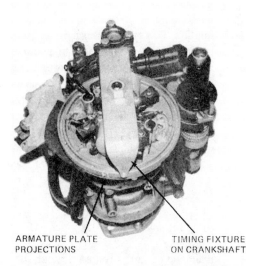

ARMATURE PLATE
PROJECTIONS

TIMING FIXTURE
ON CRANKSHAFT

Figure 8–6 OMC timing fixture used for setting points.

lands, and cylinder walls. Setting the points is setting the timing of the powerhead, and rotation of the armature plate advances the timing.

The condenser should be replaced as part of the normal maintenance. However, there doesn't seem to be much trouble with today's condensers. If they are to be reused, they should be tested on a Stevens or Merc-O-Tronic test instrument for capacity, series resistance, and shorts. Without these instruments, an ohmmeter and an eyeball-check of point condition is all you can do. A normal set of points will be gun-metal grey in color and show very little pitting. Black carboned points indicate lubrication is getting between the points. This is improper cam lubrication; oil vapor or oil may be coming from the upper crankshaft seal. If oil comes out, air goes in, leaning the fuel mixture in the crankcase.

The ignition coil can be checked with an ohmmeter. Check the primary winding for the specified resistance on the low ohm scale. The secondary winding may also be checked with an ohmmeter. Check for the specified resistance on the high ohm scale. These two tests are static tests and are only partly valid. The best way to test the coil is to put it through a power test using Stevens or Merc-O-Tronic test instruments. These machines supply the rated amperage, and show the high voltage spark at 1/4-inch.

Some older motors do not have an alternator-type charging system, but rather a belt-driven generator. The flywheel magnets cannot be tested by making a charging system output test. To test the strength of the magnets, use a large screwdriver. Grip the handle with two fingers and let it swing. Place the screwdriver tip one inch from the magnet. The magnet is strong enough if it attracts the heavy screwdriver tip to the magnet.

8.2   Low-Tension Magneto Ignition System (OMC)

The low-tension magneto ignition system consists of a driver coil, two sets of ignition points, and two condensers located under the flywheel (Figure 8–7). The flywheel has two magnets, 180 degrees apart. Bolted to the powerhead are two ignition coils with secondary leads going to the spark plugs. A key switch or push-button kill switch is used to stop the ignition. This system uses a driver coil, which delivers a higher primary voltage into the ignition coil than the standard magneto did. As magnet number one passes the driver coil, the magnetic circuit is routed through the center core to 1 heel as current is induced in the winding. Current flows through the closed point set 1 into the armature plate, then through closed point set 2 back to the other driver coil lead (Figure 8–8).

As the magnet continues to rotate with the flywheel, the magnetic circuit will be from heel of 2 through the laminations trying to go up the center core. The magnetic field cannot establish immediately because it is momentarily repelled by the attendant magnetic field. This forces the driver coil voltage to peak. The cam, rotating with the crankshaft, opens ignition point set 1, which is timed to open at the peak voltage. Driver-coil current is now forced through the primary winding of the ignition-coil 1, ballooning a magnetic field across the secondary winding. This induces a quick-rise voltage into the secondary circuit, firing the spark plug in cylinder 1 (Figure 8–9).

Magnet 2 is 180 degrees away and of reverse polarity. As it passes the driver coil, the magnetic circuit is routed through the center core to heel 1, as current is induced in the winding. Current flows through the closed point set 2 into the armature plate. It continues

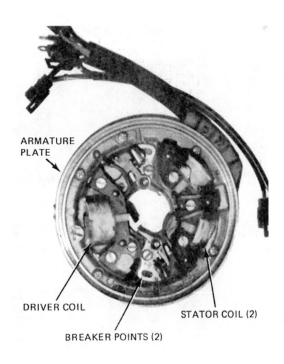

ARMATURE
PLATE

DRIVER COIL

BREAKER POINTS (2)

STATOR COIL (2)

Figure 8–7 Low-tension magneto assembly.

up through ignition point set 1 to the other driver coil lead. As the magnet continues to rotate with the flywheel, the magnetic circuit will be from the center core to heel 2 through
the laminations. Once again, the magnetic field cannot establish immediately because it is
repelled by the attendant magnetic field. This forces the driver coil voltage to peak. The
cam rotating with the crankshaft opens point set 2, which is timed to open at the peak
voltage. Driver coil current is now forced through the ignition-coil 2 primary winding, ballooning a magnetic field across the secondary winding. This induces a very high voltage of
quick rise time in the secondary circuit, firing the spark plug in cylinder 2. As both point
sets open, the condenser helps step up primary voltage by speeding up the collapse of the
magnetic field (Figure 8–10).

The simplified circuit illustrations that support the theory of operation are intended for
instructional purposes only.

8.2.1   Troubleshooting the Low-Tension Magneto Ignition System

Install a spark checker and set the gap for 7/16-inch. If you obtain a spark, the ignition
system is okay. Then check or replace the plugs. If there was no spark on either cylinder,
check the kill switch and then the driver coil. Using a flywheel puller that attaches to the
bolt pattern at the flywheel hub, pull the flywheel. Disconnect the driver coil and test it for

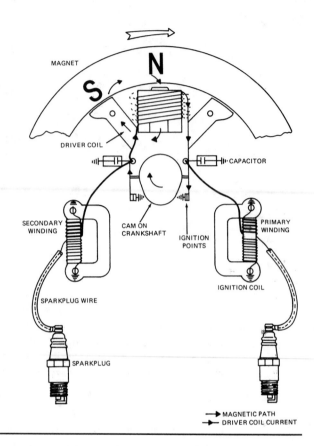

Figure 8–8   Driver current flow, low-tension.

→ MAGNETIC PATH
→ DRIVER COIL CURRENT

resistance on the low ohm scale. (A dealer would check it using a peak voltmeter or M-90.) Also check the ignition point condition and note the color. Gun-metal grey with some pitting is normal. Black is an indication of lubricant between the points. Improper cam lubricant has been used or oil vapor is coming from the upper crankshaft seal. The point gaps should be set and timed to open 180 degrees apart. The ignition coils should be checked with an ohmmeter for resistance in both the primary and secondary windings. If Merc-O-Tronic or Stevens ignition testers are available, the ignition coils should be power tested and an insulation test made. For details about ignition coil testing refer to "Section 8.1.4 Troubleshooting the Standard Magneto Ignition."

If the powerhead is an electric start, there probably is an alternator charging system. The alternator system uses the same magnets for induction, so this system can be used to check the strength of the flywheel magnets. You will be checking the alternator system as well. To determine if the magnets are weak, an alternator output test is performed. Place an ammeter in series, or place an induction type ammeter next to the rectifier battery lead or positive battery cable. Load-test the battery to lower the battery voltage or use a Stevens Load

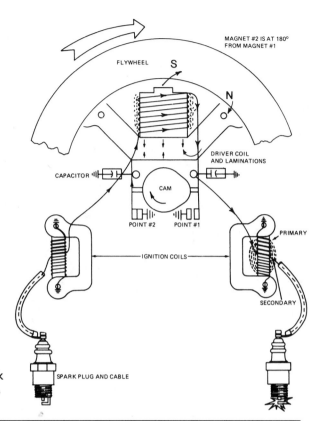

MAGNET #2 IS AT 180°
FROM MAGNET #1

FLYWHEEL   S   N

DRIVER COIL
AND LAMINATIONS

CAPACITOR

CAM

POINT #2   POINT #1

PRIMARY

IGNITION COILS

SECONDARY

SPARK PLUG AND CABLE

Figure 8–9 Driver coil peak voltage is routed through ignition-coil 1.

Bank instrument. With the outboard in the water, increase RPM to full throttle and read the ammeter. If full output is obtained, the strength of the flywheel magnets is okay. If full output is not obtained, see Chapter 7, *Alternator Charging System,* before replacing the flywheel. The stator, rectifier, battery, and wiring will have to be checked before the flywheel is replaced. If there is no charging system on a small fishing motor, use the screwdriver test as outlined under Section 8.1.4 "Troubleshooting the Standard Magneto Ignition."

8.3   **Magneto Capacitor Ignition System (OMC)**

Six major components make up the capacitor discharge (CD-2) ignition system: Flywheel, flywheel hub, charge coil, power pack, ignition coils, and spark plugs. The flywheel contains two oppositely charged magnets, placed 180 degrees apart. The flywheel hub may have a magnet used to generate a charge in the sensor coils, which are used to trigger the gate of the silicon-controlled rectifier (SCR). Under the flywheel, a charge coil generates alternating current used to charge the capacitor. Mounted to the side of the powerhead is the power pack, which houses the capacitor and other necessary electronics to produce a

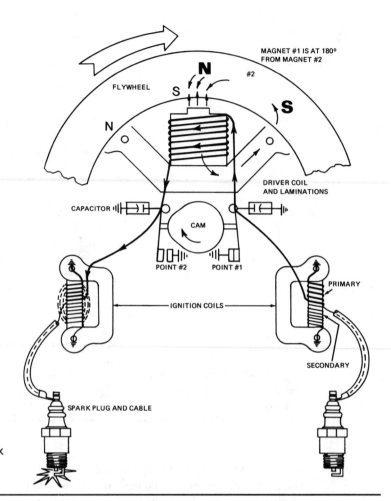

Figure 8–10   Driver coil peak voltage is routed through ignition-coil 2.

timed ignition. The ignition coils are also mounted to the side of the powerhead and spark plugs ignite the fuel mixture.

8.3.1   Capacitor Charge Circuit

The CD-2 ignition system will be used as a basic study system. Three of the circuits, capacitor charge, capacitor discharge, and ignition stop, are basic in all the OMC capacitor discharge ignition systems. The CD-2 ignition system on the older units is for OMC 4.5 through 60 models. It uses the basics of magnetic induction to generate an AC voltage at an approximate 250V in the charge coil. A group of four diodes is used to rectify the voltage to DC. A capacitor momentarily stores the DC voltage. Two SCRs are used to trigger the

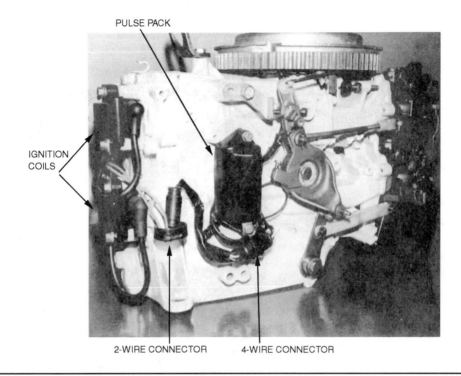

PULSE PACK

IGNITION COILS

2-WIRE CONNECTOR   4-WIRE CONNECTOR

Figure 8–11   **CD-2 ignition.** *(Courtesy of Outboard Marine Corporation)*

capacitor voltage, and ignition coils step up the voltage to the spark plugs (Figures 8–11 and 8–12).

Installed in the flywheel are two magnets of opposite polarity. As shown in Figure 8–13, the magnets, with their accompanying north-south magnetic fields, pass the charge coil (8), and alternating current is induced. Current flows out of the charge coil (8) through the wire (7), which becomes positive and enters the power pack. Current then flows through the diode (1), which applies a positive charge to the ground side of the capacitor (5). Current is blocked by diodes (2 and 3). Returning to the charge coil (8), current flows from the capacitor (5) through the diode (4) and wire (6), completing the circuit.

Current flow is reversed as the opposite magnet (10) (which was at 180 degrees) passes by the charge coil. The wire (6) is now positive. This positive current enters the power pack and flows through the diode (2), which applies a positive charge to the ground side of the capacitor. Current flow is blocked by the diode (1) and passes by another diode (4). Returning current flows from the capacitor (5) through the diode (3) and the wire (7) completing the circuit to the charge coil (8).

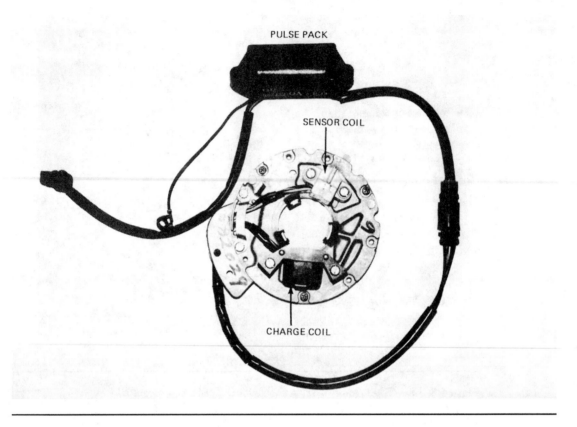

Figure 8–12   CD-2 ignition components

The charge coil delivers an alternating current to the power pack that is rectified by the diodes. The capacitor receives a positive charge that is maintained by the diodes on the ground side of the capacitor. This happens regardless of the constantly changing charge coil output.

8.3.2   Triggering Circuit

The capacitor is now charged and needs a means of releasing this charge at the correct time. This is the job of the triggering circuit, which uses the same flywheel magnets and has developed 3V in the sensor coil that flows into four diodes and two SCR. (The standard ignition points are replaced by the SCRs and sensor coil.) The ignition timing is controlled by magnetic field polarity moving with the flywheel, which induces a positive pulse to a particular SCR gate, discharging the capacitor (Figure 8–14).

The sensor is mounted to a metal plate called a timer base. The sensor is precisely positioned on the timer base so that it turns on an SCR at the correct time.

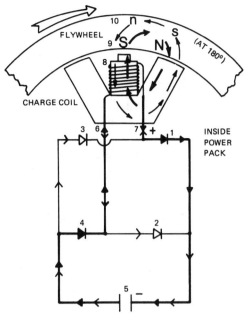

Figure 8–13 Capacitor
charge circuit at peak voltage.

MAGNETIC NORTH/SOUTH INDUCTION–HEAVY LINES ➡
MAGNETIC SOUTH/NORTH INDUCTION–FINE LINES ➡

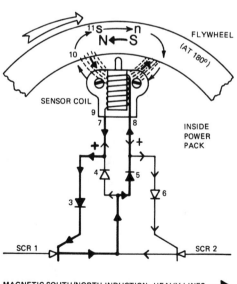

Figure 8–14 Triggering cir-
cuit. *(Courtesy of Outboard Marine
Corporation)*

MAGNETIC SOUTH/NORTH INDUCTION–HEAVY LINES ➡
*MAGNETIC NORTH/SOUTH INDUCTION HAPPENS
180° FROM SOUTH/NORTH MAGNET–FINE LINES* ➡

The timer base can rotate several degrees to vary the spark timing. The timer base is usually connected to the throttle linkage so that when the throttle opens and closes, spark timing is automatically adjusted.

As shown in Figure 8–14, as the magnet (10), with its accompanying south-north magnetic field, passes the sensor coil (9), it induces an alternating current. Current flows out of the sensor coil (9) through the wire (7), which becomes positive, and enters the power pack. Current then flows through a diode (3), which applies a positive pulse to the gate of SCR 1, turning on SCR 1. Current is blocked by diodes (4 and 6). Returning to the sensor coil (9), current flow bypasses a diode (4), going through another diode (5) and wire (8), completing the circuit.

Current flow is reversed as the opposite magnet (11) (which was at 180 degrees) now passes by the sensor coil (9). The wire (8) is now positive. This positive current enters the power pack and flows through a diode (6), which applies a positive pulse to the gate of SCR 2, turning on SCR 2. Current flow is blocked by diodes (3 and 5). Returning to the sensor coil, current flow bypasses a diode (5), going through another diode (4) and wire (7), completing the circuit.

The sensor coil has delivered an alternating current to the power pack, which was rectified by the diodes. The positive pulse was felt at the gate of SCR 1, which triggered the capacitor discharge. Then the magnet (11) of opposite polarity passed the sensor coil; a positive pulse was felt at the gate of SCR 2, which again triggered the capacitor discharge for a different cylinder.

8.3.3   Capacitor Discharge Circuit

As the gate of SCR 1 is pulsed with a positive current from the sensor coil, it is turned on. The positive charge stored in the capacitor is released. Starting at the capacitor, the current flows to the power pack ground (6) (as indicated by the heavy arrows in Figure 8–15) to ignition-coil A's primary ground (3). It then flows through the coil-A primary winding and flows to the other side of the capacitor (5) through SCR 1 until discharged. At this time, SCR 1 turns itself off. The capacitor is now ready to be recharged.

When the flywheel magnet of opposite polarity has rotated 180 degrees, the sensor coil sends a positive pulse to the gate of SCR 2, turning it on. The positive charge stored in the capacitor (5) is released (as indicated by the fine arrows). It flows to the power pack ground (6) to the ground (4), passing through coil-B primary winding flowing through SCR 2 to the other side of the capacitor (5). This continues until the capacitor is discharged and SCR 2 turns itself off. The capacitor is now ready to recharge again.

As the capacitor discharges through the primary winding, a magnetic field instantly balloons across the primary and secondary windings. This induces a quick-rise voltage in the secondary winding to create a spark across the spark plug electrodes, igniting the fuel for combustion. The amount of voltage required to fire the spark plug is dependent upon spark plug condition, electrode gap, fuel ratios, compression, and cylinder temperature.

 Safety

If the spark plug wire was removed from the plug and held in the air, voltage could rise to 40,000 volts. It packs a wallop, and therefore should be respected by using a spark tester to determine that the circuit is working.

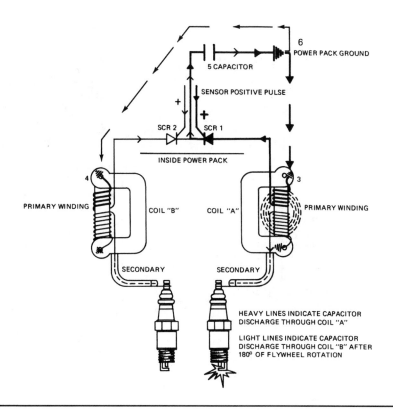

Figure 8–15   The CD (capacitor discharge) circuit. *(Courtesy of Outboard Marine Corporation)*

8.3.4   Ignition Stop Circuit

In order to stop the powerhead, the capacitor must be prevented from receiving a charge. A quick look at the CD circuit shows that the capacitor is already grounded on one side at 6 (see Figure 8–16). Therefore, to prevent the charging of the capacitor, the other side must be grounded by the stop button or key switch at 7. With both sides of the capacitor grounded, the capacitor is bypassed, and there is no spark. Note that the push button or key switch completed the stop circuit to ground. To start the powerhead, the circuit is opened by the switch.

On smaller outboards the steering handle contains a combination stop switch/emergency stop device. When the clip and lanyard assembly are *removed,* the emergency stop device is in the engine stop position. When the clip and lanyard assembly are *installed,* the emergency stop device is in the *run* position. To stop the engine when the clip and lanyard assembly are in place during normal operation, press the stop button *inward.* This momentarily grounds the ignition system.

To test the stop circuit, connect an ohmmeter between a clean engine ground and the appropriate colored wire (E terminal on OMC 9.9 HP) at the 5-pin connector. The ohmmeter

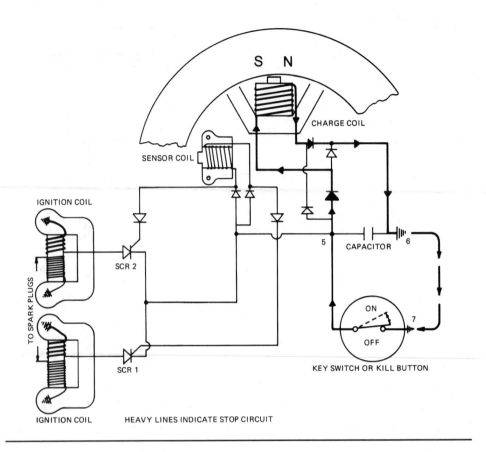

S N

CHARGE COIL

SENSOR COIL

IGNITION COIL

SCR 2

TO SPARK PLUGS

SCR 1

IGNITION COIL

5   CAPACITOR   6

ON

7

OFF

KEY SWITCH OR KILL BUTTON

HEAVY LINES INDICATE STOP CIRCUIT

Figure 8–16 The ignition stop circuit. *(Courtesy of Outboard Marine Corporation)*

must indicate a high reading. *Momentarily* press inward on the stop button. The ohmmeter should now show a low reading indicating a ground contact is made. Next *remove* the lanyard clip; the ohmmeter should show a low reading, indicating a ground contact. Both ground contacts short out the ignition system.

8.4   Capacitor Discharge-3 Ignition System (OMC)

The CD-3 ignition is used on OMC 70–75 HP, three-cylinder powerheads. It functions basically the same as CD-2 except for the triggering circuit. Also there are six inner rim flywheel magnets, three ignition coils, and the location of the sensor magnets changes (see Figures 8–17 and 8–18). Review the previous section if you are not familiar with CD-2, because the capacitor charge, discharge and ignition stop circuits of CD-3 are the same as CD-2, and will not be repeated here. This section only explains the differences between the CD-2 and CD-3 systems (see Figures 8–13, 8–15, and 8–16).

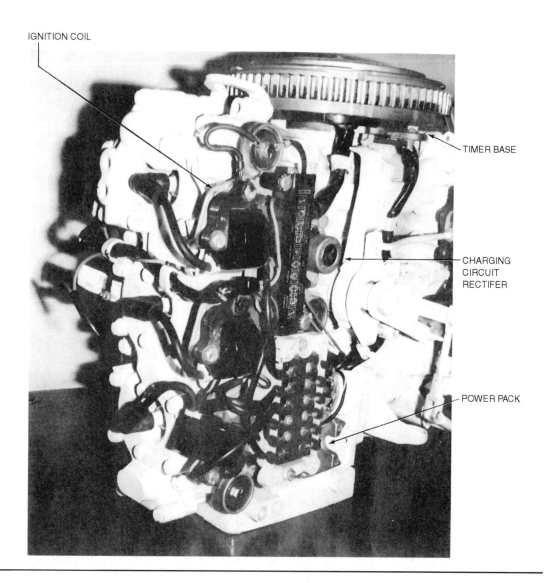

IGNITION COIL

TIMER BASE

CHARGING
CIRCUIT
RECTIFER

POWER PACK

Figure 8–17   CD-3 ignition.

8.4.1   Triggering Circuit

The flywheel *hub* is equipped with two magnets of opposite polarity, placed 180 degrees from each other (see A in Figure 8–19).

This is in addition to six magnets in the inner rim circumference of the flywheel. As the flywheel magnets in the *hub* turn, the first pole slot passes sensor coil 1, and one end of the

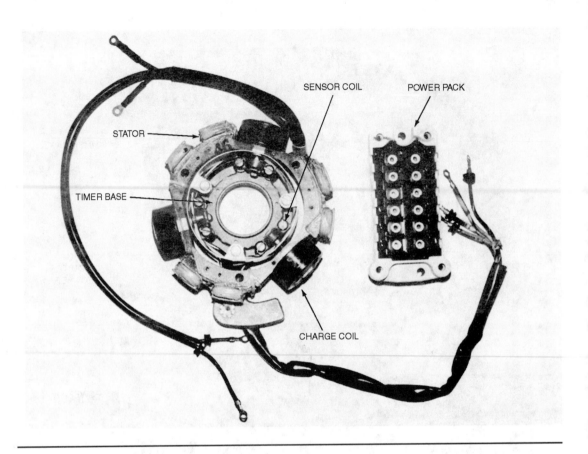

Figure 8–18   CD-3 ignition components.

sensor-coil winding becomes positive, while the other end becomes negative (see B in Figure 8–19).

This positive electrical pulse triggers the gate of SCR 1, and the capacitor discharges through ignition-coil-1 primary winding, ballooning a magnetic field across the secondary winding, returning back through SCR 1 to the capacitor. This induces a quick-rise voltage in the ignition-coil secondary winding sufficient to break down the spark plug gap and fires the spark plug in cylinder 1. After the capacitor discharge, SCR 1 turns itself off.

One hundred-twenty degrees of rotation later, the first pole slot passes sensor-coil 2, which induces a positive pulse, and in turn triggers the gate of SCR 2 (see C in Figure 8–19).

This discharges the capacitor-stored electrical charge through ignition-coil-2 primary winding, ballooning a magnetic field across the secondary winding, back through SCR 2. This induces a quick-rise voltage in the ignition-coil secondary winding sufficient to break down the spark plug gap. This fires the spark plug on cylinder 2. After the capacitor discharge, SCR 2 turns itself off.

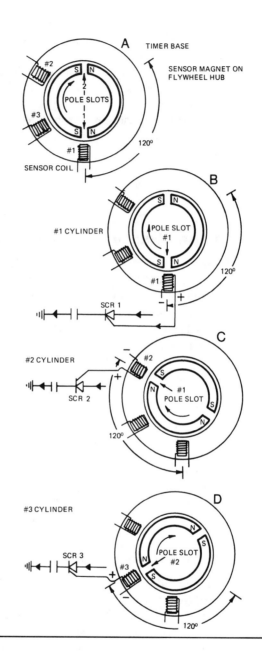

Figure 8–19 CD-3 ignition trigger circuit.

The second pole slot on the flywheel hub is located 180 degrees away, and is of the opposite polarity (see A in Figure 8–19). It travels 120 degrees, passing sensor-coil 3, and inducing a positive charge to the gate of SCR 3 (see D in Figure 8–19). The charge in the capacitor is released into the primary winding of ignition-coil 3, ballooning a magnetic field across the secondary winding returning through SCR 3. This induces a quick-rise voltage in the secondary winding sufficient to break down the spark plug gap. This fires the spark plug in cylinder 3. SCR 3 turns itself off after the capacitor is discharged.

One power pack is used to fire three cylinders. It has taken one revolution of the crankshaft, and each cylinder was fired exactly 120 degrees apart from each other. Timing is advanced by rotation of the armature plate, which advances the sensor coils into the magnetic field.

Magnets are still used in the inner circumference of the flywheel to induce a current flow in the charge coil to pump up the capacitor. These same magnets are also used to induce current in six stator coils, which is rectified for charging the battery.

The simplified circuit illustrations that support the theory of operation are intended for instructional purposes only.

8.5   Capacitor Discharge-4 Ignition system (OMC)

The CD-4 ignition system is used on older 90-, 115-, and 140-HP powerheads. The basic components of the system are the flywheel assembly consisting of six charging magnets cast in the inner circumference 60 degrees apart, and sensor magnets with two poles 180 degrees apart and placed on the flywheel hub. The timer base houses two sets of sensor coils that trigger two power packs (one on later models) containing electronic components necessary to produce a timed ignition spark (Figure 8–20).

Two charge coils generate alternating current that charges the capacitors. These coils are located in the stator assembly along with the stator windings for the alternator system. Four ignition coils are mounted on the cylinder heads and transform the capacitor voltage stored in the capacitor into high secondary voltage sufficient to break down the spark plug gaps and fire the spark plugs.

If you are not familiar with the CD-2 ignition, capacitor charge, capacitor discharge, and ignition stop circuits, review the complete CD-2 ignition system in this chapter, before studying this system. The capacitor charge, capacitor discharge, and ignition stop circuits are basically the same in the CD-4 as they are in the CD-2 system and need not be repeated here. However, there are two charge coils used to charge capacitors located in two power packs, which are mounted on the port and starboard sides.

8.5.1   Triggering Circuit

The real difference in the CD-4 system is the triggering circuit, which uses two sensor coils 90 degrees apart. Each sensor has two windings within the sensor coil (see Figure 8–21). Sensor-coils A and C are used to trigger the capacitor in the starboard-side power pack. The capacitor in the port-side power pack is triggered by sensor-coils B and D. The ignition for each cylinder is controlled and timed by the sensor circuit. The sensor magnets located in the flywheel hub present a north-south pole slot to the sensor coil. This induces a positive polarity at one end of the sensor winding and a negative polarity at the opposite end. As the crankshaft rotates, the charge coils charge the capacitor and then the sensor magnet north-south pole slot sweeps sensor coil A, producing a positive signal at the gate of SCR 1, and returning to the negative side of the sensor coil (Figure 8–22).

Figure 8–20   CD-4 ignition.

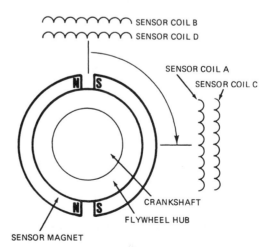

Figure 8–21   CD-4 sensor
coils and magnet.

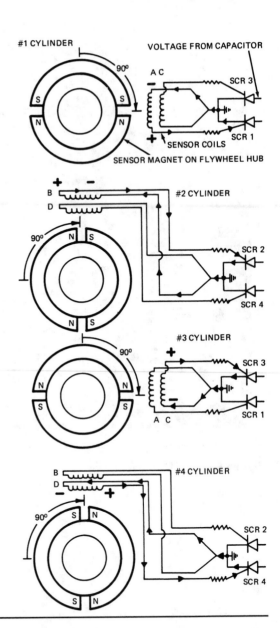

Figure 8–22 CD-4 ignition trigger (sensor) circuit.

This turns on SCR 1, which discharges the starboard power pack capacitor to ground through ignition-coil-1 primary winding. A magnetic field is instantly ballooned across the primary and secondary windings as the current returns back to the capacitor. This ballooning action induces a quick-rise voltage in the secondary winding to create a spark across the electrodes of spark plug 1, igniting the fuel. The conducting SCR turns itself off, and the capacitor can be recharged again for the firing of cylinder 3.

To fire cylinder 2 the sensor magnet travels 90 degrees as a south-north magnetic field sweeps sensor-coil B. This produces a positive signal in the port power pack at the gate of SCR 2, and then returns to the negative side of the sensor coil. (see Figure 8–22, #2). This turns on SCR 2, which discharges the port power pack capacitor to ground and through the ignition-coil primary winding. A magnetic field is instantly ballooned across the primary and secondary windings as the current returns back to the capacitor. This ballooning action induces a quick-rise voltage in the secondary winding, firing spark plug 2, which ignites the fuel. The conducting SCR turns itself off and the capacitor can be recharged for firing of cylinder 4.

To fire cylinder 3 the sensor magnet travels another 90 degrees as a south-north magnetic field sweeps sensor coil C, producing a positive signal in the starboard power pack at the gate of SCR 3, and then returns to the negative side of the sensor coil (see Figure 8–22, #3). This turns on SCR 3, which discharges the starboard power pack capacitor to ground and through ignition-coil-3 primary winding. A magnetic field is instantly ballooned across the primary and secondary winding as the current returns back to the capacitor. This ballooning action induces a quick-rise voltage in the secondary winding, firing spark plug 3 ,which ignites the fuel. The conducting SCR turns itself off and the capacitor can be recharged for firing of cylinder #1.

To fire cylinder 4 the sensor magnet travels still another 90 degrees as a north-south magnetic field sweeps sensor coil D, producing a positive signal in the port power pack at the gate of SCR 4, and returns to the negative side of the sensor coil (see Figure 8–22, #4). This turns on SCR 4, which discharges the port power pack capacitor to ground and through ignition-coil-4 primary winding. A magnetic field is instantly ballooned across the primary and secondary windings as the current returns back to the capacitor. This ballooning action induces a quick-rise voltage in the secondary winding, firing spark plug 4, which ignites the fuel. The conducting SCR turns itself off and the capacitor can be recharged for the firing of cylinder 2. This all takes place in one revolution of the power-head crankshaft and then the sequence begins all over again for the next revolution.

The simplified circuit illustrations that support the theory of operation are intended for instructional purposes only.

8.6   Capacitor Discharge-6 Ignition System (OMC)

The CD-6 ignition system is used on 150-, 200-, and 235-HP powerheads. The basic components of the system are the flywheel assembly consisting of six charging magnets cast in the inner circumference 60 degrees apart, and sensor magnets with two poles 150 degrees apart placed on the flywheel hub. The timer base houses six sensor coils that trigger two power packs containing electronic components necessary to produce a timed ignition spark. Two charge coils generate alternating current, which is rectified to DC to charge a capacitor in each power pack. These coils are located in the stator assembly along with the stator windings for the alternator charging system. Six ignition coils are mounted on the cylinder heads and transform the capacitor voltage into high secondary voltage sufficient to break down the spark plug gap and fire the spark plugs (see Figures 8–23 and 8–24).

If you are not familiar with the CD-2 ignition, capacitor charge, capacitor discharge, and ignition stop circuits, review the complete CD-2 ignition system (earlier in this

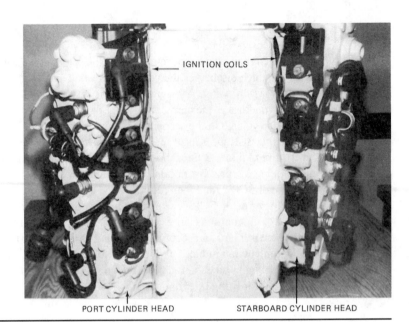

Figure 8–23   OMC V-6 ignition.

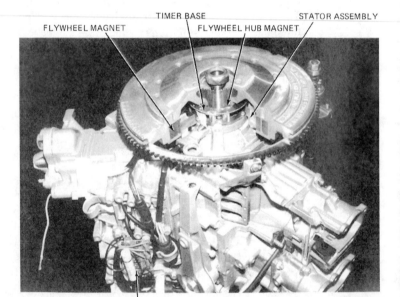

Figure 8–24   CD-6 ignition components. *(Courtesy of Outboard Marine Corporation)*

chapter) before studying this system. The capacitor charge, capacitor discharge and ignition stop circuits are basically the same in the CD-6 as they are in the CD-2 system and need not be repeated here. However, the magnets on the flywheel hub are placed at different degrees, and the two charge coils are used to charge capacitors located in the power packs which are mounted on the port and starboard sides. The real difference in the CD-6 system is the triggering circuit which uses six sensor coils in a timer base under the flywheel (Figure 8–25).

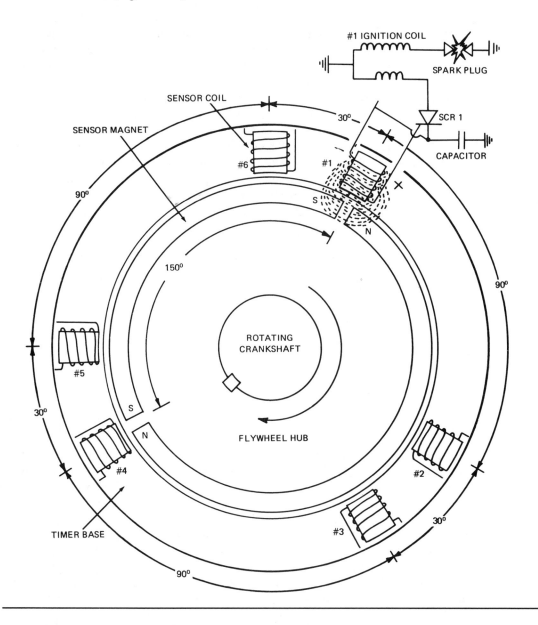

Figure 8–25   CD-6 sensor coil location.

8.6.1   **Triggering Circuit**

The ignition timing for each cylinder is controlled by the sensor circuit (see Figure 8-26). The sensor magnets located in the flywheel hub present a north-south pole slot to the sensor coil. This induces a positive polarity at one end of the sensor winding and a negative polarity at the opposite end. As the powerhead runs, the sensor magnet turns 150 degrees further, presenting a south-north pole to the sensor coil. This induces a reverse polarity in the sensor coil. The SCR gates in the power packs are *arranged to trigger only upon receiving a positive pulse from the sensor coils*. The south-north pole slot will not trigger the SCR gate because the polarity is wrong.

As the powerhead runs, the sensor magnet presents a north-south pole slot to sensor-coil 1 (see Figure 8–26, #1). This induces a positive pulse in the sensor coil lead attached to the gate of SCR 1. This turns on SCR 1, which discharges the previously charged starboard power pack capacitor to ground and through ignition-coil-1 primary winding. A magnetic field is instantly ballooned across the primary and secondary windings as the capacitor voltage returns to the capacitor. This induces a quick-rise voltage in the secondary winding sufficient to break down the spark plug gap, and creates a spark across the electrodes of spark plug 1, igniting the fuel. The conducting SCR turns itself off, and the capacitor can be recharged again for firing the next cylinder on the starboard side.

To fire cylinder 2 the sensor magnet travels 90 degrees from sensor coil 1 and presents the north-south pole slot to sensor-coil 2 (see Figure 8–26, #2). This induces a positive pulse in the sensor coil lead attached to the gate of SCR 2. This turns on SCR 2 which discharges the previously charged port power pack capacitor to ground and through ignition-coil-2 primary winding. A magnetic field is instantly ballooned across the primary and secondary windings as the current returns to the capacitor. A quick-rise voltage in the secondary winding is induced, firing spark plug 2 which ignites the fuel. The conducting SCR turns itself off and the capacitor can be charged again for firing the next cylinder on the port side.

To fire cylinder 3, the sensor magnet travels 30 degrees from sensor-coil 2 and presents a north-south pole slot to sensor-coil 3 (see Figure 8–26, #3). This induces a positive pulse in the sensor coil lead attached to the gate of SCR 3. This turns on SCR 3, which discharges the previously charged starboard power pack capacitor to ground and through ignition-coil-3 primary winding. A magnetic field is instantly ballooned across the primary and secondary windings of ignition-coil-3 as the voltage returns to the capacitor. This induces a quick-rise voltage in the secondary winding, firing spark plug 3 which ignites the fuel. The conducting SCR turns itself off and the capacitor can be recharged again, for firing the next cylinder on the starboard side.

To fire cylinder 4 the sensor magnet travels 90 degrees from sensor-coil-3 and presents a north-south pole slot to sensor coil 4 (see Figure 8–26. #4). This induces a positive pulse in the sensor coil lead attached to the gate of SCR 4, which turns on SCR 4, discharging the previously charged port power pack capacitor to ground and through ignition-coil-4 primary winding. A magnetic field is instantly ballooned across the primary and secondary windings of ignition-coil 4 as the voltage returns to the capacitor. This induces a quick-rise voltage in the secondary winding, firing spark plug 4 which ignites the fuel. The con-

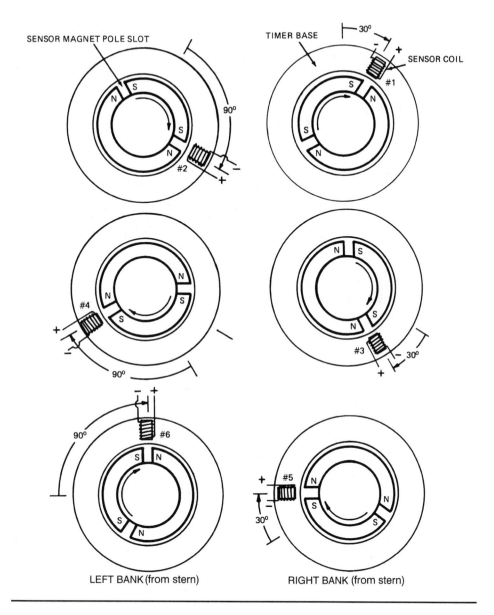

Figure 8–26   OMC CD-6 sensor coils.

ducting SCR turns itself off and the capacitor can be recharged again for firing the next cylinder on the port side.

To fire cylinder 5 the sensor magnet travels 30 degrees from sensor-coil 4 and presents a north-south pole slot to sensor-coil 5 (see Figure 8–26, #5). This induces a positive pulse in the sensor coil lead attached to the gate of SCR 5, and turns on SCR 5, which discharges the previously charged starboard power pack capacitor to ground through ignition-coil-5 primary winding. A magnetic field is instantly ballooned across the primary and secondary windings of the ignition-coil 5 as the voltage returns to the capacitor. This induces a quick-rise voltage in secondary winding firing spark plug 5 which ignites the fuel. The conducting SCR turns itself off and the capacitor can be recharged again for firing the next cylinder on the starboard side.

To fire cylinder 6, the sensor magnet travels 90 degrees from sensor-coil 5 and presents a north-south pole slot to sensor-coil 6 (see Figure 8–26, #6). This induces a positive pulse in the sensor coil lead attached to the gate of SCR 6. This turns on SCR 6, which discharges the previously charged port power pack capacitor to ground and through ignition-coil 6 as the voltage returns to the capacitor. This induces a quick-rise voltage in the secondary winding, firing spark plug 6 which ignites the fuel. The conducting SCR turns itself off, and the capacitor can be charged again for firing the next cylinder on the port side.

As the sensor magnet continues to rotate another 30 degrees from sensor-coil 6, the sequence is started all over again with numbers 1 cylinder. The cylinders have been fired at 30-90-30-90-30-90 degrees for a 360-degree rotation (which is one revolution of the crankshaft).

The simplified circuit illustrations that support the theory of operation are intended for instructional purposes only.

8.6.2 Troubleshooting the Capacitor Discharge Ignition System (OMC)

The magneto capacitor discharge ignition system is easy to check out. Approach the problem systematically, so a test or components will not be overlooked.

When a problem is suspected in the ignition system, a visual inspection of the total system should be made. Check for broken, pinched, or loose leads. Check the two-, three-, and four-prong connectors for correct mating, corrosion, and for secure connections. If the connector is hard to get apart or hard to go back together, squirt some isopropyl alcohol into the connector at both ends. This will lubricate the rubber connector and will not leave a mess inside the connector, easing disassembly. A silicone electrical compound (grease) can be used when reinstalling connectors.

Check all ignition coils for proper bonding to ground. A stainless steel star washer should be between the electrical lead and the cylinder head (powerhead). This will bond the connection by biting into the electrical lead and the cylinder head (powerhead). Without a good ground in the system, the power pack may not be able to discharge the charge in the capacitor.

A variety of ignition testers is available for testing ignition components. Popular testers are manufactured by Merc-O-Tronic Instruments Corp. and Stevens Instrument Co. However, for our troubleshooting procedure, the ohmmeter will be used. This is not the best instrument for the job, but is more common with the outboard owner. Probably the best instrument for troubleshooting is a *Peak* voltmeter, found in many dealerships. This instrument is capable of giving voltage output readings, which make the test results very accurate (Figure 8–27).

After visual inspection is made, a spark checker should be hooked up to determine if there is a spark. Set the tester gap for 1/2-inch. Voltage output of the secondary circuit is

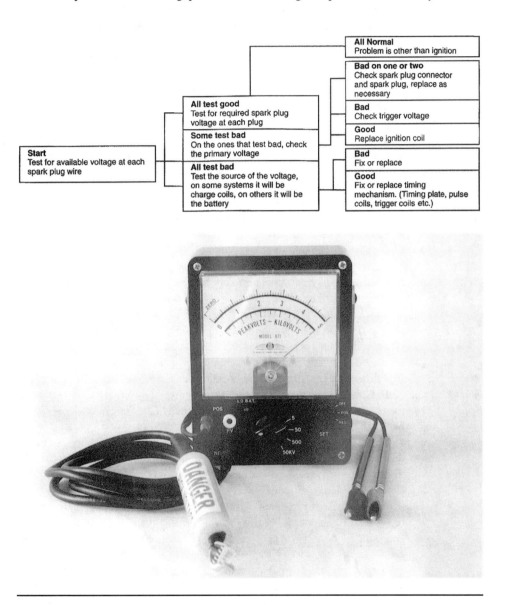

		All Normal Problem is other than ignition
	All test good Test for required spark plug voltage at each plug	**Bad on one or two** Check spark plug connector and spark plug, replace as necessary
		Bad Check trigger voltage
	Some test bad On the ones that test bad, check the primary voltage	**Good** Replace ignition coil
Start Test for available voltage at each spark plug wire	**All test bad** Test the source of the voltage, on some systems it will be charge coils, on others it will be the battery	**Bad** Fix or replace
		Good Fix or replace timing mechanism. (Timing plate, pulse coils, trigger coils etc.)

Figure 8–27　Kilo and peak voltmeter, Model 871. *(Courtesy of Merc-O-Tronic Instruments Corporation)*

about 40,000V, so be careful. Observe the spark. Is there a spark on all cylinders? Is there spark just on one bank? Is there no spark on one cylinder? Is there no spark at all? Let us assume that there was no spark on any cylinder. First consider what components would affect all cylinders, i.e., the charge coil(s), power pack ground, the power pack on 2- and 3-cylinder powerheads, and the ignition stop circuit. Always check out the easiest, most likely circuit first. In this case, start by checking the ignition stop circuit. The procedure given in the service manual for manual-start powerheads is something like the following: Start by separating the three-prong connector. Place the ohmmeter on high scale and the red lead in the A terminal (D terminal on V-4 and V-6 electric) lead going to the ignition coil and connect the black lead to a clean powerhead ground.

The reading should show an open circuit if the stop button circuit is okay. Now push the stop button in and it should show continuity, a closed circuit to ground. You should see an opening and closing of the ignition stop circuit as the stop button is pushed in and released. The ignition stop circuit grounds the capacitor and will not allow the capacitor to charge up. A malfunction of this circuit will be indicated by a no-spark condition on any cylinders, or you will not be able to use the button to shut off the powerhead. The larger powerheads use a key switch so you should turn the key on and off, and the ohmmeter should register continuity when the key is in the off position. To prove the wiring to the key switch, disconnect the wire at the M terminal on the back of the key switch. If the circuit now functions properly, the key switch is at fault.

The power packs cannot be tested using an ohmmeter, so the diagnostic procedure is used to test all other components and, by process of elimination, prove the power pack bad. A dealership can test output of an installed power pack by using a peak voltmeter.

Now consider the condition of spark only on some cylinders. The following permit spark on some cylinders but not on others: the sensor coils, a charge coil on V-4 and V-6 models, ignition coils, and spark plug leads. The sensor circuit times the spark to each cylinder, so it should be checked first. The service manual indicates that there is a specific procedure and there is a specific resistance on a V-6: jumper wires are installed into the four-wire connector A, B, C, and D terminals leading to each sensor coil (see Figure 8–28). By placing the ohmmeter leads from A to D, then B to D, and finally C to D, resistance of each sensor coil on one bank can be determined. A normal reading is 15 + or – 6 ohms on each sensor.

These tests determine if the sensor coils and leads are within specifications. Therefore, they are able to send a pulse to the gate of a particular SCR. Repeat these tests on the other three sensor coils for the other bank. If the readings are all normal, then the question arises, are they shorted to ground? Place the ohmmeter on high scale, with the black lead attached to the timer base for a ground. Touch the red lead to A, B, C, and D terminals. There should be no reading. Upon doing this, you determine if the insulation around the coil windings and leads is good. If you get a reading, then that sensor coil or lead is shorted to ground and cannot trigger the gate of the SCR to which it is attached. If there are any abnormal readings, then the total timer base must be replaced. Individual sensor coil repair is not possible. Along with the four-wire connector, the isolation diode used on V-4 and V-6 models should also be checked. If it is bad, the powerhead will be a rough runner. Place the ohmmeter leads into both D terminals on one bank. There should be a reading in one direction and no reading when the ohmmeter leads are reversed.

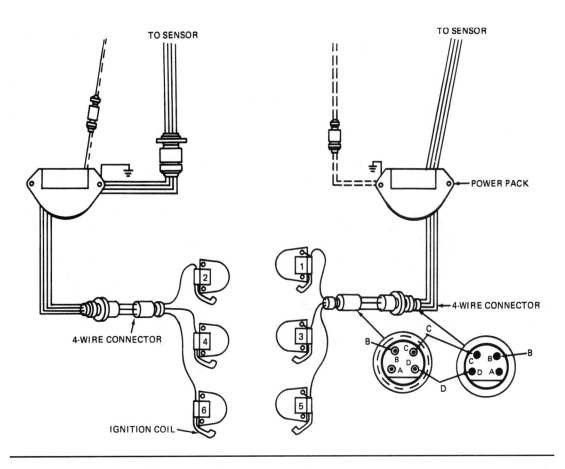

Figure 8–28   Partial schematic of V-6 ignition. *(Courtesy of Outboard Marine Corporation)*

The sensor coils on other models are checked in the same way. There are just fewer coils to check. Follow the procedure given in the service manual and refer to specifications.

Next, consider the problem of no spark on one bank of a V-4 or V-6 model. This could be a problem with a charge coil or power pack. The service manual instructions are to separate the charge coil, two-wire connectors, and installing jumper leads into the stator and at A and B terminals. The charge coil can be checked for resistance, shorts to ground, and for opens. Connect the ohmmeter across A and B terminals and take a reading. On the high scale the resistance should be 560 + or – 75 ohms. With the ohmmeter black lead attached to ground at the timer base and touching terminals A and B with the ohmmeter red lead, you can tell if the charge coil is shorted to ground. There should be no reading. When making these tests on the various coils, move the lead wires around and see if the reading varies. A lead wire may be broken and open up the circuit in a particular position. Again, the power pack cannot be tested with an ohmmeter.

If there is a problem with an ignition coil, it will only affect that one cylinder. Use the ohmmeter to check the secondary and the primary windings. Again, test for resistance in each of the windings. The service manual instructions are to disconnect the two-, three-, or four-wire connection (depending upon model) between the ignition coil and the power pack. To power-test the coil, a Merc-O-Tronic or Stevens instrument has to be used.

The ohmmeter black lead is placed on a clean powerhead ground. Set to the low ohm scale and attach the red lead to a jumper wire in A terminal (B terminal on 4.5–60 HP). This tests the primary winding. Resistance should be between 0.1 + or − 0.05 ohms. Set the ohmmeter on the high scale. Place the black lead on the secondary terminal and the red lead into the A terminal (B terminal on 4.5–60 HP). Resistance reading should be 275 + or − 50 ohms for the secondary winding. Both of these windings, tested in an inactive condition, may have tested well, but the coil still will not work. The unknown factor here is the condition of the coil insulation. Will it hold up against an approximate 40,000V when the powerhead is running? We do not know this when the ohmmeter is used for testing. The ignition coils should be power tested and the insulation stressed. To do this, use one of the special testers that the dealership has (see Figure 8–29).

What about the spark plug wires or the spring terminal in the boot end? If the wire leaks voltage to ground or the spring terminal is corroded, there may not be sufficient voltage for spark at the spark plug. Using the ohmmeter on high scale, attach the meter leads to each end of the spark plug wire when removable, or connect to the coil primary ground wire. There should be a reading from 0 to 300 ohms. If there is no reading, clean the terminal ends and retest. To test the insulation on the wire, a spark plug wire tester (see Figure 8–38) or a special tester at the dealership will have to be used.

Next, consider the problem of a weak spark on all cylinders. After making a full series of tests on the ignition stop circuit, sensor coils, charge coils, ignition coils, connections, and grounds, you may be ready to throw in the towel! But wait! Stop and think what causes this spark. Go back to the basics! A magnetic field sweeps the charge coil and the sensor coils to charge and trigger the capacitor charge into the ignition coil primary winding. This induces a voltage in the secondary winding firing the spark plug. In all these tests we have not considered the magnets in the flywheel. A weak spark in all cylinders can be caused by weak magnets. They may be strong enough to give some spark, but not strong enough to induce maximum voltage.

To test the magnets, an alternator output test is made. If the alternator can produce a maximum amperage output, there is nothing wrong with the flywheel magnets. If the alternator output is low and there is apparently nothing wrong in the alternator circuit, the flywheel magnets are weak. In this case, the flywheel would have to be replaced. Testing magnets with a screwdriver attraction is not an accurate enough test in this case. For theory instruction on the alternator output test, refer to Chapter 7, *Alternator Charging System*.

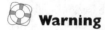 **Warning**

A word of caution when removing the flywheel—use the recommended puller that *pulls from the flywheel hub*. A puller that grips the outside of the flywheel will bend the flywheel. Also, do not strike the puller when it has been tightened down. The shock of the blow may very well break or crack the sensor magnets located in the flywheel hub. If the flywheel magnets are cracked or loose, the flywheel must be replaced.

You have now tested or checked the connections, grounds, stop circuit, sensor coils, charge coils, ignition coils, spark plug wires, and magnets. You have tested to isolate the power pack. Now if you have no spark at all or no spark on one bank, then replace the power pack, after careful attention is given to the power pack leads. The leads may be broken or shorted out to ground. At times the powerhead may stop after it is warmed up. The power pack can be temperature sensitive and shut down. It runs fine when cold. The reason for this type of problem is hard to find. If all the other components check out, then replace the power pack.

An OMC V-4 or V-6 may be equipped with an economixer, which is an electronic variable ratio oil injection system. This system functions upon receiving signals from the alternator, ignition spark, battery voltage, and the position of the throttle. If one of these signals fails to reach the microprocessor on top of the oil tank, the economixer will go into back-up mode. This will sound a beep and turn on an indicator light at the helm. The back-up mode limits the powerhead to an approximate 1500 RPM. The powerhead will act very erratically with this limited RPM. The back-up mode may be mistaken for an ignition problem, which it truly is, but it is caused by the malfunctioning economixer circuit. If you suspect that this is the case, follow the troubleshooting procedures given for the economixer system in the service manual.

Figure 8–29   ST-75 Ignition analyzer. *(Courtesy of Stevens Instrument Company)*

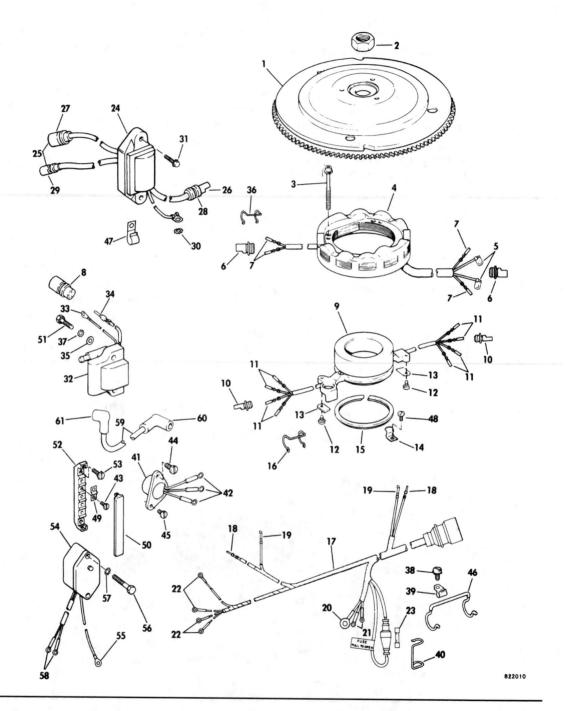

Figure 8–30   Ignition system (continued). *(Courtesy of Outboard Marine Corporation)*

OP0919C

Ref. No.	P/N	Name of Part	Qty.	Ref.No.	P/N	Name of Part	Qty.
*1	582403	FLYWHEEL ASSY.	1	31	328726	SCREW, Power pack to cylinder	4
*2	316828	NUT, Flywheel	1	32	582106	COIL ASSY.	6
3	325135	SCREW, Stator mounting	4	33	510767	. RING TERMINAL	1
4	582497	STATOR ASSY., 10 amp.	1	34	511469	. PIN	1
5	511399	. TERMINAL	2	35	304051	WASHER, Ignition coil mounting ...	12
6	511989	. CONNECTOR	2	36	511618	RETAINER, Stator connector	2
7	581656	. SOCKET	4	37	303480	LOCKWASHER, Ground	6
8	511449	CONNECTOR, Coil to power pack	2	38	303311	SCREW, Cable spring	1
9	582394	TIMER BASE & SENSOR ASSY.	1	39	326365	CLAMP, Spring, motor cable	1
10	511991	. CONNECTOR	2	40	321077	RETAINER, Fuse holder	1
11	581656	. SOCKET	8	41	582399	RECTIFIER & LEAD ASSY.	1
12	307019	. SCREW	2	42	511733	. RING TERMINAL	3
13	511387	. CLAMP, Leads	2	43	327785	SCREW, Terminal block	5
14	326439	CLAMP, Timer base	4	44	314576	SCREW, Rectifier mounting	1
15	313666	RETAINER, Timer base	1	45	552421	SCREW, Rectifier mounting	1
16	322549	RETAINER, Connector	4	46	315182	CLAMP	1
17	389452	MOTOR CABLE ASSY.	1	47	310493	CLAMP, Power pack	1
18	511469	. PIN	2	48	302812	SCREW, Retainer clamp	4
19	308808	. KNIFE DISCONNECT	2	49	315004	JUMPER	1
20	204053	. RING TERMINAL	1	50	390318	COVER ASSY.	1
21	204036	. RING TERMINAL	2	51	328725	SCREW, Coil mounting	12
22	510780	. RING TERMINAL	4	52	326255	TERMINAL BLOCK	1
23	510884	. FUSE, 20 amp.	1	53	510195	SCREW, Block to bracket	2
24	582138	POWER PACK, Port side	1	54	173640	VOLTAGE REGULATOR ASSY	1
24	582712	POWER PACK, Starboard side	1	55	510818	. TERMINAL	1
25	511469	. PIN	6	56	304913	. SCREW, Mounting	2
26	581656	. SOCKET	4	57	306488	. LOCKWASHER	2
27	511449	. CONNECTOR	1	58	510780	. TERMINAL	2
28	511991	. CONNECTOR	1	59	582365	LEAD ASSY., Ignition coil	6
29	511616	. CONNECTOR	1	60	581027	. TERMINAL & COVER	1
30	303480	LKWASH., Power pack to ground ..	2	61	580339	. TERMINAL & COVER	1

• **Note** Lap fit the flywheel taper to crankshaft taper with valve
lapping compound. Use OMC Ultralock P/N 388517 on taper.
Use OMC Nutlock P/N 384849 on crankshaft thread. Use a
new locknut whenever service requires removal of flywheel.

Figure 8–30   Continued. *(Courtesy of Outboard Marine Corporation)*

The spark plug must come in for some consideration as well. All too often it is replaced; the symptom goes away and then later reappears. This in itself vindicates the spark plug. Problems that affect the spark plug performance are stale fuel, improper oil mix, carburetor adjustments, preignition, detonation, synchronization of timing to carburetion, and cooling system problems. The spark plug is installed in the combustion chamber, so therefore it becomes a good indicator of powerhead performance and problems. See Section 10.8 High Performance for information on combustion—normal and abnormal, spark plug analysis, and outboard troubleshooting.

8.7   QuikStart System (OMC)

Some late models (125C to 250 HP) from OMC incorporate *QuikStart electronic starting* in the ignition system. When the powerhead temperature is below 96°F (36° C) the timing is automatically advanced for approximately 5 seconds each time the powerhead is started, regardless of temperature. Once the outboard has been at operating temperatures the QuikStart is canceled. It will not function again until the key is shut off and the outboard is restarted, which recycles the QuikStart electronic ignition. Another feature of QuikStart is the canceling of advanced timing when the powerhead speed tops approximately 1100 RPM.

The QuikStart system can be checked by making a temporary top dead center (TDC) mark of the flywheel for each cylinder. Run the outboard in a test tank with a test propeller. Check synchronization and linkage adjustments and that idle RPM is correct. With the white/black lead between the port temperature switch and the power pack disconnected, hook up the timing light to #1 spark plug lead (Figure 8–31). With the powerhead running and idle speed below 900 RPM the timing light must show #1 TDC near your temporary timing mark. Next disconnect the white/black lead and the #1 should shift approximately 1 inch left to the factory #1 TDC. This indicates that the ignition timing has been released and is now at the normal timing setting. After this test, the powerhead is shut off and the other cylinders are tested in sequence. If each cylinder test does not show a retard to the left, the timer base will have to be replaced.

8.8   S.L.O.W. Warning System

When you are out boating and the powerhead temperature exceeds 203° F (95° C) on a V-4 or 212° F (100° C) on a V-6 model, the powerhead RPM will automatically slow down to 2500. This warns the operator that an overheat condition has occurred. The operator should shut down the outboard and let it cool. (Check for blockage of the water inlet on the lower unit.) When the temperature drops below 162° F (72° C), the outboard can again be operated at normal speed. Information on powerhead temperature comes from the port or starboard temperature switch and tan leads.

To test the warning system, disconnect the starboard and port tan leads at the temperature switches. With the outboard in a test tank and a test propeller installed, run it at 3500 RPM. Attach the starboard tan test lead to a clean powerhead ground and the powerhead should slow to 2500 RPM. Perform the same test on the port side. If RPM drops in both tests to 2500 RPM the response is normal. Next, remove the temperature switches and test the warning system while heating in a container of automotive oil raising the temperature. Watch an ohmmeter for changes in resistance while the temperature increases. Follow the listing of changing temperature/resistance guide in the service manual (open at 170° F [77° C] and closed at 203° F [95° C]).

8.9   Breakerless Ignition System (Mercury)

The breakerless ignition system is a capacitor discharge ignition system. Components in the system are a battery, switch box, one ignition coil, tilt switch, key switch, and a distributor. The distributor is belt-driven from the crankshaft, and distributes a timed spark to each cylinder.

Battery voltage is applied to the red terminal of the switch box. Once inside the switch box, the DC voltage is converted to AC, stepped up, and rectified back to substantially higher DC voltage to charge the capacitor (see Figures 8–31 and 8–32).

When the key switch is placed in the start or on position, battery voltage is routed through a resistor in the switch box. The lowered voltage continues on to the trigger unit located in the distributor. The trigger unit consists of two oppositely faced coils and a slotted (window) disc. One coil produces a magnetic field and is considered a sender coil. The second coil is a receiver coil and pulses the SCR. The solid part of the slotted disc interrupts the magnetic circuit between the trigger coils. When the magnetic circuit is once again ballooned out by an open

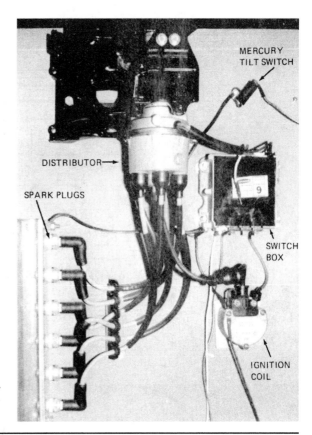

Figure 8–31 Breakerless ignition. *(Courtesy of Mercury Training Center)*

window in the disc, the oppositely faced coils function to pulse an SCR in the switch box. This releases the high primary voltage from the capacitor through ground, through the primary winding of the ignition coil, and back to the capacitor. This instantly balloons a magnetic field across the secondary winding inducing a very high secondary voltage to create the spark across the electrodes of the spark plug. The conducting SCR returns to the nonconductive state and the solid disc once again is between the trigger coils. There is a window in the disc for each cylinder of the powerhead. As each window passes the trigger coils and the capacitor is discharged, secondary voltage from the coil is routed to the distributor cap center terminal. A rotor picks this voltage up and directs it to the particular cylinder to be fired. To advance the timing, the distributor is turned by linkage to advance with the carburetor opening.

There is a mercury switch connected to the trigger circuit that conducts to ground if the outboard tilts enough for the propeller to come out of the water. This grounds the trigger circuit momentarily killing the ignition. The action of the switch prevents excess RPM that would damage the powerhead. As the outboard returns to normal trim, the mercury switch opens and ignition resumes. A tachometer circuit is also connected to the switch box. The tachometer counts ignition pulse and gives a reading in revolutions per minute (RPM).

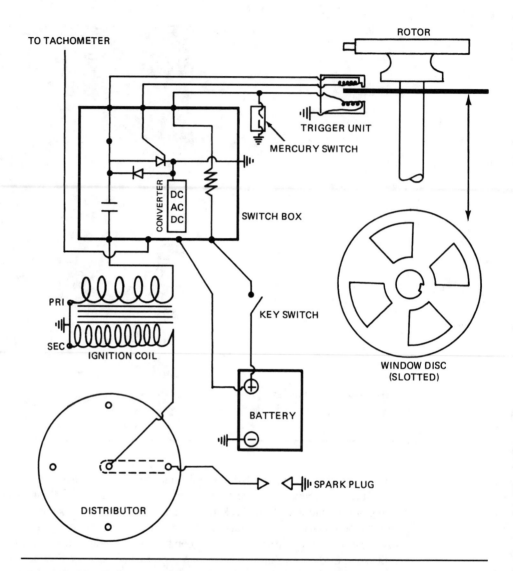

Figure 8–32   Breakerless ignition schematic. *(Courtesy of Mercury Marine)*

The simplified circuit illustrations that support the theory of operation are intended for instructional purposes only.

8.9.1   Troubleshooting the Breakerless Ignition System

The breakerless ignition is a durable system if properly serviced. Improper service procedures will damage the switch box. Observe the normal electrical service precautions of correct polarity: no sparking of wires, no disconnecting wires when in operation, and ground only to the switch box and its mounting. Resistor-type spark plugs should not be used; the Mercury tachometer or one compatible to the system should only be used.

The reason for the ignition system is to create a spark. To check for spark, install a multi-cylinder spark tester with the gap set at 1/2-inch (Figure 8–35). Crank the powerhead and observe the spark. If there is a spark on all cylinders, there is nothing wrong. Check the timing and spark plug condition.

If there is spark only on some cylinders, examine the distributor cap for cracks. These cracks will leak secondary high voltage to ground. Using an ohmmeter, test the spark plug wires. To do this, touch the red lead to the distributor cap insert, and the black lead to the end of the wire. With the ohmmeter on high scale, the reading should be a zero. This test determines if resistance has developed, which will prevent the plug from firing.

If there is no spark at all, check if battery voltage to power the system is present at the switch box red and white terminals. With the key on, a voltmeter should read battery voltage at both terminals. If not, check the battery voltage and/or voltage at the back of the key switch. If the two switch box terminals have battery voltage, power to the switch box and other circuits will need to be checked.

If there is a faulty mercury switch, the trigger circuit may not function. Testing this switch will determine if the trigger circuit is grounded. Place an ohmmeter black lead to ground and the red lead to the switch. Tilt the outboard through the maximum tilt range and back down. The switch should have made continuity when the outboard came up, and should have opened as the outboard came down. This switch grounds the ignition if the propeller comes out of the water.

What has been determined so far? You have determined that there was, or was not, spark on all cylinders. If there wasn't, you know about voltage to the switch box and through to the ignition switch. You also know the condition of the distributor cap, that the mercury switch is operational, and that it is not grounding the trigger circuit.

Connections to the switch box are always in question when troubleshooting. Check each connection for corrosion, tightness, and proper hookup. Check the ground wire where the switch box mounts to the front cover of the bottom cowl. It would not hurt to bond this connection and all ground straps. If the trigger circuit fails, there will be no spark at any cylinder. To test the trigger circuit, position the powerhead on #1 TDC and substitute trigger action by attaching a jumper to the blue terminal and ground. With the key on, a spark should occur when the jumper is removed. If a spark occurs using the jumper wire procedure and there is no spark otherwise, the trigger circuit is faulty. What did we do? We created our own trigger circuit in order to prove the trigger circuit faulty in the distributor. This test is run *only* on Mercury 800, 3051041 through 3052380, and Mercury 800, 3144219 through 3192962.

To test all other trigger circuits with the thunderbolt ignition and distributor, substitute the trigger assembly with a variable speed pulser. Disconnect the wires to the trigger at the switch box and install a substitute trigger. Pass a feeler gauge through it and watch for a spark at the spark tester. If there is a spark, the installed trigger is faulty. No spark on any cylinder may also be caused by the ignition coil. To test it, use an ohmmeter on low scale and attach to the primary terminals. It should read within specifications. Shift the ohmmeter to high scale and attach one lead to the coil secondary terminal and one to the primary terminal. The reading should be within specifications. This is all the testing that can be done with an ohmmeter. By testing the coil with the ohmmeter, you have determined that there is continuity and specified resistance in each winding. If the coil is within specifications and there is no insulation breakdown, the coil should work. For best results, the ignition coil should be power-tested and an insulation test run on it and the distributor rotor. This uses the special tester shown in Figure 8–29.

You have tested around the switch box. If all other components of the system checked out fine, then substitute the switch box and test for spark again.

 Warning

Remember the caution about the switch box: *improper testing procedures will damage the unit.* Follow the service manual instructions exactly.

If the distributor has been disassembled and the disc removed, it is possible to install the disc upside down. This will cause the SCR to be triggered at the wrong time. When setting the timing follow the service manual instructions for timing/synchronization procedures.

8.10   Alternator Driven Ignition System (Mercury)

This ignition system does not use a distributor. Instead, it is alternator-driven, charging a capacitor. The ignition's major components are the flywheel charging magnets, flywheel hub magnet, trigger ignition, stator assembly, switch box, ignition coils, and the spark plugs (Figure 8–33).

The alternator driven ignition (ADI) system was popular in the Mercury outboard line. Only the ADI system used on the 70 HP model will be used for the theory explanation. There are differences in other ADI systems, such as one box and capacitors on the four cylinder, and a bias circuit in the six cylinder model.

The stator is mounted under the flywheel, housing two stator ignition coils and the alternator charging coils. Charging magnets positioned in the outer circumference of the flywheel pass the ignition stator coils inducing an AC electrical flow, which is rectified in the switch box to DC. It is then routed to the positive side of the capacitor (Figure 8–33).

At low RPM (under 2000), the low-speed stator ignition coil provides a high primary ignition voltage for starting and low-speed operation. When powerhead RPM exceeds 2000, the low-speed coil is near the saturation point. An operating voltage detector circuit disconnects the low-speed coil. At this time, the high-speed coil picks up the primary ignition load and supplies primary voltage up through the maximum operating range. The primary ignition voltage is used to charge the capacitor with a high DC voltage, which is stored in the capacitor until released by a triggering SCR.

The sensor part of the trigger circuit is also under the flywheel, and is located in the trigger assembly. Three coils are located in this assembly, mounted 120 degrees apart. There is a split ring (sandwiched) magnet attached to the flywheel hub. The magnet poles are located 180 degrees apart, and are reversed from each other (see Figure 8–36). As the flywheel rotates, a north-south pole is presented to sensor-coil 1, producing a timed positive pulse in the lead con-

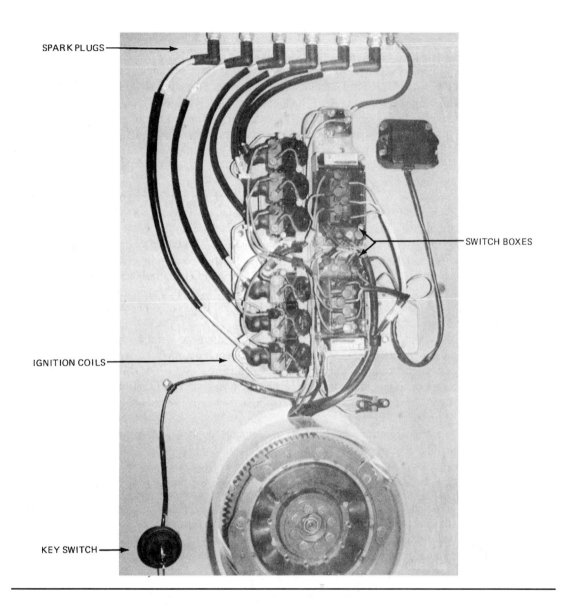

Figure 8–33   **Alternator driven ignition.** *(Courtesy of Mercury Training Center)*

nected to SCR 1 in the switch box (see Figure 8–34). When the south-north pole passes the trigger coil, the generated negative pulse will not trigger the SCR. The positive pulse at the SCR gate turns on the SCR as the trigger coil voltage returns to the negative side of the trigger coil. The triggered SCR conducts and the capacitor voltage is routed to the positive terminal of ignition-coil 1. The capacitor voltage enters the primary winding, ballooning a magnetic field across the secondary winding while returning to the other side of the capacitor through ground. This magnetic field induces a stepped-up quick-rise voltage in the ignition coil secondary winding, sufficient to break down the gap at the spark plug electrodes. The secondary circuit is completed back to the ignition coil, and the conducting SCR turns itself off.

What happened for cylinder 1, happened within 120 degrees of crankshaft rotation. The next cylinder in sequence to be fired is cylinder 2. The process of sensor-coil 2 (pulsing SCR 2, releasing the capacitor voltage into the primary winding of ignition-coil 2) takes place again in the next 120 degrees of rotation. Again, in the next 120 degrees, trigger-coil 3 will pulse SCR 3, and the capacitor voltage will be released into the primary winding of ignition-coil 3, firing the spark plug in cylinder 3.

In one revolution of the crankshaft and flywheel hub, trigger magnets have pulsed each of the three coils at 120 degrees. These positive pulses triggered an SCR for each cylinder, discharging the capacitor and firing a particular cylinder.

The ignition timing can be advanced or retarded by rotating the trigger assembly. This happens when the throttle linkage changes the location of the trigger coils in relation to the permanent trigger magnets on the flywheel hub. Timing is initially advanced, then the carburetor throttle plate starts to open. Both continue to advance/open until maximum timing is reached. At this point, timing advance stops and the throttle plates continue to open to the wide-open throttle position.

Three-cylinder switch box electronics are used on the V-6 models. The trigger circuit and assembly are changed with a bias circuit added between the switch boxes. This is a common circuit controlled by diodes between the two switch boxes. It is very important that the ground(s) on the switch box(es) be clean and bonded. This is done by using a stainless steel washer between electrical leads and ground. If the ground connection fails there will be no spark.

The simplified circuit illustrations that support the theory of operation are intended for instructional purposes only.

8.10.1   Troubleshooting the Alternator Driven Ignition System

The comments made here are intended for theoretical guidance. The exact service procedures given in the Mercury service manual should be followed.

Improper testing of an electrical circuit can create all types of problems. The switch box in particular can be damaged if the wrong procedures are used. Remember the following safety procedures when making service repairs:

- *do not reverse* battery connections; do not spark battery connections, including jumper cables
- *do not disconnect* battery cables when powerhead is *not grounded*.

Voltage travels through the ground circuit just as surely as it does through the wires on the ignition system. For this reason, all grounds should be bonded by using a stainless steel star washer *between the electrical lead and the powerhead.* This is not the normal location to use a lock washer to retain a bolt. The purpose of placing the stainless steel star washer *between the electrical lead and the powerhead* is to provide a bond by biting into the lead and into the powerhead. This will reduce corrosion buildup at the connection and insure a

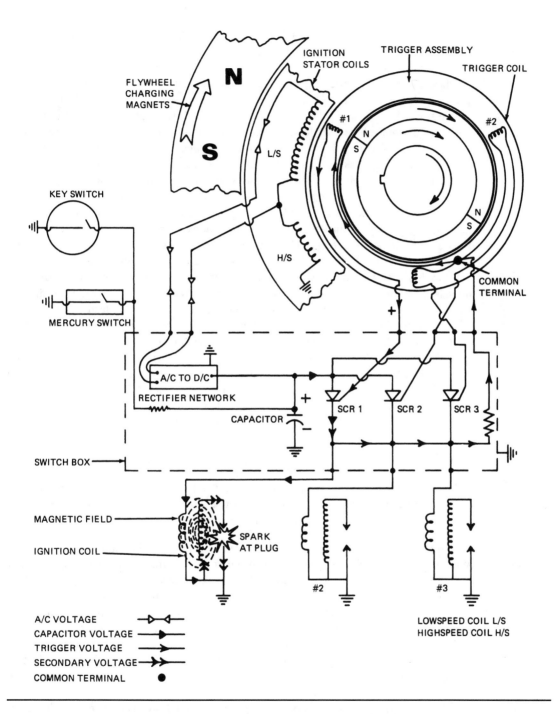

Figure 8–34   Alternator driven ignition system schematic. *(Courtesy of Mercury Marine)*

ground. If there is a poor ground, there will be no spark and it is possible to damage the switch box. For a first check of the ignition system the grounds should be checked and bonded if necessary.

The spark should be checked by using a spark tester set to the proper gap (Figure 8–35). Keep your hands clear of the tester, as voltage can approach 35,000–45,000V. During the spark test watch to see if there is spark on all cylinders. Is the spark weak, is it intermittent, or is it just on some of the cylinders? If there is no spark, the stop switch (key switch) circuit needs to be checked using an ohmmeter. With the orange wire (on older models, yellow on new models) disconnected at the switch box, and the ohmmeter connected to a good ground and the orange wire, the circuit should show continuity when the stop switch is pressed in (key switch off), and should read an open circuit when the stop switch is released (key switch on). This check is made to determine if the switch(es) is (are) functioning, and that there is no ground being made by broken insulation exposing a bare wire somewhere between the switch box and the switch. Also, check that the mercury switch is not grounded. The mercury switch grounds only when the propeller comes out of the water.

If the stop circuit is okay, the ignition stator coils or their leads can cause a no spark condition. An ohmmeter can be used to check these coils. With the blue and red leads disconnected from the switch box, place the ohmmeter between each of the leads. There should be a reading to specifications. This tells us that the coils are whole and complete and should function unless there is an insulation breakdown. A reading lower or higher than specifications indicates problems within the ignition stator coils. Check to see if the coils are grounded by placing an ohmmeter lead to the red wire and to ground. A reading other than specification indicates that the ignition stator coils are grounded. The complete stator assembly would have to be replaced if any test failed. A dealer can test these coils using a peak voltmeter (Figure 8–27). This is a more accurate test.

 ## Safety

A word of caution about temperature: powerhead temperature can affect readings taken. When an ohmmeter reading is taken, the powerhead should be at room temperature to receive an accurate reading.

If the spark is weak at high (low) speed, the low (high) speed ignition stator coil could be the problem. The checks given above would apply for either high- or low-speed stator coils. A weak spark can also be caused by slow cranking (caused by either lazy rope pull or a problem in the starter motor circuit). This could mean a battery in a low state of charge, or starter motor or cable problems. If the flywheel magnets are weak, spark will be weak. To check the magnets, an alternator output test should be made (see Chapter 7).

An intermittent spark can be caused by poor connections. Inspect all connections for tightness and be sure that the plug connectors are fully engaged. The terminals must be free of corrosion. If corrosion of electrical terminals is a problem, Mercury offers a corrosion guard to waterproof and insulate various electrical connections.

If there is a cylinder without spark, the trigger circuit could have a problem. The trigger circuit pulses the gate of the SCR at the proper time to release the capacitor charge for the spark.

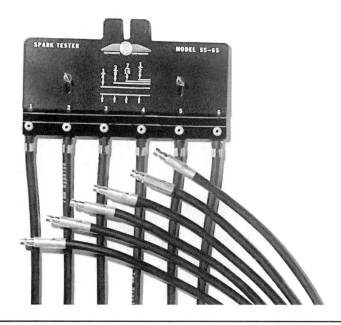

Figure 8–35   Spark tester. *(Courtesy of Merc-O-Tronic Instruments Corporation)*

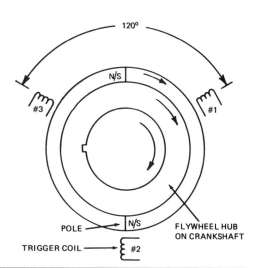

Figure 8–36   Alternator driven ignition, three-cylinder assembly.

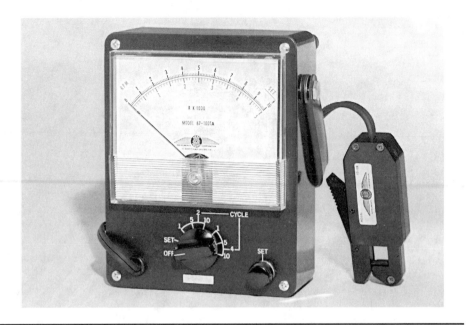

Figure 8–37 Tachometer operates on magneto, battery, or CD ignition systems. *(Courtesy of Merc-O-Tronic Instruments Corporation)*

If the circuit does not work because of bad leads or a bad trigger coil, then there will be no spark for that particular cylinder. The trigger coils can be checked using an ohmmeter. Place the ohmmeter between the disconnected brown, white/black trigger leads, then between the white, white/black trigger leads, and again between the violet and white/black leads. There should be a reading to specifications for each coil checked, indicating that the trigger coils and leads are whole and complete. This checks all three trigger coils. If a test on any coil in the trigger assembly is out of specification, the trigger assembly will have to be replaced.

A no spark or weak spark condition on one cylinder may be caused by the ignition coil. Check the primary and secondary windings by using the ohmmeter. Leads are placed on the disconnected positive and negative terminals and between either the positive or negative terminal, and coil tower (secondary) and readings should be to specifications. This gives an indication of the condition of the primary and secondary windings. If the specified reading is not obtained, the coil should be replaced. However, the ohmmeter test on the ignition coil is not conclusive. You do not know if the insulation is breaking down under high voltage. The ignition coil should be power tested using the special ignition analyzer at the dealership. At this time, the insulation can be tested while the coil is power tested.

Having tested the trigger assembly, ignition stator coils, ignition coils, and stop circuit, you have effectively tested around the switch box by using the process of elimination. However, if the switch box itself is suspect, it will have to be checked using the Quicksilver Multimeter DVA Tester at the dealership.

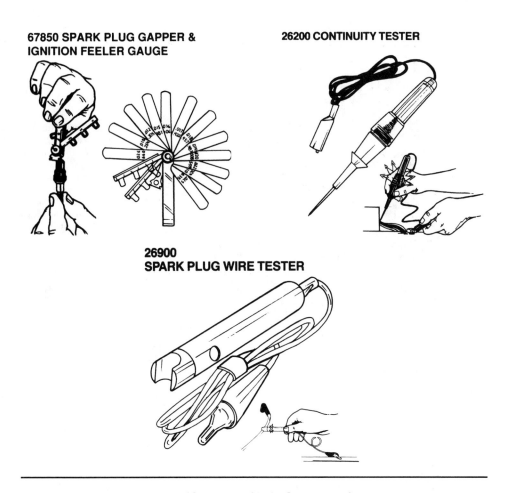

**67850 SPARK PLUG GAPPER &
IGNITION FEELER GAUGE**

26200 CONTINUITY TESTER

**26900
SPARK PLUG WIRE TESTER**

Figure 8–38   Specialty tools. (Courtesy of Lisle Corporation)

Other models that use the ADI system are tested in the same basic way. There will be more trigger coils and ignition coils to check on the 4-cylinder and V-6 models. (Check also the bias circuit of the V-6.) There will be fewer checks for the smaller horsepower models.

The key to good troubleshooting is a knowledge of the circuit and a systematic approach to your testing.

8.11   Two-Cylinder Capacitor Discharge Module (CDM) Ignition System (Mercury)

For this section, examine Figure 8–39 on the following pages and read the text that appears in each diagram. Figure 8–40 shows a typical testing device used in the procedures described in this and other sections of this chapter.

308   *Chapter 8*   *Ignition Systems*

2 CYLINDER CAPACITOR DISCHARGE MODULE (CDM) IGNITION

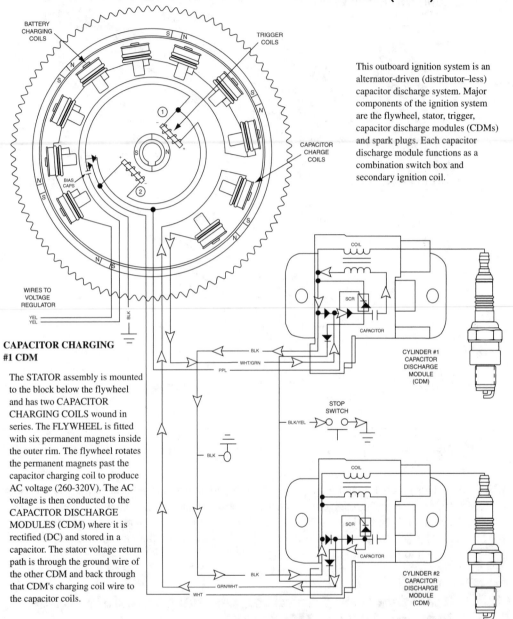

This outboard ignition system is an alternator-driven (distributor–less) capacitor discharge system. Major components of the ignition system are the flywheel, stator, trigger, capacitor discharge modules (CDMs) and spark plugs. Each capacitor discharge module functions as a combination switch box and secondary ignition coil.

CAPACITOR CHARGING #1 CDM

The STATOR assembly is mounted to the block below the flywheel and has two CAPACITOR CHARGING COILS wound in series. The FLYWHEEL is fitted with six permanent magnets inside the outer rim. The flywheel rotates the permanent magnets past the capacitor charging coil to produce AC voltage (260-320V). The AC voltage is then conducted to the CAPACITOR DISCHARGE MODULES (CDM) where it is rectified (DC) and stored in a capacitor. The stator voltage return path is through the ground wire of the other CDM and back through that CDM's charging coil wire to the capacitor coils.

Figure 8–39   Two-cylinder capacitor discharge module (CDM) ignition (continued). *(Courtesy of Brunswick Corporation)*

2 CYLINDER CAPACITOR DISCHARGE MODULE (CDM) IGNITION

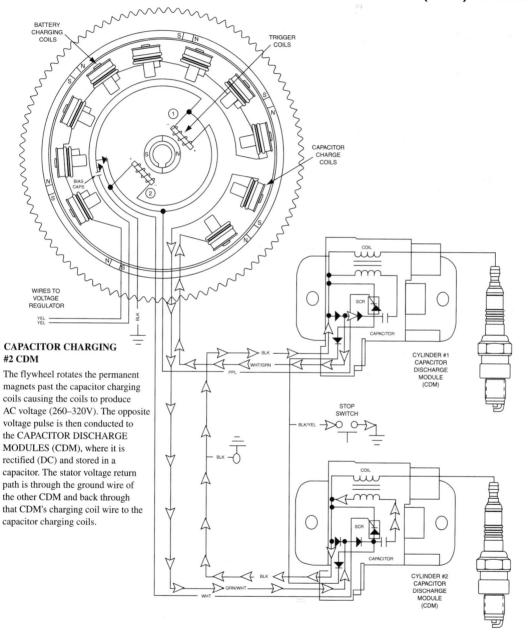

**CAPACITOR CHARGING
#2 CDM**

The flywheel rotates the permanent
magnets past the capacitor charging
coils causing the coils to produce
AC voltage (260–320V). The opposite
voltage pulse is then conducted to
the CAPACITOR DISCHARGE
MODULES (CDM), where it is
rectified (DC) and stored in a
capacitor. The stator voltage return
path is through the ground wire of
the other CDM and back through
that CDM's charging coil wire to the
capacitor charging coils.

Figure 8–39   Two-cylinder capacitor discharge module (CDM) ignition (continued). *(Courtesy of
Brunswick Corporation)*

2 CYLINDER CAPACITOR DISCHARGE MODULE (CDM) IGNITION

IGNITION COIL CIRCUIT

As the capacitor voltage flows through the primary windings of the ignition coil, a voltage is induced into the ignition coil secondary windings. This secondary voltage rises to the level required to jump the sparkplug gap and return to ground. This secondary voltage can, if necessary, reach approximately 40,000V. To complete the secondary voltage path, the released voltage enters the ground circuit of CDM module.

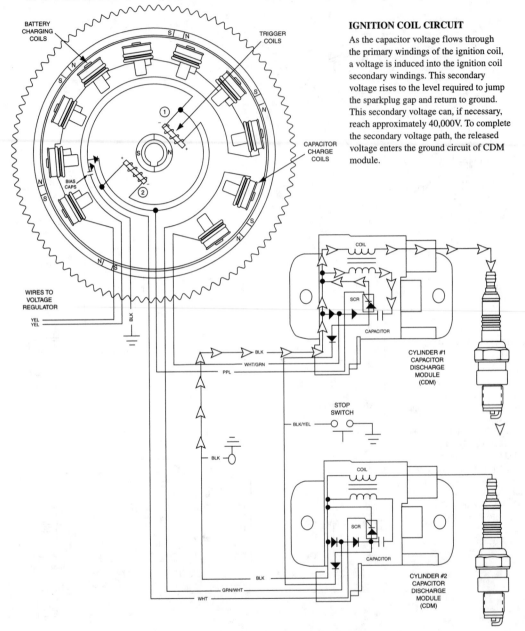

Figure 8–39 Two-cylinder capacitor discharge module (CDM) ignition (continued). *(Courtesy of Brunswick Corporation)*

2 CYLINDER CAPACITOR DISCHARGE MODULE (CDM) IGNITION

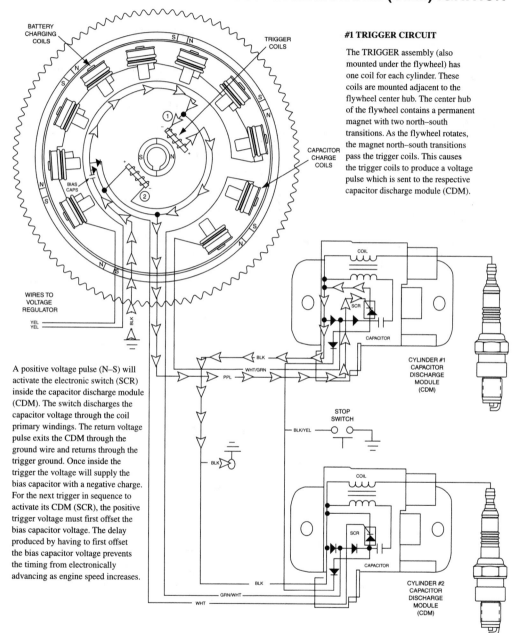

#1 TRIGGER CIRCUIT

The TRIGGER assembly (also mounted under the flywheel) has one coil for each cylinder. These coils are mounted adjacent to the flywheel center hub. The center hub of the flywheel contains a permanent magnet with two north–south transitions. As the flywheel rotates, the magnet north–south transitions pass the trigger coils. This causes the trigger coils to produce a voltage pulse which is sent to the respective capacitor discharge module (CDM).

A positive voltage pulse (N–S) will activate the electronic switch (SCR) inside the capacitor discharge module (CDM). The switch discharges the capacitor voltage through the coil primary windings. The return voltage pulse exits the CDM through the ground wire and returns through the trigger ground. Once inside the trigger the voltage will supply the bias capacitor with a negative charge. For the next trigger in sequence to activate its CDM (SCR), the positive trigger voltage must first offset the bias capacitor voltage. The delay produced by having to first offset the bias capacitor voltage prevents the timing from electronically advancing as engine speed increases.

Figure 8–39   Two-cylinder capacitor discharge module (CDM) ignition (continued).   *(Courtesy of Brunswick Corporation)*

2 CYLINDER CAPACITOR DISCHARGE MODULE (CDM) IGNITION

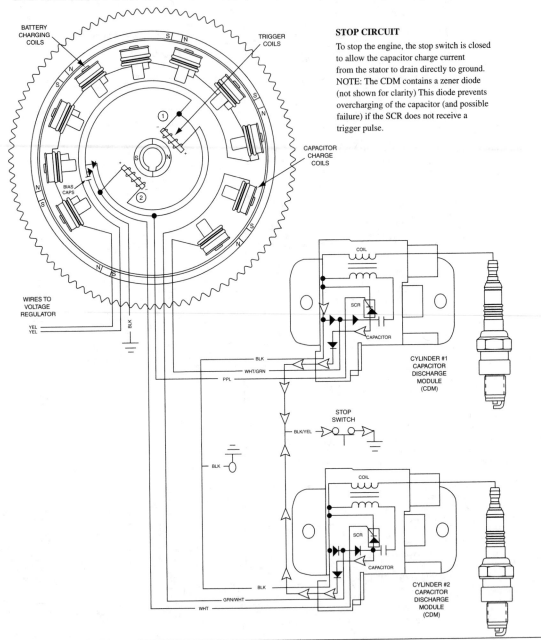

STOP CIRCUIT

To stop the engine, the stop switch is closed to allow the capacitor charge current from the stator to drain directly to ground. NOTE: The CDM contains a zener diode (not shown for clarity) This diode prevents overcharging of the capacitor (and possible failure) if the SCR does not receive a trigger pulse.

Figure 8–39 Two-cylinder capacitor discharge module (CDM) ignition (continued). *(Courtesy of Brunswick Corporation)*

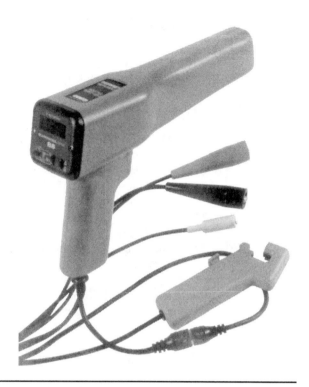

Figure 8–40   Model ST-86
Ferret eluminator timing light.
*(Courtesy of Stevens Instrument
Company)*

KNOW THESE PRINCIPLES OF OPERATION

- Magnetic fields.
- Primary and secondary coil windings.
- Function of SCR and diodes.
- Low-tension magneto ignition system.
- Magneto capacitor ignition system.
- Alternator driven ignition system.
- Operation of charge coils and trigger circuits.
- Capacitor discharge.
- Capacitor discharge module ignition.

REVIEW QUESTIONS

Standard Magneto Ignition System

1. To advance the timing the armature plate is rotated.
   a. true
   b. false

2. The magnetic field of the flywheel magnet flows through the coil laminations.
    a. true
    b. false
3. The ignition coil is mounted on the _____ of the laminations.
    a. center core
    b. first heel
    c. second heel
4. The primary winding in the ignition coil consists of several
    a. thousands of turns of heavy copper wire
    b. hundreds of turns of fine copper wire
    c. hundreds of turns of heavy copper wire
5. The ignition points are opened by the cam on the crankshaft.
    a. true
    b. false
6. The condenser is charged just as the breaker points open.
    a. true
    b. false
7. When does the condenser discharge?
    a. after the spark plug fires
    b. immediately after the points open
    c. when the points close
8. The stop kill switch is wired
    a. to the primary circuit.
    b. to the secondary circuit.
    c. directly to coil ground.
9. The ignition coil acts as a step up transformer.
    a. true
    b. false
10. The secondary winding in the ignition coil is grounded at one end and then wrapped over the primary winding and is insulated from it.
    a. true
    b. false
11. Current is induced in the coil primary winding as the magnet passes the
    a. first heel.
    b. center core.
    c. Both a and b are correct.
12. What causes the primary voltage to peak?
    a. The attendant magnetic field temporarily opposes the polarity change in the center core.
    b. The condenser is charged.
    c. The magnetic path continues through center heel and the first heel.
13. The breaker points are made to open
    a. when the condenser is charged.
    b. at peak voltage in the primary winding.
    c. when the magnetic path is past the second heel of the laminations.

14. What induces a very high voltage surge in the secondary winding to fire the spark plug?
    a. The attendant magnetic field overcomes the flywheel magnetic field.
    b. The points open, and there is a heavy magnetic surge through the secondary and primary windings as the attendant magnetic field collapses.
    c. The points open, and the attendant magnetic field is sustained.
15. The condenser discharge helps speed up the collapse of the attendant magnetic field.
    a. true
    b. false
16. Why is the spark checked using a spark tester set for 1/4-inch gap?
    a. to develop maximum coil output
    b. to stress the coil insulation
    c. Both a and b are correct.
17. Excessive metal transfer at the breaker points indicates a
    a. shorted secondary winding in the coil.
    b. condenser problem or loose connections.
    c. weak spring on the breaker points.
18. On a two-cylinder powerhead, cylinder 2 breaker points should open _____ degrees from number one cylinder.
    a. 90
    b. 120
    c. 180
19. A normal set of points is gun-metal grey in color. Black breaker points indicate that
    a. condenser capacity is high.
    b. lubrication is on the points.
    c. condenser capacity is low.
20. To test the flywheel magnetic strength on powerheads not equipped with an alternator
    a. make a generator.
    b. exchange flywheels and check performance.
    c. use a large screwdriver.

Low-Tension Magneto Ignition System (OMC)

21. The ignition coils are mounted under the flywheel.
    a. true
    b. false
22. The driver coil delivers high primary voltage to the ignition coils.
    a. true
    b. false
23. Induction of current into the driver coil is the same as induction into the ignition coil in the standard magneto ignition system.
    a. true
    b. false

24. Breaker point 1 opens at driver coil peak voltage.
    a. true
    b. false

25. What induces the quick-rise voltage in the secondary winding of ignition coil 1?
    a. driver coil voltage is forced through the secondary winding, ballooning a magnetic field across the primary winding.
    b. the points close, causing condenser discharge through the primary winding.
    c. driver coil current is forced through the primary winding, ballooning a magnetic field across the secondary winding.

26. Primary circuit current flow is from the driver coil, through the ignition coil primary winding, armature plate, closed ignition points, and back to the other end of the driver coil.
    a. true
    b. false

27. The condenser acts to increase arcing at the points to keep them clean.
    a. true
    b. false

28. The flywheel puller attaches to the hub of the flywheel.
    a. true
    b. false

29. The color of normal ignition points is
    a. blue
    b. black
    c. grey

30. The alternator charging magnets are different than the magnets used with the low-tension magneto ignition system.
    a. true
    b. false

31. Two methods used to test flywheel magnets on outboards equipped with an alternator are the use of a large screwdriver and the alternator output test.
    a. true
    b. false

Capacitor Discharge Ignition System (OMC)

32. Silicon-controlled rectifiers are used to
    a. trigger the charge coil.
    b. control high voltage.
    c. trigger capacitor discharge.

33. The charge coil induces DC voltage into the rectifier.
    a. true
    b. false

34. The capacitor receives a positive charge that is maintained by the diodes on the ground side of the capacitor.
    a. true
    b. false
35. The job of the sensor circuit is to _____ the gate of the SCR.
    a. start
    b. pulse
    c. block

Capacitor Discharge-2 Ignition

36. In the CD-2 ignition system, the standard breaker points are replaced by the _____.
    a. SCR
    b. rotor
    c. condenser
37. Ignition timing is controlled by the sensor circuit.
    a. true
    b. false
38. The capacitor is discharged when the sensor circuit sends a _____ signal to the SCR.
    a. positive
    b. negative
    c. DC
39. The ignition stop circuit is effective because it completes a ground to positive side of the capacitor, preventing a charge.
    a. true
    b. false

Capacitor Discharge-3 Ignition System (OMC)

40. Draw the schematic symbol for the following components used in the CD ignition system.

    A diode   A capacitor
    A silicon-controlled rectifier   A ground
41. How many coils are required for the CD-3 Ignition System?
    a. 1
    b. 3
    c. 6
42. What are the three basic circuits found in the CD-2 and CD-3 ignition circuits? (Write them on your answer sheet.)
    a.
    b.
    c.

43. The sensor magnets for the CD-3 ignition are placed on the _____ _____ and are _____ degrees apart.
44. The gate of SCR 1 is triggered by _____ _____ .
45. The capacitor discharges through ignition-coil-1 primary winding, ballooning a magnetic field across the secondary winding, inducing a quick-rise voltage to break down the plug gap and fire the spark plug.
    a. true
    b. false
46. Cylinder 2 fires 180 degrees after cylinder 1 on an OMC 70–75 HP three-cylinder powerhead.
    a. true
    b. false
47. How many power packs are used in the CD-3 ignition system?
    a. 1
    b. 2
    c. 3
48. What magnets are used to pump up the capacitor on the CD-3 ignition system?
    a. flywheel hub magnets
    b. flywheel inner circumference magnets
    c. neither; the battery charges the capacitor

Capacitor Discharge-4 Ignition System (OMC)

49. What is the main difference between the CD-3 and CD-4 ignition systems?
    a. the trigger circuit
    b. the way the capacitor discharges
    c. the way the capacitor is grounded
50. The CD-4 ignition system uses two sensor coils placed 90 degrees apart.
    a. true
    b. false

Capacitor Discharge-6 Ignition System

51. The CD-6 ignition system uses six sensor coils and flywheel hub magnets placed 150 degrees apart to trigger the SCR within the power pack.
    a. true
    b. false
52. The SCR gates in the power pack are arranged to trigger only upon receiving a negative pulse from the sensor coils.
    a. true
    b. false

53. What lubricant is used to ease the separation of the connectors?
    a. 50/1 oil
    b. WD 40
    c. isopropyl alcohol
54. All ignition coils are bonded to powerhead ground by the use of a
    a. regular lock washer.
    b. stainless steel star washer.
    c. secondary lead.
55. For testing the spark on a CD-6 ignition system the spark checker gap should be set at _____ inch.
    a. 1/4
    b. 1/2
    c. .035
56. The best meter for testing the output of the sensor coil is a
    a. peak voltmeter.
    b. ohmmeter.
    c. Merc-O-Tronic 7100.
57. When the stop circuit is checked for a fault, you should disconnect the connector on the powerhead and then use the _____ to test the circuit.
    a. ammeter
    b. voltmeter
    c. ohmmeter
58. Power packs can be tested with an ohmmeter.
    a. true
    b. false
59. The ohmmeter may be used to check the output of the charge coil.
    a. true
    b. false
60. To determine if the sensor coil is shorted to ground, you should place the ohmmeter leads with
    a. both leads into the connector.
    b. one lead in the connector and one lead to the timer base ground.
    c. both leads on the ignition coil terminals to check for pulse.
61. If a short is found in sensor-coil 3 on a V-6, the complete timer base would be replaced.
    a. true
    b. false
62. Consider the problem of no spark on one bank of a V-4 or V-6 model.
    a. it could be a bad ground at the power pack.
    b. it could be a defective power pack.
    c. check the charge coil against specifications, check all grounds, and prove the power pack with a peak voltmeter.
    d. a systematic check of the total system is the correct method of repair.
    e. all of the above are correct.

63. If there is a problem with the ignition coil, it will affect more than one cylinder.
    a. true
    b. false
64. When the ignition coil is checked with an ohmmeter, what is the unknown factor?
    a. resistance of the secondary winding
    b. resistance of the primary winding
    c. the ability of the insulation to contain the electricity to the secondary winding
65. Consider the problem of weak spark, on all cylinders.
    a. both power packs are bad on a V-6
    b. the sensor circuit is out of time
    c. alternator output is low and the stator and rectifier have tested okay, so the flywheel magnets are weak
66. When pulling the flywheel, the puller is placed
    a. on the flywheel ring gear
    b. on the flywheel hub
    c. a puller is not used; a knocker is used
67. The OMC economixer may upset the ignition system if the oil tank is low on oil.
    a. true
    b. false

Alternator Driven Ignition System (Mercury 70 HP)

68. How many coils make up the ignition stator?
    a. 1
    b. 2
    c. 3
69. The split ring trigger magnets are located on the flywheel hub.
    a. true
    b. false
70. The trigger magnet poles are located 90 degrees apart from each other.
    a. true
    b. false
71. In one revolution of the crankshaft and flywheel hub, trigger magnets have pulsed the three trigger coils each at 120 degrees.
    a. true
    b. false
72. Ignition timing is advanced or retarded by rotating the trigger assembly.
    a. true
    b. false
73. When the stop switch is depressed or the key is turned off, the stop circuit is _____.
    a. completed (closed)
    b. open

74. An intermittent spark may be caused by a
    a. loose connection.
    b. loose flywheel hub magnet.
    c. grounded stop switch.

Two-Cylinder Capacitor Discharge Module Ignition (Mercury)

75. Two capacitor charge coils are located under the flywheel and are wound in
    a. parallel.
    b. series.
    c. series-parallel.
76. Where is the voltage from the capacitor charge coils rectified to DC?
    a. voltage regulator
    b. capacitor discharge module
    c. bias capacitor
77. How much voltage is produced by the charging coils?
    a. 260–320V
    b. 325–425V
    c. 100–200V
78. The capacitor voltage flows through the secondary winding to fire the spark plug.
    a. true
    b. false
79. The trigger coils are located under the flywheel and are influenced by
    a. outer flywheel magnets.
    b. flywheel hub magnets.
80. The trigger coil positive voltage pulse (N-S) will activate the
    a. diode.
    b. SCR.
    c. bias capacitor.
81. The delay produced by having to first offset the bias capacitor voltage prevents the timing from ____ ____ as engine speed increases.
    a. manually advancing
    b. electronically retarding
    c. electronically advancing
82. To stop the engine, the stop switch is closed allowing capacitor charge current from the stator to drain directly
    a. to ground.
    b. to the capacitor.
    c. to bias capacitor.

CHAPTER 9

Fuel System Operation

Objectives

After studying this chapter, you will know

- Carburetion fundamentals.
- Carburetor circuitry.
- Primer operation.
- Recirculation of puddled fuel.
- Function of reed valves.
- Fuel pump operation.
- Portable and fixed fuel tank safety.
- Speed control system.
- Fuel injection.
- Variable oil ratio.

9.1   Carburetor Fundamentals

9.1.1   Function

A carburetor is a device that meters and mixes the fuel with air in correct proportions, and delivers it into the intake manifold or, design permitting, directly into the crankcase as a flammable mixture. The source of power for the two-stroke outboard is gasoline with two-cycle TC-W3 oil mixed together (see Figure 9–40), hereafter referred to as the *fuel mix*. The four-stroke outboard uses straight gasoline. In its liquid state, this fuel is of little use. The energy locked up in the fuel can only be released by combustion when combined with the proper air ratio. This air-fuel ratio is then delivered to the cylinder as a flammable mixture.

9.1.2   Carburetor Purpose

The basic purpose of the carburetor is fourfold: to meter, mix, atomize, and distribute (control) the fuel throughout the airstream moving into the powerhead (see Figure 9–1). These functions are automatically carried out over a wide RPM range.

To give the operator control over powerhead speed, the carburetor controls air-fuel mixture flowing into the powerhead. Regardless of powerhead RPM or load, automatic control over these four basic functions is always maintained. The carburetor is a precision device and, when studied one circuit at a time, it is easily understood. (Go ahead and take a look at Figures 9–15 and 9–17 before you begin.)

9.2   Basic Definitions and Laws of the Fuel System

1. Combustion of gasoline will occur in an air-fuel ratio as rich as 8 to 1, or as lean as 20 to 1.
2. Air-fuel ratio is a measure of weight of air to weight of fuel in the mixture. For instance, 14 lbs. of air to 1 pound of fuel = 14 to 1.
3. Air-fuel ratios depend upon powerhead design.

Figure 9–1   Multi-carbure-
tors, OMC 75 HP.

4.   ***Octane rating*** is a measurement of the ability of the gasoline to resist detonation (knock). Octane required for a given powerhead is dependent upon compression ratio and adjustments.

5.   *End gas* is the unburned fuel ahead of the burning flame front within the combustion chamber.

6.   Spontaneous combustion of the end gas is detonation, during the otherwise normal combustion.

7.   ***Preignition*** is caused from hot spots (glowing carbon deposits), and happens before normal ignition.

8.   When gasoline vapor is compressed on the compression stroke, heat is added.

9.   Cooler surfaces of the powerhead crankcase will condense fuel, which puddles in the lower part of the two-stroke crankcase, where it is known as ***puddled fuel***.

10.   Atomization is the process of breaking up or frothing of the gasoline into a fine mist, which aids in vaporization as heat is applied.

11.   The rate of vaporization depends upon volatility of the gasoline, temperature, pressure, and the surface area the gasoline presents to the air flow.

12.   Air passing through the carburetor venturi speeds up and creates a low pressure (vacuum) area within that venturi. This vacuum is not to be confused with the manifold or crankcase vacuum.

13. Gasoline within the passageways and the float chamber of the carburetor, when under equal pressure, will seek its own level.
14. Fuel moves within the carburetor because of pressure differential (atmospheric and manifold or venturi low pressure).
15. For lubricant use the manufacturer's recommendation of NMMA-certified oil for service—TC-W3 (two-cycle water-cooled) in properly mixed ratios. Four-stroke powerheads should use SAE30 SG, SH motor oil.

9.3　Air-Fuel Metering

Performance with economy demands good combustion. To obtain good combustion there must be a correct mixture ratio between fuel and air. If the mixture is *heavy* on fuel, there is a *rich* condition. If the mixture is *light* with fuel, there is a *lean* mixture, and possible powerhead damage may occur. To obtain good combustion, there must be a proper air-fuel ratio, so the released energy will develop evenly and progressively across the combustion chamber (see Figure 9–2). The carburetor's metering function gives proper ratios under all powerhead loads and conditions. Properly set up, the carburetor will deliver correct air-fuel ratios that are neither too lean nor too rich for the power requirements.

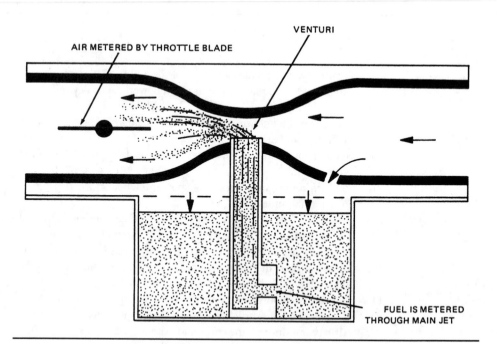

Figure 9–2　Air-fuel metering.

9.4   **Atomization**

The purpose of atomization is to have more air in contact with the fuel (see Figure 9–3). To accomplish this, the fuel in liquid form has to be broken into very small particles so that it can be readily vaporized and mixed with air. Atomization is accomplished within the carburetor in two ways. Air is bled into the fuel passages of the carburetor by the use of small orifices or tubes called *air bleeds*. This air within the passages causes a turbulence that breaks the solid stream of heavy fuel into smaller particles (similar to a frothed-up mixture). Secondly, where the narrowest point of restriction is located within the carburetor venturi, there is a low-pressure area established by the speed-up of air flow (air velocity) at open throttle. Atomized fuel is emitted into this low pressure area by the main discharge nozzle. As the air speeds by the nozzle tip, the air actually tears or pulls apart the frothed (atomized) fuel into a fine mist that has entered the airstream flowing into the powerhead.

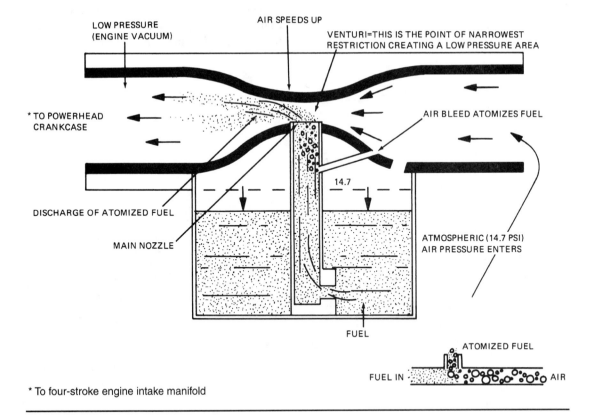

* To four-stroke engine intake manifold

Figure 9–3   Atomization.

9.5   **Float Circuit**

Fuel comes to the carburetor by gravity flow (on small fishing outboards), or by a flow created by fuel-pump pressure. Controlling this flow of fuel is the job of the needle and seat, along with the float action. As fuel enters the float chamber via the needle and seat, the float rides up on the fuel as it pivots on the hinge (axle) pin. A tang on the float pushes the needle into the seat closing off the flow of fuel at a predetermined level (see Figure 9–4).

Atmospheric pressure at 14.7 PSI is vented from the air horn to the float chamber. With atmospheric pressure applied to the fuel and a low pressure area (vacuum) applied at either the idle port or main nozzle, the *fuel is forced to flow from the high-pressure area.* As the fuel level lowers in the float chamber, because of consumption, the float lowers and pressure on the fuel in the fuel line forces the needle down, metering incoming fuel to maintain a specified fuel level. This specified float level has an effect on the metering of the fuel through the main jet, idle passage, and main nozzle. Therefore, it is essential that the float level be set to the manufacturer's specifications. If not, the metering of all the carburetor circuits will be either too rich or too lean. You may think that by turning in the mixture screw(s), the condition can be corrected. This is not so, because the circuit operation is dependent upon the low pressures applied to the idle or main circuit ports in the throat of the carburetor, therefore, controlling the lift of the fuel. It requires little lift, regardless of mixture screw setting, when fuel level is high, and it requires more lift if the level is low. To

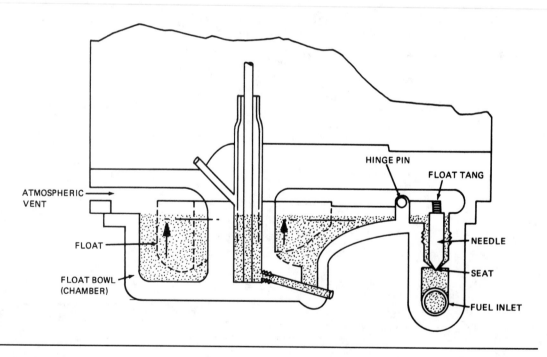

Figure 9–4   Float circuit.

make this lift easier, air is bled into the fuel to atomize it and make it lighter.

The primary responsibility of the float circuit is to maintain the correct level of fuel in the float chamber (bowl) of the carburetor. The ability to do this is dependent upon the float setting the mechanic has made, and on being able to keep dirt or water from the fuel system. Different materials are used in the construction of floats, such as copper, cork, and space-age plastics, and they all take special considerations for handling and cleaning. Check the copper floats for leaks, large dents, and for wear at the pivot points at the hinge area. To check for leaks, shake the float near the ear and listen for the slosh of fuel inside the float, or immerse the float in warm water and watch for bubbles around the seam. The old cork floats have a coat of varnish applied to the float at the factory, and are glossy in appearance and light in color. Don't mistake this for the varnished-up condition caused by stale gasoline. Many kits have a replacement float in them as these cork floats load up during service; therefore, it should be replaced whenever carburetor repairs are made. The plastic float can also load up with fuel. To check this, place the float in a container of water and check the floating height (buoyancy) of the float in comparison to a new float. If it is loaded with fuel, the old float will ride lower in the water. The hinge pin (axle) pivot is a wear-area on all floats, and is caused by vibration and normal wear from pivoting. If excessive looseness of the hinge pin is noted, the float and pin should be replaced.

 Warning

Cork and plastic floats *should not* be put in carburetor cleaner.

9.5.1   Troubleshooting the Float Circuit

The fuel level in the float chamber is controlled at a specific setting, and is a very important function in the calibration of the carburetor. If the fuel level is higher than specified, the mixture will be rich, and premature main circuit fuel delivery will occur. Spillage from the main nozzle will also occur in rough water, and in maneuvering of the boat. The result is engine carbon buildup and poor economy. When the float level is set low, a lean condition will occur. Lean mixtures burn hotter than normal mixtures, and can result in piston damage and loss of power.

Whenever fuel runs externally from the carburetor, the needle and seat are not seating and shutting off the flow of fuel. This normally is caused by water or dirt in the fuel. Foreign material between the needle and seat hold the needle off the seat allowing gasoline to continue to run (drip). The repair is twofold: get rid of the dirt, and stop it from coming into the carburetor. One very subtle form of dirt is rust from a metal tank. Once the fuel tank begins to rust internally, discard the tank. It is no longer serviceable. You may clean up the tank, but once the plating is gone, rust will form again and interfere with the ability of the needle to seat; thus, external flooding occurs. Check the fuel system for alcohol damage.

The squeeze bulb used to prime the carburetor float chamber can overpower the float and cause fuel to run externally. This is not a normal situation and the float needle and seat should be replaced.

9.6   **Backdraft Circuit (Mercury)**

The backdraft circuit uses a play on pressures applied to the fuel in the float chamber. Atmospheric pressure is emitted through the backdraft jet through a passage to the float chamber. The passage continues on to the carburetor lower venturi area where there is a low-pressure area at moderate to maximum throttle (see Figure 9–5). As the venturi low pressure gets stronger, air flow capacity through the backdraft jet is reached. At this point, the air pressure of the fuel in the float chamber is lowered slightly, leaning out the main circuit for economy. This circuit allows for a certain richness at mid-range for good acceleration, yet gives some economy at higher speeds. All carburetors do not have this circuit. It is a special feature applied to a carburetor as deemed necessary by the engineers. When repairing carburetors with this circuit be careful and keep a particular backdraft jet with a given carburetor. Other carburetors on the same outboard motor may have a different size backdraft jet.

9.7   **Carburetor Idle Circuit**

When the throttle blade is in the closed or nearly closed position, the engine is throttled. There is slower movement of air within the venturi of the carburetor. This then applies a high pressure in the venturi around the main nozzle, air bleeds and the fuel cannot come out, but air can enter the air bleeds because of the higher pressure. At this slow-speed operation, the low-pressure area is behind the throttle plate. Therefore, any port (orifice) in

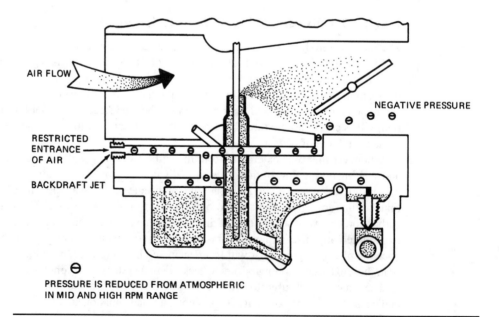

Figure 9–5   Backdraft circuit.

that area will discharge fuel. A 14.7 atmospheric pressure at sea level is applied to the fuel in the float chamber, so the pressure differential between atmospheric and engine vacuum moves the fuel from the float bowl (chamber) to the idle primary discharge port.

As the fuel moves along the idle passage, air is bled into the fuel by the idle air bleed to lighten, atomize, and froth up the fuel (see Figure 9–6). This mixture, which is now lighter, can be easily lifted to the idle primary discharge port and metered through the base of the carburetor beyond the throttle plate. Air is also bled into the frothed fuel by the secondary (slow speed) idle port, which is on the atmospheric side of the throttle plate. When in the idle position, this further atomizes the fuel (this is the second stage of atomization). Thus, the fuel moves from the main well, up the idle tube, past the air bleed, then past the mixture screw (where metering is accomplished), past the secondary (slow speed) idle discharge port, picking up more air and is discharged at the idle primary discharge port. You will notice that there is no idle RPM screw to increase idle RPM, such as you have on an automotive carburetor (Japanese outboards have an idle screw). The throttle blade is constructed either with holes in it, or there is an area cut out of the blade. Air going past the throttle blade through the cut out or holes is

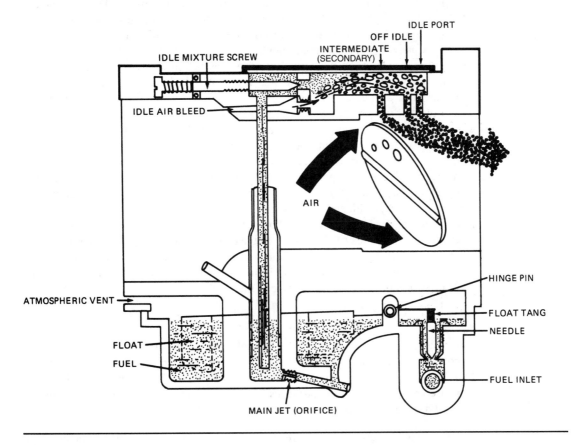

Figure 9–6　Idle circuit.

enough to provide idle RPM. To control the exact ratio of fuel, a mixture screw or jet is provided. Turning the mixture screw in leans the mixture, and turning the mixture screw out enriches the mixture. On some models, adjustment screws will lean the mixture out by admitting more air, and turning the mixture screw in enriches the mixture by decreasing air.

As the throttle valve is opened during slow-speed operation, the secondary or off idle (low speed) port is now exposed to the low-pressure area, and fuel starts to discharge from it. This additional fuel and air give an increase of speed (RPM). Therefore, the secondary (low speed) port has a dual function: it supplies air as an air bleed at idle RPM, and discharges fuel in the off idle mode.

9.7.1   Troubleshooting the Carburetor Idle Circuit

Troubleshooting the idle circuit is simple. First make sure that all powerhead conditions are normal (this means ignition and compression in each cylinder), and that there are no air or vacuum leaks (oil leaks in two-stroke powerheads) in the crankcase. If the powerhead will not idle, but will run at high speed, then the idle circuit is clogged. Disassemble and clean out the carburetor, paying special attention to the idle passages, idle tube, and air bleed in the idle circuit. Using a syringe, pass solvent (cleaner) through each passage, tube, and air bleed, and make sure, without a doubt, that the passage is open. You can't always be certain that air bleeds are open, for there may be a restriction in a passage caused by loosened gum and varnish. Therefore, trace each circuit with an aerosol can of carburetor cleaner, forcing it through the passage until you can see it coming through—then you know it is open. If you don't know where it is supposed to come out, find the passage in the service manual. Once the passages and air bleeds are cleaned and the carburetor is reassembled, make the proper adjustments and it will work.

The use of a syringe (or compressed air) forcing solvent into and through each air bleed and passage may seem strange. There are modern reasons for this procedure. There is porosity in the carburetor casting, and during the manufacturing process this carburetor casting may be sealed with an externally applied sealer. This sealer can be damaged when soaked in carburetor cleaner, even for short periods of time. With the sealer removed, air can enter through a porous casting, and then the carburetor will not perform correctly.

These carburetors do not get as greasy and grimy as automotive carburetors, but through neglect they do have lots of varnish deposits in the bowls when gasoline is not run out or a fuel stabilizer was not used before storage of the outboard.

 Note

Remember, *never* clean a carburetor by soaking it in carburetor cleaner. A good method to clean gum and varnish deposits is to use a clean bristle brush. You may have once overhauled a carburetor, soaked it, and all circuits checked out with air pressure and it looked good, but it just didn't run right. You put on a new carburetor and solved the problem. Now you know the rest of the story!

There are three wear-areas that directly affect the idle circuit and should be checked. These are the two bearing areas of the throttle shaft, where it goes through the base or carburetor. If there is considerable play between the shaft and base (body), the carburetor will have to be replaced. This is because air will enter around the shaft, upsetting the fuel ratio. Also, check the tapered end of the idle mixture screw. If it is grooved replace it; it will not meter fuel correctly. The idle screw seat may also be damaged. A poor idle may also be caused by the air bleed system malfunctioning.

9.8   Carburetor High-Speed Circuit

As there is progressive opening of the throttle blade beyond the idle secondary (off idle) port, air flow increases through the venturi of the carburetor. At a predetermined flow of air, a strong low-pressure area develops in the venturi due to the air velocity. This low-pressure area is strongest at the point of narrowest restriction in the venturi, which is the location for the main discharge nozzle, so the tip of the main (high speed) nozzle will be right in the center of this low-pressure area. With atmospheric pressure in the bowl above the fuel, we have established the pressure differential necessary to move the fuel to the venturi of the carburetor. Fuel in the bowl is pushed (by atmospheric pressure) through the main jet located in the bottom of the main well. This meters the flow of fuel to the main well as it travels up the main discharge tube heading for the venturi low-pressure area. As the fuel moves up the discharge tube air is bled into the fuel, atomizing and making the fuel lighter and therefore easier to lift to the main (high speed) nozzle tip for delivery to the airstream passing through the venturi. This atomized fuel is torn from the tip of the main nozzle by the passing airstream, and is carried on into the manifold or crankcase of the powerhead (see Figure 9–7).

Once inside the powerhead, heat is added by compression and from the running powerhead, making it a very flammable mixture ready for ignition. As long as the throttle blade is held in the open position, the low-pressure area remains at the venturi, and fuel will continue to flow from the main nozzle. As the throttle blade is returned to the area around the secondary idle port, the fuel will stop flowing from the main nozzle (because reduced air flow will not maintain a low-pressure area), and the secondary slow speed (off idle) port will be flowing, supplying fuel for a slower speed. As the blade closes below the secondary port, fuel will be delivered from the primary idle discharge port, and air once again is emitted through the secondary (slow speed) idle, atomizing this idle fuel. Understand that the fuel followed the low-pressure area within the carburetor wherever it, with control of the fuel flow metered by the strength of the low-pressure area, the size of the main jet, and the idle mixture screw setting. If the outboard is to be operated at high altitudes, then the atmospheric pressure will be less (air is less dense), and this will necessitate a change of the main jet and propeller pitch. One manufacturer starts listing jet sizes at 2,500 feet elevation. The jet size gets smaller as the altitude increases. Consult the service or parts manual for a specific size jet for a given altitude. The powerhead will run rich at high altitudes if the jetting is not corrected. This will lead to hard starting and fouled spark plugs with poor economy.

 **Warning**

Do not go to a hotter tipped plug, but correct the basic problem of improper jetting for the altitude and also change the propeller pitch.

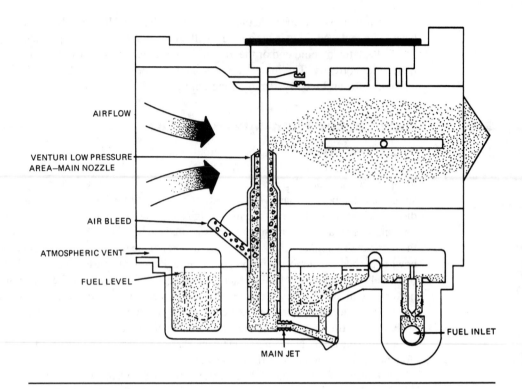

Figure 9–7   High-speed circuit.

9.8.1   Troubleshooting the Carburetor High-Speed Circuit

Troubleshooting the high-speed circuit is easy. First, make sure that the powerhead is in satisfactory condition, that the compression variation is acceptable, ignition is good, and the synchronization between the throttle linkage and the armature plate (distributor) is correct. Lastly, check that there are no air (oil) leaks at the crankcase of the two-stroke powerhead (or vacuum leaks in the four-stroke). With these items checked, if there is still a problem in the cruising or wide-open throttle range, then you may suspect the carburetor, fuel distribution line, tank vent, anti-syphon valve, fuel pump, and filters in the fuel line. (See Section 9.15 Fuel Pump Operation and Section 9.25 Fixed Fuel Tanks.)

With maximum air flow through the venturi, fuel is delivered to the tip of the main nozzle. Now, if too little fuel gets there, the passage is restricted, or if no fuel gets there, then it is clogged. This means disassembly of the carburetor. Wash it in carburetor cleaner but *do not soak it*. Then wash it in solvent or hot water and blow-dry with low air pressure. Using a syringe, put solvent into the main jet. It should exit at the nozzle tip in the venturi. Also, check the high-speed air bleed with solvent. Check the main jet for proper size for the altitude at which the powerhead is running. With no porosity and these passages open, the circuit will work. We know that the float setting will affect this circuit, so make sure it is set to specifications. Most of the newer outboards use a fixed main jet in the high-speed circuit but many of the older ones have an adjustable main jet mixture screw. Check the tapered end of the mixture screw for a ring (groove) around it, and if one is found, replace the mixture screw. Check the service manual for the setting procedure for the high-speed mixture screw for start-up and for instructions to set it for high-speed operation.

9.9   **Choke Circuit**

The choke is used because the powerhead is cold and the fuel in the powerhead quickly leaves the atomized form and condenses back into a liquid fuel on the cold surfaces of the intake manifold or crankcase. Therefore, mixture enrichment is necessary to combat this tendency (see Figure 9–8). As the choke is applied, air is restricted from entering the powerhead. A low-pressure area is created throughout the base and venturi of the carburetor. Thus, all the fuel passages, idle and main, are exposed to this low-pressure area at the same time. This permits delivery of fuel from the primary and secondary idle ports, as well as from the main discharge nozzle. With all ports feeding in fuel there is a very rich mixture

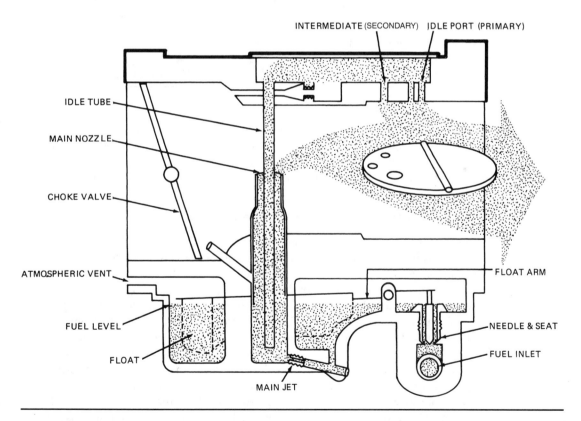

Figure 9–8   Choke circuit.

Figure 9–9 Three carbure-tors mounted on OMC 75 HP powerhead.

indeed. It is very easy to flood two-stroke powerhead crankcase/cylinders and four-stroke cylinders with fuel when the choke is overused. Some choke systems are activated by hand and others by electric solenoid, controlled from the helm. Older models may have an automatic choke closed by spring and opened by temperature and vacuum (Figure 9–9).

9.9.1   Troubleshooting the Choke Circuit

There are generally few problems with the manual choke system. Occasionally, there are problems with the linkage. A spring may come off or the plastic pull handle may break. Probably the worst problem is the overuse of the choke by the operator causing a flooded condition. This may force the removal and cleaning of the spark plugs. Some choke linkages have a detent to hold the choke in a selected position. The operator may forget to open the choke fully, causing a very rich mixture and possible fouling of the spark plugs.

The electrically controlled choke may develop problems in the electrical circuit, or the solenoid plunger linkage may break or come loose. Electricity is supplied through a fused circuit from the battery to the choke switch (key switch) and on to the choke solenoid. If the fuse is blown out or corroded, it will probably be evident from the loss of electricity to other circuits as well. If there is no choke operation, check for electricity at the solenoid connector, with the switch (key switch) depressed. There should be 12V. If there is voltage, check the solenoid with an ohmmeter for a short, open, and continuity to ground. Check for physical binding of the choke shaft and blade. If there was no electricity to the solenoid, then check for electricity at the battery terminal and solenoid terminal of the choke switch (key switch). The switch could be inoperative. The wire leading to the solenoid should be checked out using a test light or ohmmeter.

9.10   **Enrichment Circuit (Mercury)**

The enrichment valve circuit is found on some mid-1980s and later model-year carburetors. The circuit is only in the top carburetor. This circuit replaces the conventional choke circuit, which is not the best means of enriching the two-stroke crankcase. When the choke is applied, there is little air intake. All ports in the low-pressure area behind the choke valve discharge fuel, and the crankcase/cylinders are quickly flooded with excessive fuel. To help control this condition the enrichment system was created. It can be manually operated or electrically controlled from the helm. When the solenoid is energized it rotates and opens the enrichment valve to unmetered idle circuit fuel. Atmospheric pressure applied on the fuel in the float chamber pushes fuel through the idle and enrichment circuit to the low-pressure area at the idle discharge ports (see Figure 9–10).

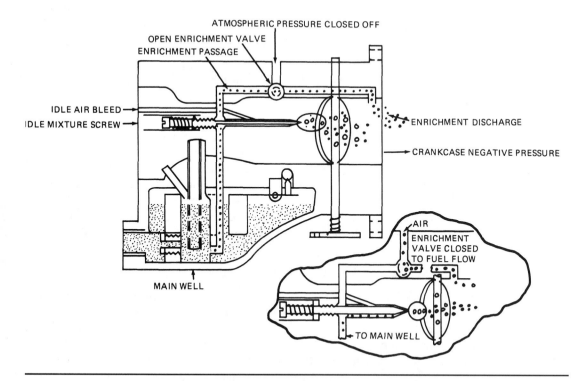

Figure 9–10   Enrichment circuit.

Low pressure in the crankcase pulls the fuel into the two-stroke crankcase. As the piston moves into the crankcase, the fuel is pressurized and pumped through the transfer passage and through the intake port into the cylinder for an enriched start. Because this fuel is metered partially through the idle circuit and only one additional port is opened, the two-stroke crankcase does not flood as easily. Air flow for normal starting is maintained, thus providing a quicker start. As the powerhead starts, the solenoid is deactivated by release of the switch at the helm. The enrichment circuit is closed to fuel flow and opened to air, which is bled into the idle circuit.

9.11   Manual Primer Operation (OMC)

On some mid-1980s and later model-year outboards there is no choke, but there is a primer, which can be operated either electrically or manually (Figure 9–11). The primer is a means of putting fuel into the intake ports or the intake manifold, therefore reducing puddled fuel in the two-stroke crankcase. Fuel is squirted into the intake port or into the manifold, which is drawn directly into the cylinder. This is a more effective means of starting a cold powerhead. There are more parts involved to pump and route the fuel to the intake port or manifold, but the cold start-up is better. In case of the manual primer, with the powerhead cranking, the primer handle is pushed and pulled, which draws fuel in from the bowl of the carburetor. Fuel is then forced past a check valve at the primer and routed to the check valve tee. Here the fuel splits, passing two other check valves and is then routed out to the nipples where the fuel squirts at the intake ports or intake manifold.

9.11.1   Troubleshooting the Manual Primer

The object of the primer is to squirt fuel. To check it, push the hose off the nipples at the crankcase and crank the powerhead while activating the primer. Fuel should squirt from the hoses. If it does, check the nipples in the bypass cover or intake manifold for a clog. If there is no fuel from the hoses, then move back up the circuit and pull the hose going into the tee. Again crank the powerhead and activate the primer. If there is fuel from the hose, check the filter screen and check the valves at the tee. If there is no fuel delivery, repair the primer, providing there is fuel coming from the carburetor bowl or fuel inlet line. Primer hoses should be routed so they won't kink. Use the OEM hose and there will be no problem with the hose ends splitting and coming off.

9.12   Electric Primer Operation (OMC)

The electrically operated primer is activated by turning on and pushing in the key switch. Twelve volts are routed to the primer solenoid creating a magnetic field that activates the valve plunger (see Figure 9–12).

Fuel pumped from the fuel manifold by the fuel pump is forced through the primer hoses to the nipples which spray fuel into the intake manifold. This prime fuel is continuous, as long as the primer valve is open and the powerhead is cranking. Fuel is routed to each cylinder through primer hoses. Equal fuel is delivered to each cylinder giving a very

A. **Plunger shown in warm-up position**
B. **Plunger shown in off position**
 1. **Nipple at intake port**
 2. **Check valve tee assembly**
 3. **Filter screen**
 4. **Gasket**
 5. **Check valves**
 6. **Check valve retainer**
 7. **Duck-billed check valve**
 8. **Fuel in from bottom carburetor**
 9. **Prime stroke**
10. **Fuel flow to check valve tee assembly**
11. **Check valve**
12. **Washer**
13. **End cap**
14. **Brass plunger**
15. **Connector-prime pump**
16. **Wire end cap retainer**
17. **O-rings**

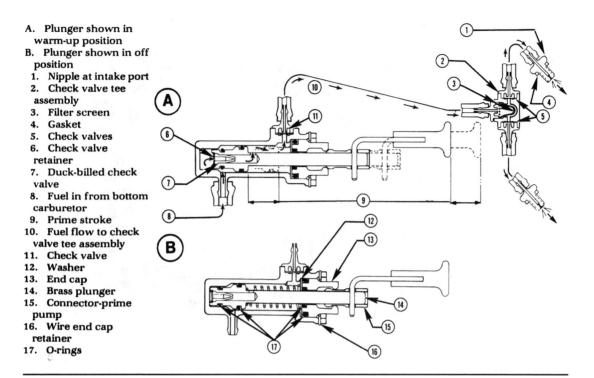

Figure 9–11   Manual primer. *(Courtesy of Outboard Marine Corporation)*

effective prime for cold starting. The system will not deliver fuel unless the key is in and the powerhead is cranking or running. There is a manual override for use with the squeeze bulb and rope starting in emergency situations. For later V-8 models and built-in fuel systems, there is an electric primer kit that installs near the fuel supply and forces fuel through the distribution lines and to the engine fuel system.

9.12.1   Troubleshooting the Electric Primer

Because the primer is electrically operated by battery voltage, the first check is to load test the battery. (See Chapter 5, *The Marine Battery*.) Voltage from the battery goes to the key switch. By turning on and pushing in the key switch, voltage should go to the solenoid, activating the primer solenoid and then to ground. Check the solenoid to insure that it is functioning. A poor ground connection to the powerhead will cause the solenoid not to operate. If there appears to be corrosion at the ground connection, clean it up with sandpaper and install a stainless steel washer between the electrical ground lead and the powerhead. This will bond the lead to the powerhead.

If the portable fuel tank is pressurized by the sun (vent closed), and the key switch is pushed in or the manual valve is open, fuel will flow through the opened plunger valve into

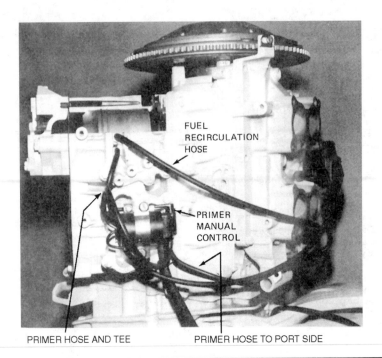

Figure 9–12   OMC electric primer.

the cylinder without the powerhead being cranked and may cause a hydrostatic lock. Manual priming is affected by pushing in the key switch and squeezing the primer bulb which forces fuel into each intake port or manifold. The system can be checked with the powerhead running at 2000 RPM. Push in the key switch and watch the tachometer for a 1000 RPM drop. This RPM drop indicates a rich fuel condition and that the system is working fine. If there is an insufficient RPM drop, pull the hoses and check for fuel delivery. If there is no delivery, check for blocked, kinked primer lines and/or a clogged filter in the tank or primer assembly. Also check for correct hookup to the upper nipples (OMC). Remember that the primer hoses are special hoses and should not be substituted with automotive fuel or vacuum hoses.

9.13   Recirculation/Puddled Fuel (Bleed) System

There are inherent fuel delivery problems with the two-stroke design. Fuel flow in atomized/vaporized form is stopped in the crankcase momentarily as the piston comes into the crankcase during the power/exhaust stroke. During lower engine speeds, the heavy fuel is settling out and is being pressurized. This is caused by the changing displacement (volume) of the crankcase as the piston comes into it. There is some turbulence and agitation as the piston moves in, but fuel is momentarily going nowhere. Reed valves and intake ports are closed. During this time, the heavier fuel droplets fall out to the bottom of the crankcase and reed valve area. As the intake port opens, the heavier fuel residue remains in the lower reaches of the crankcase and reed valve (block) area, instead of being pressurized (pumped) out. If not removed through a recirculation system (bleed system) the puddled fuel will foul the plugs in the affected cylinders. This fallout of fuel is also caused by cold operation, with fuel condensing on the crankcase walls running into the lower part of each crankcase. In some designs, puddled fuel may also form in the intake manifold. Now the problem is, what to do with this puddled fuel? Several methods have been used to recirculate the fuel to lubricate the crankshaft bearings or to put the fuel into the transfer passage and have it pumped into the cylinders. Wherever the fuel is taken to, it is pumped by crankcase pressure through one-way check valves and tubing (Figures 9–13 and 9–14). At high engine RPM, fuel quickly passes through the crankcase and there is little puddled fuel.

RECIRCULATION HOSES

RECIRCULATION CHECK VALVE
AND NIPPLE ASSEMBLY

Figure 9–13   Recirculation system (puddled fuel).

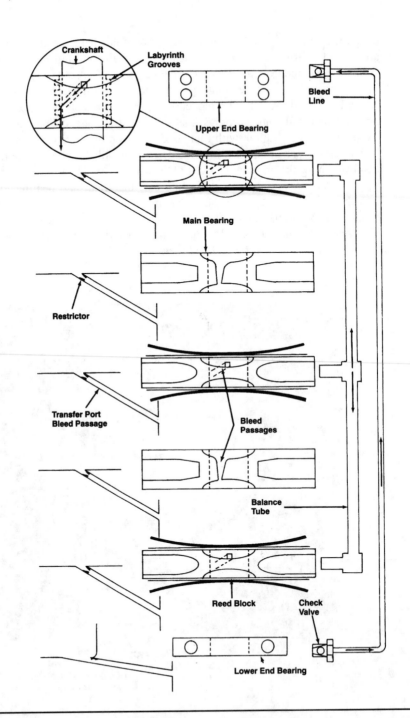

Figure 9–14   In-line engine—new type bleed system. *(Courtesy of Brunswick Corporation)*

9.13.1   Troubleshooting the Recirculation/ Puddled Fuel (Bleed) System

Indications that the system is not working at idle speed include the following: rough running, carbon buildup, fouled plugs in lower cylinders (because of a rich condition in the cylinders), and excessive smoke when accelerated after extended idle. To check the check valves in this system apply solvent using a syringe, first in one direction and then in the other. No solvent should pass in one direction, and reversing the valve, solvent should pass through. If this does not happen, replace the valve. Do not attempt to clean the valve with carburetor cleaner, a wire tool, or air pressure because of damage to the valve diaphragm. Also, check to see if the plastic tubing and nipples are open. The nipples can be cleaned using a welding tip cleaner. Check the service manual for the proper routing of the tubing. The system can be checked by placing the outboard in the trimmed-out position for shallow-water operation and removing the recirculation lines from the nipples. With the powerhead running, fuel should pulse from each line. If it does not, check the check valves, tubing, and proper routing of the tubing (see Figures 9–15, 9–16, and 9–17).

9.14   Two-Stroke Reed Valves

Reed valves may be located just after the carburetor mounting, at the reed block inside the crankcase (around the crankshaft), or mounted between the cylinder block and the intake manifold. This depends upon the model of the engine, such as a two-cylinder, V-4, or V-6.

The ***reed valve*** is a flat spring that flutters and is activated by crankcase atmospheric pressure. As the piston moves toward the spark plug end of the cylinder, a negative pressure is felt on the underside of the reed valve, and vacuum/atmospheric pressure forces the valve to flex open, allowing fuel-oil mixture and air to enter the crankcase. As the piston reverses direction, the increasing crankcase pressure closes the reed valve, prohibiting escape of the fuel, air, and oil mixture back through the carburetor (see Figure 9–18). The maximum amount that the reed valve can open is dependent upon the total induction system. Some of the larger powerheads have a reed valve stop to limit opening of the reed. There is a reed valve for each cylinder. A carburetor may share cylinders, but the reed valve never shares cylinders.

9.14.1   Troubleshooting the Reed Valves

The reeds should be inspected for pitting, cracks, rust, holding open, breakage, seal (seat) deterioration, and reed stop measurement. To do this, the reeds have to be exposed by removing the crankcase portion of the cylinder assembly (on some models). If the powerhead is operational, run it at high RPM. Put your hand very close to the front of the carburetor and see if your hand is sprinkled with fuel. If it is, the fuel is escaping the crankcase as the piston moves in and the reed is unable to completely close off. There are specifications given to measure the closed reed and reed stops. If the reed is damaged you should replace the entire set rather than an individual reed in the reed plate or block.

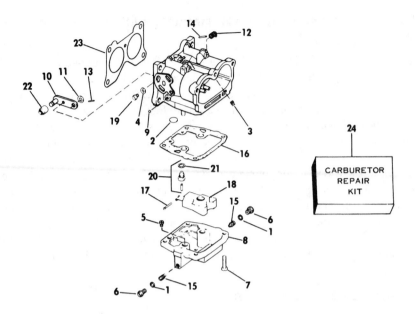

832033

OP0920C

Ref. No.	P/N	Name of Part	Qty.	Ref.No.	P/N	Name of Part	Qty.
*	394433	CARB ASSY., Upper 2.6 litre	1	12	314887	. SPRING, Throttle return	1
*	394434	CARB ASSY., Middle 2.6 litre	1	13	314575	. ROLL PIN, Return spring	1
*	394433	CARB ASSY., Lower 2.6 litre	1	14	303261	. ROLL PIN, Shaft	2-1
††1	321117	. WASHER, Screw plug	2	15	322810	. ORIFICE PLUG, 2.6 litre, No.	
††2	202310	. CORE PLUG	2			65C	2
3	324698	. ORIFICE, Idle air bleed, 2.6		††16	323784	. GASKET, Float bowl	1
		litre, No. 37	2	††17	302661	. PIN, Float	1
•4	328735	. WASHER, Follower screw	1	††18	383437	. FLOAT ASSY.	1
5	318832	. ORIFICE, 2.6 litre, No. 35	2	•19	328714	. SCREW, Cam follower	1
6	323783	. SCREW, Plug	2	††20	387658	. FLOAT VALVE ASSY.	1
7	307175	. SCREW, Bowl to body	4	††21	301996	. . WASHER	1
8	392375	. FLOAT CHAMBER ASSY.	1	•22	308199	. ROLLER, Cam follower	1
9	304201	. LEAD SHOT	10	††23	327707	GASKET, Carburetor mounting	3
•10	382370	. LEVER & PIN	1	24	392550	CARBURETOR REPAIR KIT	1
11	307596	. WASHER, Throttle shaft	1				

  *   Not Shown
 ††   Also available in Carburetor Repair Kit
  •   Middle Carburetor Only

Figure 9–15   Carburetor. *(Courtesy of Outboard Marine Corporation)*

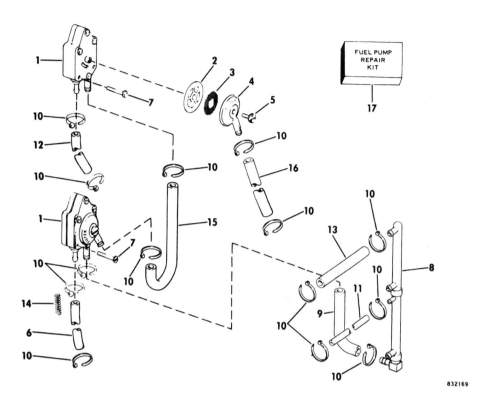

OP0898C

Ref. No.	P/N	Name of Part	Qty.	Ref.No.	P/N	Name of Part	Qty.
1	393648	FUEL PUMP ASSY., Upper	1	10	320107	CLAMP STRAP, Hose	17
1	393650	FUEL PUMP ASSY., Lower	1	11	•#	HOSE, Fuel manifold to primer	
2	312679	. GASKET, Cap to pump	1			solenoid	1
3	312675	. SCREEN, Filter	1	12	322762	HOSE, Pump to crankcase	1
4	322759	. CAP, Filter	1	13	325663	HOSE, Carburetor to manifold	3
5	312677	. SCREW, Cap	1	14	315942	SPRING	1
6	**	HOSE, Pump to crankcase	1	15	324866	HOSE, Upper to lower pump	1
7	308777	SCREW, Pump mounting	4	16	324370	HOSE, Connector to pump, upper .	1
8	390466	FUEL MANIFOLD ASSY.	1	17	393867	FUEL PUMP REPAIR KIT	1
9	324078	HOSE, Lower pump to manifold	1				

** Use 309364 Cut to Length 5-1/4" Long
•# Use 319733 Cut to Length 6-3/4" Long

Figure 9–16   Fuel pump. *(Courtesy of Outboard Marine Corporation)*

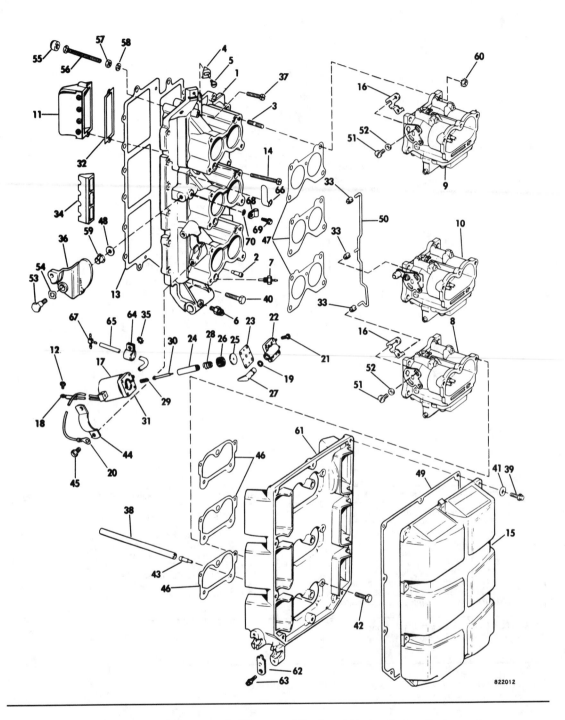

822012

Figure 9–17 Intake manifold (continued). *(Courtesy of Outboard Marine Corporation)*

OP0921C

Ref. No.	P/N	Name of Part	Qty.	Ref.No.	P/N	Name of Part	Qty.
1	390440	INTAKE MANIFOLD ASSEMBLY ...	1	36	326170	THROTTLE CAM	1
2	325659	. NIPPLE	1	37	133451	SCREW, Leaf plate to manifold,	
3	314956	. STUD	12			short	6
4	318174	. TIMING POINTER, Manifold	1	38	#	HOSE, Air silencer	1
5	303737	. SCREW, Timing pointer	1	39	308742	SCREW, Cover to base	10
6	319253	. LUBE FITTING	1	40	306418	SCREW, Manifold to c'case	16
7	392988	. CHECK VALVE & FILTER	6	41	304051	WASHER, Cover to base	10
8	†	CARB. ASSY., Lower	1	42	323480	SCREW, Air silencer base	6
9	†	CARB. ASSY., Upper	1	43	314513	NIPPLE, Air silencer base	1
10	†	CARB. ASSY., Middle	1	44	321218	CLAMP, Primer solenoid	1
11	394430	LEAF PLATE ASSY., 2.6 litre	6	45	308762	SCREW, Clamp primer solenoid	2
12	319603	SCREW, Terminal to primer solenoid	1	46	319771	GASKET, Air silencer	3
13	321198	GASKET, Manifold to c'case	1	47	327707	GASKET, Carb. to manifold	3
14	313217	SCREW, Leaf plate to manifold,		48	313079	WASHER, Cam to manifold	1
		long	6	49	321183	GASKET, Cover	1
15	320930	COVER, Air silencer	1	50	387378	LINK AND PIN, Throttle arm	1
16	320575	ADJ. LEVER, Throttle link	2	51	328714	SCREW, Lever to carb.	2
17	582111	SOLENOID ASSY., Primer	1	52	328735	WASHER, Lever screw	2
18	510780	. TERMINAL	1	53	316037	SCREW, Cam to intake	1
19	305739	. O-RING, Manual valve	1	54	204445	WAVE WASHER, Cam to intake	1
20	510818	. TERMINAL	1	55	314416	BUMPER, Adj. screw	1
21	326186	. SCREW, Cover to housing	4	56	317615	SCREW, Spark adjustment	1
22	511807	. COVER, Primer solenoid	1	57	304753	NUT, Screw, spark adjustment	1
23	511808	. GASKET, Solenoid	1	58	306396	LOCKWASHER	1
24	511809	. PLUNGER	1	59	311015	BUSHING, Throttle cam	1
25	511810	. VALVE SEAT	1	60	316291	NUT	12
26	511811	. FILTER, Valve housing	1	61	391254	BASE, Air silencer	1
27	511812	. MANUAL VALVE	1	62	325388	. RETAINER, Air silencer base	2
28	511986	. SPRING, Valve	1	63	302750	. SCREW, Retainer to silencer	2
29	511987	. SPRING, Plunger	1	64	317000	CLAMP, Hoses, primer solenoid	2
30	511988	. VALVE, Plunger	1	65	**	HOSE, Primer solenoid system	AR
31	511806	. HOUSING	1	66	325563	RETAINER, Hose to intake manifold	2
32	321907	GASKET, Leaf plate	6	67	325569	CONNECTOR, Primer hose	1
33	313329	RETAINER, Carburetor link	3	68	325637	CLAMP, Hose, intake manifold	1
34	322722	FILLER BLOCK, 2.6 litre	6	69	315287	SCREW, Clamp, intake manifold	1
35	303480	LOCKWASHER, Ground	2	70	303480	LOCKWASHER, Clamp	1

† See page 74 for Detail Service Parts
\# Cut from 203909
** Use 326449 Cut to Length

Figure 9–17   Intake manifold continued. (Courtesy of Outboard Marine Corporation)

Reed valves can be damaged when repair work requires their removal. Set aside the reed assembly so sunlight, moisture, and grit can't get on the reeds. Sunlight can deteriorate the valve seat rubber seal on some models. Rust and grit can cause breakage of the reed valve. (See Figures 9–18 and 9–19.)

9.15   Fuel Pump Operation

The operation of the two-stroke fuel pump depends on crankcase pressures (Figure 9–16). As the piston comes into the crankcase on the power stroke, a positive pressure is developed within the crankcase of that cylinder. As the piston reaches the bottom of its travel,

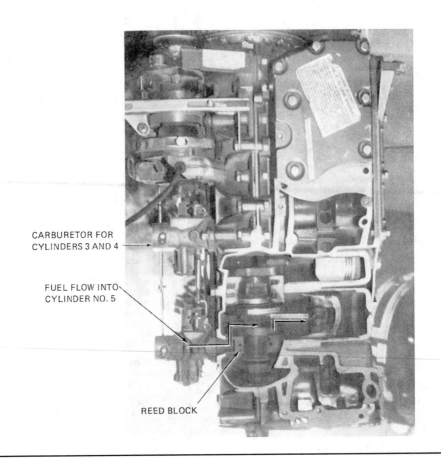

CARBURETOR FOR
CYLINDERS 3 AND 4

FUEL FLOW INTO
CYLINDER NO. 5

REED BLOCK

Figure 9–18   Fuel flow from carburetor through reed valve into crankcase.

the piston reverses itself and starts the compression stroke creating a negative pressure within the crankcase (Figure 9–20). Here are two opposite pressures that are put to work to operate the fuel pump diaphragm. The pump operation is caused from pressures ported from one cylinder. Two cylinders are used if it is a staged (booster) pump. The pump can be mounted on the side of the powerhead or it may be an internal part of the carburetor. On larger outboards there may be two fuel pumps to supply three carburetors on the power-head. Internally, in the fuel pump, the diaphragm receives the pulse through a port from the crankcase. One side of the diaphragm is exposed to fuel, while the other side feels the pulse. There are also two check valves, one to check the fuel into the pump chamber and one to check the fuel out.

As the positive pressure is exerted on the diaphragm, the outlet check valve is opened, and the inlet valve is closed, allowing fuel to flow toward the carburetor. As the piston re-

Reed Stop Measurement and Wear Sleeves

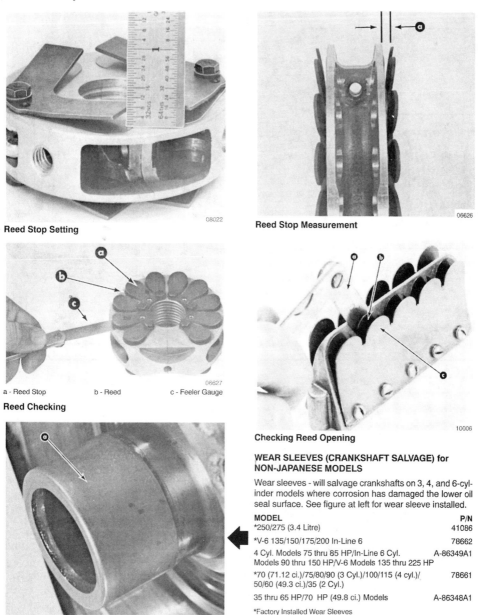

Reed Stop Setting   08022

Reed Stop Measurement   06626

a - Reed Stop   b - Reed   c - Feeler Gauge

06627

Reed Checking

Checking Reed Opening   10006

WEAR SLEEVES (CRANKSHAFT SALVAGE) for NON-JAPANESE MODELS

Wear sleeves - will salvage crankshafts on 3, 4, and 6-cylinder models where corrosion has damaged the lower oil seal surface. See figure at left for wear sleeve installed.

MODEL	P/N
*250/275 (3.4 Litre)	41086
*V-6 135/150/175/200 In-Line 6	78662
4 Cyl. Models 75 thru 85 HP/In-Line 6 Cyl. Models 90 thru 150 HP/V-6 Models 135 thru 225 HP	A-86349A1
*70 (71.12 ci.)/75/80/90 (3 Cyl.)/100/115 (4 cyl.)/ 50/60 (49.3 ci.)/35 (2 Cyl.)	78661
35 thru 65 HP/70 HP (49.8 ci.) Models	A-86348A1

*Factory Installed Wear Sleeves

a - Wear   **Wear Sleeve Installed on Crankshaft**

Figure 9–19   Reed stop measurement and wear sleeves. *(Courtesy of Brunswick Corporation)*

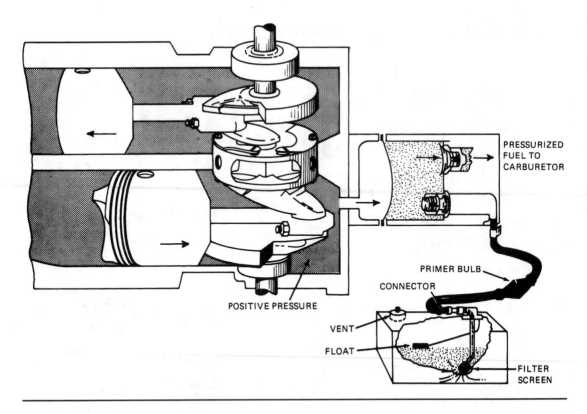

Figure 9–20 Fuel pump operation.

verses direction, a negative pressure is exerted on the diaphragm, closing the outlet check valve, opening the inlet check valve, and fuel is drawn in from the fuel line and tank. Thus, the flexing of the diaphragm pumps the fuel through check valves under an approximate 2.5 PSI (kPa 17) to the carburetor. The vent must be opened in the portable gas tank, so air can displace the fuel removed from the tank (Figure 9–21).

Mercury uses a Tillotson/Walbro carburetor (7.5 to 50 HP) that has an integral fuel pump as part of the carburetor. It also works off of crankcase pulse, with a pulse passageway running through the carburetor, through rubber and fiber check valve gaskets, to a diaphragm. When fuel is needed in the float bowl, the flexing diaphragm moves the fuel, which flows through the outlet check valve, then to the carburetor needle and seat.

On the four-cycle engine, the fuel pump is of diaphragm-displacement and is operated by a camshaft lobe. A fuel filter is located directly on the pump (OMC).

On smaller horsepower outboards, a diaphragm and reed valve are used in the pump which develops a fuel pressure of 2.5 PSI (kPa 17).

Figure 9–21   Fuel pump pressure test

9.16   Oil Injection System with Variable Ratio Oil Pump (Mercury)

An oil tank is mounted under the outboard cover for TC-W3 oil. The oil tank is equipped with an oil level sight gauge, a magnetic float, and a low oil sensor located inside the tank. If the oil level becomes low, the sensor will send a signal to the low oil/overheat warning module. The module will sound an intermittent "beep" alerting the operator of the situation (Figure 9–22).

The oil pump is gear-driven by the crankshaft and is a variable ratio pump. The variable ratio is controlled by linkage attached to the carburetor throttle shafts. As the throttle is opened, the oil ratio (flow) becomes heavier. Oil is pumped from the tank, through a check valve attached to a tee in the fuel line before the fuel pump. Therefore a prescribed ratio of oil and fuel moves through the diaphragm fuel pump into the carburetors. As throttles open and engine speed increases, so does the fuel/oil mix ratio change from lean to 50/1 at maximum RPM.

The low oil/overheat warning module also monitors engine temperature. If an electrical signal comes from the overheat switch, the module will sound a constant "beep" signaling the operator that the engine has overheated.

50/60 HP (3 Cyl. Engine w/3 Carbs.) Oil Injection System (Variable Ratio Oil Pump)

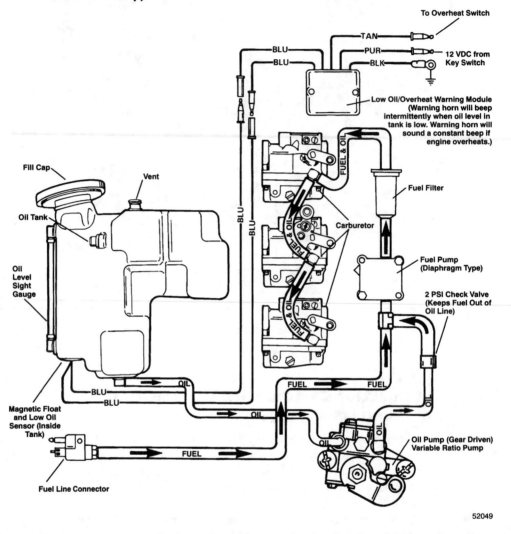

Figure 9-22 Fifty/sixty HP (3 cylinder engine with 3 carbs.) oil injection system (variable ratio oil pump). *(Courtesy of Brunswick Corporation)*

9.17   VRO (OMC)

The OMC *VRO* unit is a combination mechanical fuel pump and oil pump for OMC 40 through 55 to 125C to 250 HP (see Figure 9–23). It is *powered by pulses* from the crankcase flexing the pump diaphragms. A separate oil tank for two-cycle oil supplies the oil pump. At various engine power settings, the correct oil ratio is injected by the oil pump into the fuel system (Figure 9–24). The system has warning circuits that will sound a horn when a "No Oil," "Low Oil," or "Overheat" situation occurs, alerting the operator. OMC V-8 engines are equipped with a fuel manifold bypass valve sensitive to fuel pressures. If pressure/volume falls off from one VRO fuel pump (because of pump failure), the bypass valve will shift, permitting fuel from the second pump to maintain pressure and volume in both fuel manifolds. This is a built-in emergency feature and maximum RPM may not be possible. For cold starting, a key-activated solenoid diverts some of the VRO pump's fuel directly to the intake manifolds.

9.17.1   Troubleshooting the VRO^2

As mentioned earlier, the two-stroke fuel pump works off of crankcase pressures. If there are problems in the powerhead (e.g., like leaking seals, gaskets, O-rings, and crankcase mating surfaces), then there are problems with delivery of fuel from the fuel pump (loss or weakened pulse). When diagnosing fuel delivery problems, keep in mind a possible restriction in the distribution line, condition of filters, anti-syphon valve, and fuel tank venting. The fuel pumps (and VRO oil pump) do their job well; however, a crankcase pressure leak can cause problems as cylinders run lean on fuel and oil, scoring cylinder walls

Figure 9–23   VRO^2 is a combination fuel and oil pump.

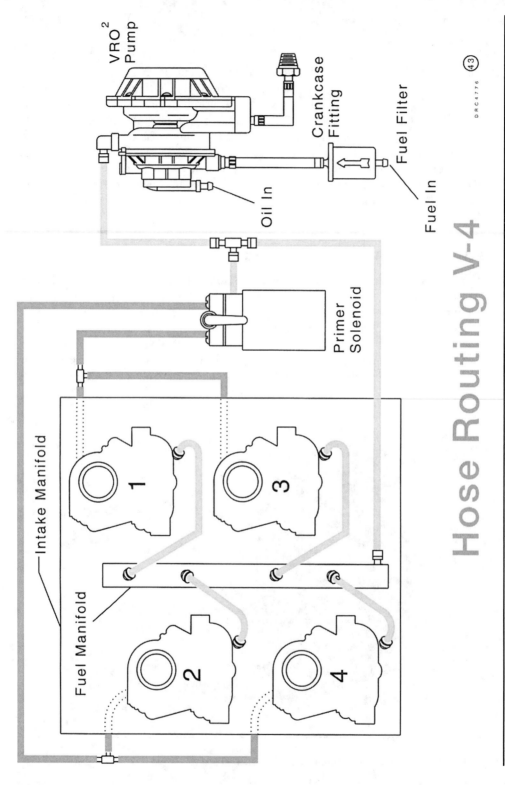

Figure 9–24 Hose routing V-4. *(Courtesy of Outboard Marine Corporation)*

and pistons. The diet of the powerhead is fuel/oil mix and this system must deliver the required amount of fuel and oil for a given RPM and horsepower. If not, the performance is off and problems are introduced. Repair kits are available to repair the pump. With today's labor costs what they are, it might be best to replace the pump when it is proven to be defective.

To determine if the fuel pump is defective, install a pressure gauge in the line between the carburetor and pump. At idle, the pressure should be 1 PSI for smaller powerheads and 2½ PSI for larger powerheads. At wide open throttle (WOT) the pressure range should be 2½ to 6 PSI with good flow (volume) and no bubbles present. These are general specifications (Figure 9–21). Check the service manual for particular powerhead specifications. The pumps will not satisfactorily lift fuel from a tank that is lower than 30 inches below the pump. If the pressures are not up to specifications there is either a clogged pulse passage/line, restriction in the distribution line, air leaks, the tank vent is not open, or it's a bad pump. To see if the trouble is in the distribution line, etc., pressurize the squeeze bulb, checking for leaks. Make a suction test or substitute the tank and line with a known good one (see Figure 9–35). Repeat the test and recheck pump pressure. Also think about one or two cylinders that foul plugs. Fouled spark plugs are usually thought of as an ignition problem, but that is not necessarily so! Do the fouled plugs correspond to the same cylinder(s) a staged fuel pump is using, or do they correspond to the cylinder that the single pulse pump is using? If so, then there probably is a pin hole in the diaphragm, allowing fuel to enter the crankcase of that (those) particular cylinder(s), causing a rich condition which is fouling the plug(s).

The VRO pump will have to be checked for oil/fuel delivery and that the "No Oil" and "Low Oil" warning circuits are functioning. The Low Oil/No Oil light (horn) will come on when the oil tank is empty or low on oil. Also, check the overheat "Water Temperature" warning light/horn circuit. It will turn on when the engine overheats.

Any time the VRO[2] oil supply line has been disconnected from the engine, you *must* run a 50/1 oil mix in the fuel tank for a period of time to verify oil consumption from the fixed tank. During that time you are watching for a lowering oil level in the reservoir. Follow the sequential checks given in the service manual.

9.18   Fuel Pump Volume Test

9.18.1   Problem: Air Bubbles

1. Rusted hole in tank pickup tube.
2. Low fuel in gas tank.
3. Fuel line connector O-ring worn out.
4. Fuel pump filter gasket or O-ring leaking.
5. Fuel pump anchor screws loose.
6. Fuel pump gaskets or fitting loose.
7. Loose gas line connection at tank or connector.
8. Crack or cut in fuel line.

9.19   Fuel Pump Vacuum Test (for Permanently Installed Fuel Systems)

1. Warm up engine.
2. Push off fuel inlet hose and tee in vacuum gauge using eight inches of clear vinyl hose (in fuel line). Do not leave this hose in the system!
3. Run engine for two minutes at full throttle while under load (test prop) watching for air bubbles (there should be none) and take a vacuum reading. It should not read higher than 4 in. If higher, check fuel distribution system for restrictions.

9.20   Fuel Pump Pressure Test

Install pressure gauge between carburetor(s) and fuel pump. Refer to service manual for specifications.

9.20.1   Problem—Poor Fuel Pressure

1. Fuel tank vent not open.
2. Air bubbles showing in clear line during volume test (see above).
3. Check for restricted flow of fuel as above.
4. Fuel tank pickup tube filter restricted or clogged.
5. Inline fuel filter dirty, anti-syphon valve dirty.
6. Problems within the fuel pump.
7. Check pulse holes/lines leading from the crankcase to the pump.
8. Test RPM too slow.

 **SAFETY**

Fuel distribution lines must be USCG Type "A1" or "B1." Vinyl hose *must not* be used in a permanently installed fuel distribution system.

9.21   Ficht Fuel Injection System (OMC)

Two-stroke technology has a problem—it is environmentally dirty. This is caused mainly by the cylinder exhaust and intake ports being open at the same time, at the end of the power stroke, and the use of the same fuel mix ratio at all speeds (50/1 ratio). This allows unburned fuel to be purged from the combustion chamber along with unwanted exhaust gases. The engine not only wastes fuel in this design, but emits annoying clouds of smoke and fumes. This problem has been corrected by the precise fuel injection used in the OMC *Ficht Fuel Injection* (FFI) system (Figures 9–25, 9–26, and 9–27).

Notable among the features of the Ficht Fuel Injection system is a simple form of fuel injection that delivers precise fuel/air mixtures directly into the combustion chamber, after the exhaust port is closed. It is controlled by a sophisticated engine management system.

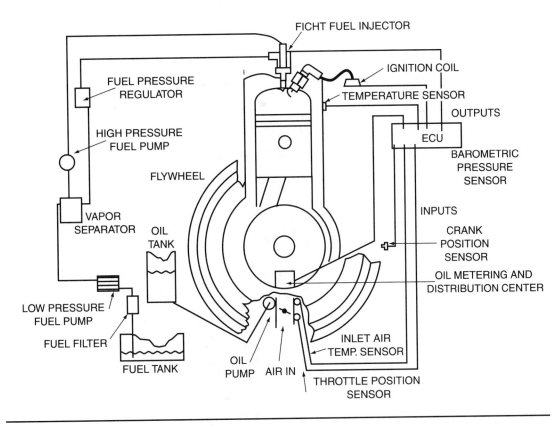

Figure 9–25   Ficht fuel injection system.

An air-cooled microprocessor-based (electronic control unit or ***ECU***) engine management system is in control of the engine temperature, throttle position, air temperature, crankshaft position, and fuel delivery system. It controls fuel blasts hammered through the injector, directly into the combustion chamber. Each injector is a combination pump and injector nozzle, with raw fuel supplied at a constant pressure (about 30 PSI). The injector pump is electromagnetic and operates like a solenoid. As the ECU sends voltage to the injector, a piston moves inside to "hammer" the fuel forward through the nozzle and into the combustion chamber. Injector (one per cylinder) function is controlled by input from sensors monitoring RPM, shifting, fuel, oil systems, water temperature, inlet air temperature, throttle position, electrical voltage, and barometric pressure. Injectors use a hydraulic impact method to inject precise amounts of fuel and air directly into the combustion chambers at a cyclic rate of up to 100 times per second and at pressures exceeding 250 PSI. The ECU rapidly turns the injector on and off. The actual injection can occur in less than 1/4 of a crankshaft revolution at an accuracy of plus or minus three degrees of crankshaft rotation. This "hammer pulsing" of the fuel atomizes it in a way that greatly enhances the fuel

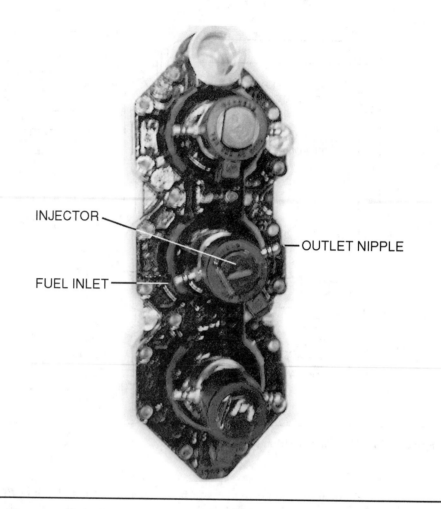

Figure 9–26   The Ficht fuel injection system uses one injector per cylinder, that is threaded directly into the cylinder.

burn. The new powerhead is 35 percent better in overall fuel economy over conventional outboards because of the closed exhaust ports and the quick fuel injection. This new system complies with the EPA requirements to clean up the outboard emissions by the year 2006.

This new powerhead has an oiling system that does not rely on the operator to mix the fuel/oil ratio. This system features an engine-pulse-driven pump that supplies oil from a remote boat-mounted oil-supply tank and is controlled by the ECU. The oil is delivered to and contained in a compartment in the powerhead that is constantly pressurized. As oil is needed, a solenoid-activated piston in the compartment pumps it into the crankcase; any unused oil is returned to the reservoir tank. There are mechanical metering blocks within an oil sump that are mounted to the bottom of the crankcase. The metering blocks meter and route the TC-W3

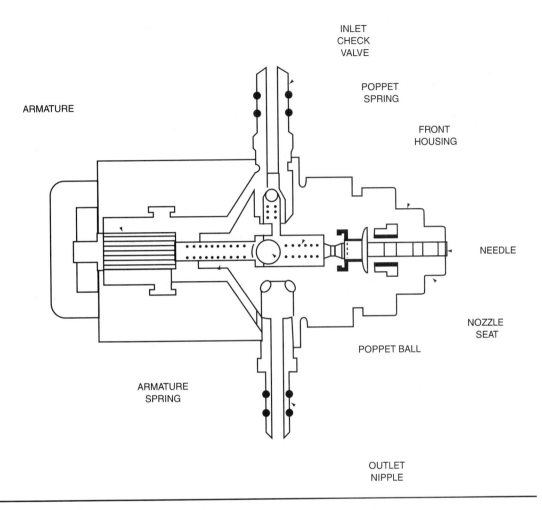

INLET
CHECK
VALVE

POPPET
SPRING

FRONT
HOUSING

ARMATURE

NEEDLE

NOZZLE
SEAT

POPPET BALL

ARMATURE
SPRING

OUTLET
NIPPLE

Figure 9–27   Ficht injector that blasts short bursts of fuel into the combustion chamber.

oil flowing to critical parts of the powerhead and are controlled by the ECU. One big differ-ence with this system is that there is no gasoline going through the crankcase.

The ECU also controls the plasma ignition system, a digital system that forms an oscil-lator circuit that drives and recharges the capacitive discharge (CD) ignition multiple times per second to provide 10–15 sparks to ignite the fuel completely, instead of using just one spark. This all but eliminates fouling of the spark plug electrodes due to misfire.

Since the Ficht system increases fuel efficiency along with power, outboards equipped with FFI burn an average of 35 percent less gas and 50 percent less oil. If no oil, over rev, or high temperatures occur, the ECU will automatically reduce engine speed to 1800 RPM and a light will be turned on at the helm.

9.21.1   How Does the Ficht Fuel Injection System Work?

Fuel coming from the fuel tank moves through a filter into a low pressure diaphragm fuel pump activated by crankcase pressures. It then passes through a vapor separator where the fuel pressure is increased by a high-pressure fuel pump, which moves it to the injector mounted on the cylinder head. Fuel not used is returned to the vapor separator after passing through the fuel pressure regulator. To inject the fuel into the cylinder the ECU system takes over. The ECU ("brain") has sensors monitoring the oil metering & distribution system, crank position, powerhead temperature, air inlet temperature, throttle position, barometric pressure, and electrical system voltage. The ECU then puts out electrical signals to control oil, ignition timing, and to activate the fuel injectors for a precise period of time.

9.21.2   Troubleshooting the Ficht Fuel Injection System

The ECU is a built-in computerized diagnostic system. The system's trouble codes can be read without a computer through the OMC system check gauge on the engine. The system takes the technician through a series of window prompts, allowing them to check the engine's ECU, timing, RPM profile, break-in, and cylinder. Tests can be done without running the engine, and all previous maintenance the engine has had is chronicled.

The technician's personal laptop computer can also be put to work. The Ficht ECU will analyze the engine, provide problem points, and list appropriate courses of action. It can instantly identify problems in the powerhead, fuel injection, oil, or ignition systems. It can present the actual condition of every sensor and switch.

When doing service work on Ficht-equipped outboards it is important to handle the sophisticated electronics and the 11 sensor circuits per service manual instructions. If not, damage to the ECU or sensor may occur.

Unlike competitive fuel-injection systems which are equipped with automotive-style fuel injectors, the Ficht injector can be easily unscrewed from the cylinder head, then quickly disassembled, cleaned, reassembled, and reinstalled. When the outboard is not going to be used in the off-season, use OMC fuel stabilizer to prevent fuel gum and varnish buildup.

9.22   Production V-6 Outboard EFI Electrical Components Description*

9.22.1   Electronic Control Unit (ECU)

The ECU is continually monitoring various engine conditions (engine temperature, engine detonation control, engine throttle opening) and climate conditions (induction air temperature, barometric pressure, and altitude level) needed to calculate fuel delivery (pulse width length) through injectors (see Figure 9–28). The pulse width is constantly adjusted (richer or leaner) to compensate for operating conditions, such as cranking, cold starting, climate

*This section is courtesy of Brunswick Corporation.

PRODUCTION V-6 OUTBOARD ELECTRONIC FUEL INJECTION SYSTEM

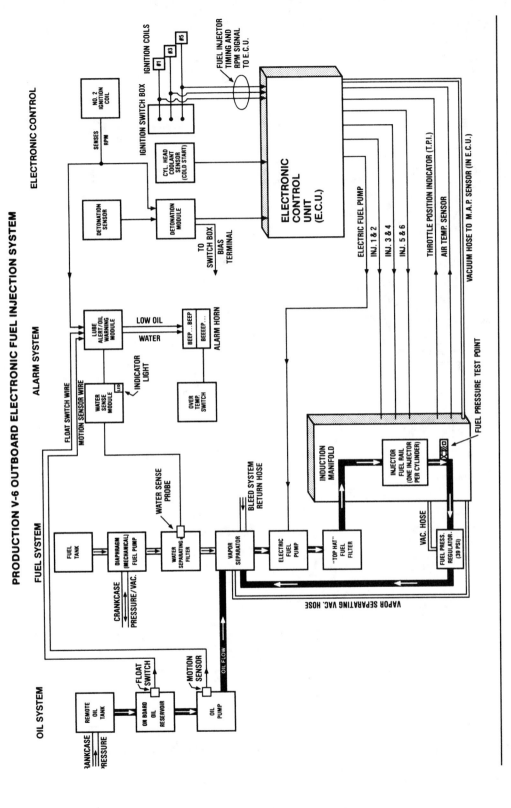

Figure 9–28 Production V-6 outboard electronic fuel injection system. *(Courtesy of Brunswick Corporation)*

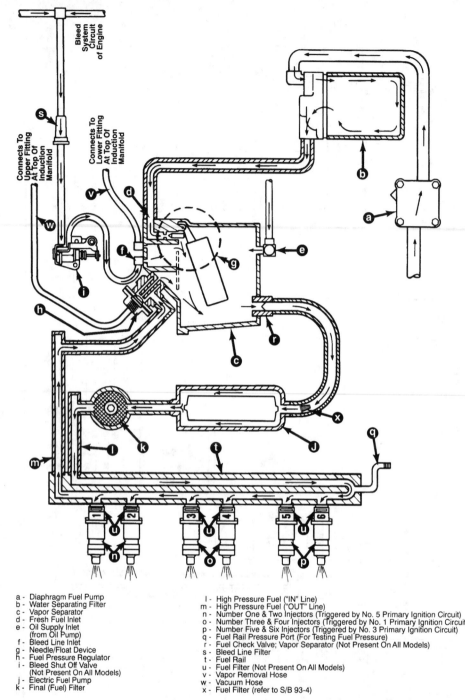

a - Diaphragm Fuel Pump
b - Water Separating Filter
c - Vapor Separator
d - Fresh Fuel Inlet
e - Oil Supply Inlet
    (from Oil Pump)
f - Bleed Line Inlet
g - Needle/Float Device
h - Fuel Pressure Regulator
i - Bleed Shut Off Valve
    (Not Present On All Models)
j - Electric Fuel Pump
k - Final (Fuel) Filter

l - High Pressure Fuel ("IN" Line)
m - High Pressure Fuel ("OUT" Line)
n - Number One & Two Injectors (Triggered by No. 5 Primary Ignition Circuit)
o - Number Three & Four Injectors (Triggered by No. 1 Primary Ignition Circuit)
p - Number Five & Six Injectors (Triggered by No. 3 Primary Ignition Circuit)
q - Fuel Rail Pressure Port (For Testing Fuel Pressure)
r - Fuel Check Valve; Vapor Separator (Not Present On All Models)
s - Bleed Line Filter
t - Fuel Rail
u - Fuel Filter (Not Present On All Models)
v - Vapor Removal Hose
w - Vacuum Hose
x - Fuel Filter (refer to S/B 93-4)

Figure 9–29 Production V-6 outboard electronic fuel injection fuel flow diagram. *(Courtesy of Brunswick Corporation)*

conditions, altitude, acceleration, and deceleration, allowing the outboard to operate efficiently at all engine speeds.

 **Warning**

Sensor Interruption with the ECU: DO NOT run the engine for extended periods of time with sensors disconnected or bypassed (shorted). Serious engine damage may result.

9.22.2   Air Temperature Sensor

The air temperature sensor (see Figure 9–28) transmits manifold absolute air temperature, through full RPM range, to the ECU. As air temperature increases, sensor resistance decreases, causing the ECU to decrease fuel flow (leaner mixture). Disconnecting the air temp sensor (open circuit) will increase fuel flow (richer) by 10 percent. Bypassing (placing a jumper wire between tan/black wires from ECU) air temp sensor (short in circuit) will cause fuel flow to decrease (leaner) 10 percent.

The air temperature sensor circuit can be tested using the EFI tester.

9.22.3   Manifold Absolute Pressure (MAP) Sensor

The MAP sensor (see Figure 9–28) is a nonserviceable sensor mounted in the ECU box. The MAP sensor is used to sense changes in manifold absolute pressure and is connected to the intake manifold by a vacuum hose. The MAP sensor is functioning through the full RPM range and is continually signaling induction manifold pressure readings to the ECU. The ECU in turn determines fuel flow as signals are received. Drawing a vacuum on the MAP sensor hose will create a lean fuel condition, altering engine operation. If no change occurs when drawing vacuum, the MAP sensor is not functioning properly.

The MAP sensor can be tested with the EFI tester.

9.22.4   Engine Head Temperature Sensor

The Engine Head Temperature Sensor provides the ECU with signals related to engine temperature, to determine the level of fuel enrichment during engine warm-up. The ECU is reviewing information at all engine temperatures but stops fuel enrichment at the engine temperature of 90° F (32.2° C). An open circuit in the temperature sensor will increase fuel flow up to 40 percent but will not have an effect on wide-open throttle operation. If no change occurs when sensor is disconnected, it probably is not functioning properly. The engine head temperature sensor can be tested.

 Note

If sensor does not make clean contact with cylinder head a rich condition may exist.

9.22.5   Throttle Position Indicator (TPI)

The TPI transmits information to the ECU, during low-speed and midrange operation, related to throttle angle under various load conditions. TPI adjustment is a critical step in engine setup. Disconnecting the TPI will increase fuel flow 40 percent at idle but does not have an effect on-wide open throttle operation.

Note

The higher the resistance the richer the fuel flow.

9.22.6   Detonation Control System (Not Present on All Models)

The detonation control system (see Figure 9–28) consists of a detonation control sensor located on the port side cylinder head and a detonation control module mounted on the powerhead. The detonation control model has seven wires:

- **White/Blue.** Connects to knock sensor, transmits knock signal to control module.
- **Green.** Connects to primary ignition wire 2. The primary ignition voltage signals the detonation control module to monitor combustion "noise" during a window of time. (See *Detonator Control System Function,* following.)
- **White/Black.** Two white/black wires connect to the switch boxes (bias circuit terminals). A third white/black wire, from the idle stabilizer module, is spliced into the white/black wire that connects to the inner switch box.
- **Gray/White.** Connects to the ECU; signals ECU to enrich fuel mixture when knock occurs.
- **Purple.** Twelve-volt power supply.

9.22.7   Detonation Control System Function

1.   Combustion noise (or vibration) excites the piezoelectric circuit located inside the detonation sensor, which transmits a voltage to the control module.
2.   When cylinder 2 ignition primary fires, it signals the control module to look at a one millisecond window of knock sensor output, which it retains as a reference level of combustion "background noise."
3.   When "background noise" reaches a measurable value, usually between 2500 and 3500 RPM (it is dependent on load), the ignition timing is advanced 6 degrees beyond what the mechanical timing is set at. Timing advance is accomplished by lowering the bias voltage in the switch boxes.
4.   The control module continues to monitor knock sensor output. If the output exceeds a predetermined threshold level over the "background noise" (which indicates knock is occurring), the ignition timing will be retarded by as much as 8 degrees and fuel flow will be enriched by as much as 15 percent, until the knock sensor output is reduced below the threshold level.

The detonation control system actually acts as an ignition advance module; when knock occurs it takes away the advance. Ignition timing will not advance if:

a.   Knock sensor fails.
b.   Blue/white wire becomes disconnected.
c.   Black wire has poor ground connection.
d.   Purple power wire becomes disconnected.

 Note

Disconnected gray/white wire will not affect ignition timing and will not allow fuel enrichment.

Other components associated with the ECU:

Rectifier: Connected to a 12V power supply. The rectifier provides power to the ECU even if the ignition switch is in the off position.

 Safety

Before disassembling the EFI system, disconnect the battery cables.

Starter Solenoid: Provides 12V signal when key is in the start position. In the start position, injector pulse widths are tripled [when engine head temperature is below 90° F (32.2° C)] to provide adequate fuel for quick start-up. When key is returned to the run position or engine head temperature is above 90° F (32.2° C), pulse width returns to normal width.

Fuel Injectors: A four-wire harness connects the fuel injectors to the ECU. The red wire is 12V and connects to all injectors. The blue, yellow, and white wires, from the ECU, each go to a pair of injectors. The ECU opens each pair of injectors by grounding either blue, yellow, or white wires.

Electric Fuel Pump: The ECU contains a fuel pump driver circuit that controls the electric fuel pump. The fuel pump positive terminal (+) red wire is 12V. The ECU causes the electric fuel pump to run by grounding the red/purple wire. The fuel pump is run at two speeds by the ECU. It is run on slow speed during slow speed engine operation, and at a faster speed when engine is operated above approximately 2000 RPM.

9.22.8   Water-Sensing System

The system consists of a water-separating fuel filter (starboard side powerhead), sensing probe (bottom of filter) and a water-sensing module (below ECU box). The water-sensing module has four wires:

Purple: Connects to 12V power supply.
Light Blue: Connects to lube alert, which sounds the warning horn when activated.

Tan: Connects to sensing probe.
Black: Connects to ground.

9.22.9   Water-Sensing System Function

1. The filter separates the accumulated water from the fuel.
2. A voltage is always present at the sensing probe. When water reaches the top of the probe it completes the circuit to ground.
3. The completed circuit activates the warning system. The warning system has a 5–10 second delay, then the module's red light illuminates and the warning horn intermittently sounds.

The system can be tested by disconnecting the tan wire from the sensor probe and holding to a good engine ground for 10 seconds. If system is functioning properly, the red light should illuminate, and the warning horn should intermittently sound.

9.23   Production V-6 EFI Outboard Fuel Components Description (from Fuel Flow Diagram, Figure 9–29)*

A description of each component as shown in Figure 9–29 will aid in the diagnosis of EFI problems. The letters in parentheses refer to labels in this figure.

9.23.1   Diaphragm Fuel Pump (a)

The diaphragm fuel pump, mounted on the crankcase, delivers fuel through the water-separating filter to the vapor separator. Typical fuel pressure (at 5000 RPM) is 6 to 8 PSI.

9.23.2   Water Separating Filter (b)

The water-separating filter protects fuel system components from water and debris. The filter contains a sensor probe which monitors water level in the filter. If water is above the sensor probe, the red light on the water-sensing module will illuminate, triggering a series of beeps from the warning horn.

9.23.3   Vapor Separator (c)

The vapor separator can be considered a fuel reservoir that continuously blends and circulates fresh fuel, oil, and unused fuel/oil from the fuel rail.

Fuel inlet: fresh fuel delivered to system.
Oil inlet: oil delivered from oil pump.
Crankcase bleed inlet: recirculated fuel (unburned) delivered from bleed lines at low RPM.

*This section courtesy of Brunswick Corporation.

Fuel pressure regulator inlet: unused fuel/oil from the fuel rail flows into the vapor separator for recirculation.

The fresh fuel delivered to the vapor separator is controlled by a needle/float device located in the vapor separator (g).

9.23.4 Bleed System (f, i, s)

Unlike carbureted engines, the unburned excess fuel from the crankcase bleed system flows into the vapor separator. The bleed system flow is shut off to the vapor separator, when operating the engine above approximately 3000 RPM, by the bleed shut-off valve. The bleed shut-off valve is activated by the throttle linkage on the induction manifold. At idle speeds the flow is approximately 1000 cc's of gasoline per hour. Also, a small filter (30 micron) is installed in the bleed line to keep contaminants from entering the vapor separator. If the filter becomes clogged, the engine will be prone to load up (run rich) at idle speeds and hesitate upon acceleration.

9.23.5 Electric Fuel Pump (j)

The electric fuel pump is continually providing fuel in excess of engine demands. The excess fuel circulates through the fuel rail back to the vapor separator. With the key in run position (engine not running), the ECU signals the pump to run for approximately 30 seconds, and then shut off. With the key in run position (engine running), the ECU determines pump speed (2 speeds) depending on RPM. During low-speed operation, the pump runs at low speed.

A 130 micron filter (x) is located just before the electric fuel pump. This filter prevents large particles from entering the electric fuel pump and causing damage or stoppage (not on all models—refer to S/B 93–33).

9.23.6 Final Fuel Filter (k)

The final fuel filter is located above the electric fuel pump. The filter collects debris flowing from the electric fuel pump to the fuel rail and can withstand blockage of up to 50 percent and still allow adequate fuel flow.

9.23.7 Fuel Injectors (n, o, p)

The fuel injectors are located inside the induction manifold on the fuel rail. The injector valve body consists of a solenoid actuated needle and seat assembly. The injectors are controlled by the electronic control unit. The ECU determines how long the injector needle is lifted from its seat (pulse width) allowing fuel to flow. The pulse width will widen (richer) or narrow (leaner) depending on various signals received from sensors connected to the ECU. The ECU receives signals from primary ignition circuits 1, 3, and 5 to open each "pair" of injectors accordingly.

9.23.8 Fuel Pressure Regulator (h)

The fuel pressure regulator is located on top of the vapor separator and is continuously regulating fuel pressure in the fuel rail. The electric fuel pump is capable of producing 90 PSI (621 kPa) of fuel pressure. The pressure regulator regulates fuel to injectors down to a usable 36 to 39 PSI (248 to 269 kPa).

9.23.9 Induction Manifold

The induction manifold is a common plenum chamber for accurate pressure measurement. It contains four throttle shutters on two throttle shafts. The shutter opening (idle air opening) can be adjusted during EFI setup procedure. The manifold contains the fuel rail, injectors, throttle position sensor, and air temperature sensor. A fuel rail pressure port (q) is located on the manifold (port side) and is used to determine fuel rail pressure.

9.24   Portable Fuel Tanks

Everybody understands that the fuel tank's purpose is to hold fuel. A simple task, but yet does the tank do this safely, during transportation, in operation, and in storage? Many of the old pressurized tanks are still around. They are easily identified by the double hose. These tanks are installed on outboards that don't have a fuel pump. A positive pressure hose is routed from the crankcase to the sealed tank, pressurizing the tank. Pressure on top of the fuel pushes the fuel out through the tank pickup screen and tube, through the squeeze bulb and line, to the connector and carburetor. The powerhead crankcase pressure is capable of quickly pressurizing this tank, and this system works well as long as there are no air leaks. To out check this tank, the lines and tank have to be pressurized, testing for leaks.

The current portable tanks are *not* pressurized, but are vented through the cap and can also be sealed by closing that vent (see Figure 9–30). When transporting the gas tank, the cap and vent should be sealed to prevent spillage of fuel. The sealed tank is subject to pressures created by the sun's heat (see Figure 9–31). This means the tank has to be constructed heavy enough to withstand the flexing and changing pressures. This also means that the cap, vent, squeeze bulb, and line have to be strong enough to take the pressure without developing a leak. A leaking tank or line in the back of your pickup or boat are definite fire hazards.

9.24.1 Troubleshooting Portable Fuel Tanks

These metal or plastic gas tanks should not be patched or repaired as automotive gas tanks are. Automotive tanks are not subject to flexing and changing pressures because they are always vented. So repairs are acceptable. Repaired outboard portable tanks may rupture again, leaking at the weakened repair area and creating a fire hazard. Pressure-test the tank at least seasonally. If the hose or squeeze bulb is replaced, the tank should be pressurized

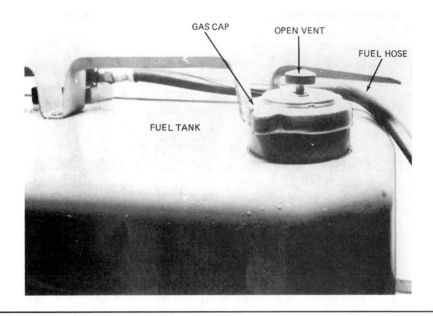

Figure 9–30   Portable tank venting system.

Figure 9–31   Fuel tanks are made of corrosion-proof, durable, lightweight plastic with special inhibitors and plasticizers to protect against ultraviolet rays and aging. *(Courtesy of Mercury Marine)*

Figure 9–32   Using air pressure to test portable fuel tank.

with the new hose and squeeze bulb attached, to determine that no leaks exist (see Figure 9–32). Use USCG approved fuel hose only.

Manufacturers have an adapter available that fits the tank opening. Short bursts of air from a hand pump can be applied through it, thereby pressurizing the tank, bulb, hose, and connector. By submerging one end of the tank at a time, bubbles indicate any leak (see Figure 9–33).

The metal tank should be looked into, and if rust is present in the tank, it will have to be replaced. There is no effective repair. The rust can be cleaned off, but rust will reappear and will again cause clogging of the fuel filter and carburetor flooding. The only effective solution to a rusty tank or leaking gas tank is to replace it with a certified ABYC plastic tank (see Figure 9–31).

9.25   Fixed Fuel Tanks

Fixed fuel tanks are generally located in one of three areas in the boat: bow, stern, or amidships. The fill is probably located in the gunnel for easy access for filling.

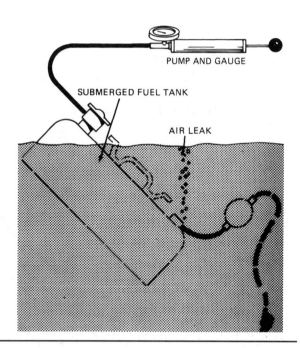

PUMP AND GAUGE

SUBMERGED FUEL TANK

AIR LEAK

Figure 9–33   Leak testing a partially submerged fuel tank.

  Warning

When filling the tank, contact should be made between the gasoline pump nozzle and the fuel tank. This is to ground the boat tank to the pump, so there will be no problems with static electricity jumping around and igniting the fuel.

The boat should also be emptied of any occupants during the fill operation, just in case there is a fire or explosion.

The tank vent is on the top of the tank and vents through the side of the boat (see Figure 9–34). There is a special cover over the vent to keep the water out. This is not always foolproof, and water might enter the vent causing fuel contamination. The vent allows atmospheric pressure into the tank to supply pressure differential when the fuel pump is pumping the fuel from the tank. The fuel distribution line (outlet) attaches to the tank at the *anti-siphon valve*. This valve prevents fuel from leaving the tank unless there is a pressure drop of 2.5 inches of Mercury. This is at a flow rate of 20 gallons of fuel per hour. This is a safety factor, so no fuel will flow if the fuel line leaks or comes off. These valves are troublesome at times, but they should *never* be taken out of the line. If there is a problem with fuel flow from the tank, replace the anti-siphon valve, or pickup tube and screen inside the tank. This screen is of stainless steel #304 wire cloth, 100 mesh, and can become clogged with debris, gum and varnish from stale gasoline.

There should be a filter in the distribution line in an accessible area near the stern. Use a marine filter (not an automotive filter) designed to work with outboard fuel pumps. The filter should have no more than 0.4 inches of Mercury drop across it, at the rate of 20 gallons of fuel per hour.

The distribution fuel lines must be USCG Type "A-1" or "B-1." Vinyl hose *must not* be used in a permanently installed fuel distribution system. The neoprene fuel line should be at least 3/8-inch for V-6 and V-8 models. This will insure that there is minimal resistance to the flow of fuel, for the larger V models and in-line 6-cylinder outboards.

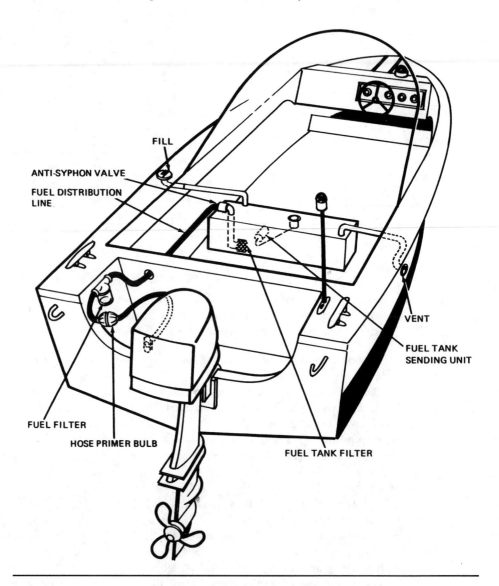

Figure 9–34 Fixed fuel tank distribution system.

9.25.1   Troubleshooting Fixed Fuel Tanks

If there are two four-cylinder outboards or larger installed, taking fuel from the same tank, two distribution lines, two filters, two anti-siphon valves, and two tank pickups should be used. The fuel pump suction test is used to determine if there is a restriction in the distribution line supplying larger outboards. Tee a vacuum gauge into the line, using tank connectors. With the powerhead warmed up, either on the dynomometer or in the water, accelerate to wide-open throttle in gear and watch the vacuum gauge. Hold the RPM long enough to insure that there is demand on the line. Specifications for a 4-cylinder should be more than 4 inches and no more than 4 inches of vacuum on a 6-cylinder (Figures 9–35 and 9–36).

If vacuum is greater than specifications, the fuel pump(s) cannot pump the required flow of fuel, and the outboard will have performance problems, such as:

- Loss of or low fuel pump pressure
- High-speed surging
- Preignition/detonation
- Hesitation upon acceleration
- Loss of power

Symptoms

1. Loss of fuel pump pressure
2. Cuts out or hesitates on acceleration
3. Loss of power
4. Powerhead runs rough
5. Powerhead stops and cannot be restarted
6. High-speed surging at wide-open throttle
7. Detonation/preignition (Erosion of piston head/dome)
8. Vapor lock
9. Powerhead will not start and ignition is OK
10. Runs and stops

Note: The above symptoms may be caused by a restricted anti-syphon valve in a fixed tank installation or a restricted pickup tube filter in a portable tank. (See sections on fixed and portable fuel tanks.)

Suction Test: Tee in a vacuum gauge at the fuel line connector. With the outboard in the water, run at wide-open throttle and read vacuum gauge.

Specifications: 4 inches for four-cylinder models, 4 inches for 6-cylinder models.

If the vacuum gauge is not available, operate the outboard with a separate portable 6-gallon fuel supply which is known to be good. Make a comparison run on the water.

Figure 9–35   Checking for restricted flow in the fuel line—tank to connector.

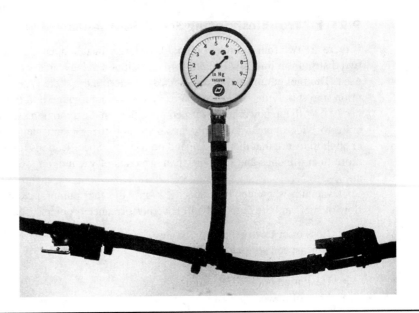

Figure 9–36   Vacuum gauge used to test for restrictions in fixed fuel tank installations.

- Running rough
- Vapor lock

To correct the problem, repair or replace the line, filters, anti-siphon valve, and/or the pickup screen in the tank. This is a check of the distribution system only, and it is not a test of the fuel pump. If you do not have a vacuum gauge, then disconnect the distribution line at the powerhead and substitute a known good 6-gallon portable tank. This tank is used because of the free flowing 3/8-inch pickup tube. Then make your performance (dyno) run at wide-open throttle, allowing time to gain a good fuel flow in the line. If this improves performance, then you need repairs in the distribution system.

There are pressure tests to be made on the fuel pump(s) to determine condition. When a pressure test is made on the fuel pump and insufficient pressure is noted, then the vacuum gauge should also be teed into the distribution line, the suction test made, and readings checked against the above specifications (this is for larger outboards). Restrictions also equal low volume (see Figure 9–37).

A quick means of checking an anti-siphon valve is to *remove* it from the fuel line. Make up a clear plastic tubing the size of the inlet. Mark the tubing at 20 inches and 25 inches. Install the anti-siphon valve inlet into the tubing. Hold the tubing in a vertical position and fill the tubing with water to the 20-inch mark. Water must not run through the valve. Increase the water level to the 25-inch mark. Water must begin to run through the valve before reaching the 25-inch mark. Replace the anti-siphon valve if your test results vary. A

Fuel System Trouble Isolation Chart

Problem: Engine Malfunction Due To Fuel System

Step 1. Check Fuel Supply

Is there fuel in tank? Is the fuel/oil mixture correct? Is the fuel fresh? Is the fuel tank clean inside? Is there water in the fuel?

Result

Fuel OK - Go to step 2.

Fuel questionable - Drain and fill tank with correct mixture.

Fuel tank dirty - Clean fuel tank.

Step 2. Check Fuel Filter

Remove filter cover and inspect screen for dirt and varnish. Inspect gasket.

See Inline Filter Instructions.

Result

Filter screen clean - Go to step 3.

Screen dirty - Clean or replace.

Gasket OK - Go to step 3. Not OK - Replace.

Step 3. Check Fuel Delivery

a. Disconnect fuel hose at motor. Hold outlet end of primer bulb up and squeeze primer bulb until definite pressure is felt. Inspect for leakage.

b. Connect fuel hose to motor. Hold up and squeeze primer bulb until definite pressure is felt. Inspect for leakage from hoses, fuel pump, filter and carburetor.

Result

a. Primer bulb shows definite pressure, no leakage - Go to step 6.

Primer bulb does not show definite pressure, no leakage present - Repair primer bulb valves.

Leakage present - Repair as necessary.

b. Primer bulb shows definite pressure with hose connected to motor - Go to step 6.

No pressure, leakage present - Repair leakage as necessary.

Primer bulb does not firm up, no leakage present - Go to step 5.

Step 4. Check Carburetor

Remove air silencer cover. Squeeze primer bulb until it firms up. Inspect throat of carburetor for flooding.

Result

Primer bulb firms up, no flooding - Go to step 6. Primer bulb does not firm up, flooding present - Repair carburetor(s).

Primer bulb does not firm up, no flooding present - Go to step 5.

Key

1. No Fuel or Old Fuel - Wrong Mixture
2. Blocked Pickup Screen
3. Dirt, Varnish, Water
4. Cover
5. Filter Screen
6. Gasket
7. Hold Outlet Upward

Figure 9–37   Fuel system trouble isolation chart (continued). *(Courtesy of Outboard Marine Corporation)*

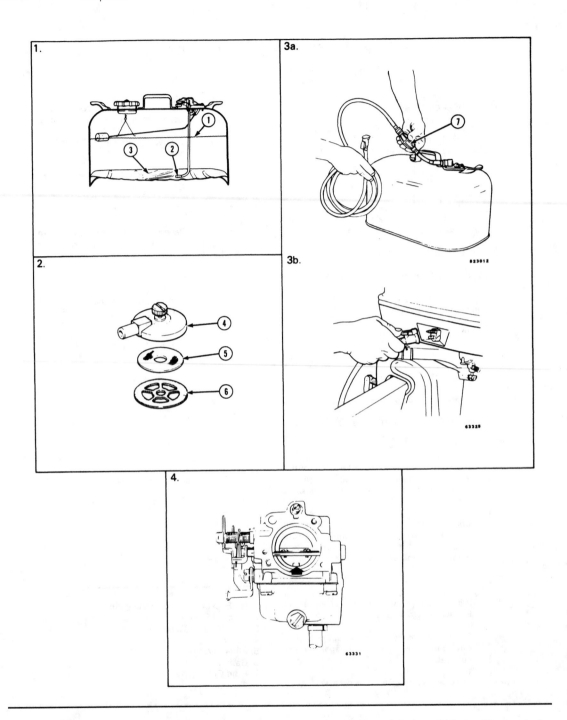

Figure 9–37 Fuel system trouble isolation chart continued. *(Courtesy of Outboard Marine Corporation)*

Fuel System Trouble Isolation Chart (Cont.)

Step 5. Test Fuel Pump

After checking the ignition system and all leaks are corrected and filters cleaned, connect a fuel pressure gauge between fuel pump and carburetor. Run engine at WOT. Pressure should read 2.5 psi (17 kPa) minimum.

Result

Fuel pump OK - Go to step 6. Not OK Replace fuel pump.

Step 6. Check Adjustments

Adjust carburetor linkage and throttle cam See Section 1, **Engine Synchronization and Adjustments.** Test motor.

Result

Motor not OK - Check timing.

Squirt fuel in carburetor's throat. Does engine start? If not, suspect ignition.

Timing OK - Remove and service carburetor and leaf valve assemblies.

Step 7. Check Primer System V-4 and V-6

With engine running approximately 2000 rpm, push in key. Engine should run rich and drop approximately 1000 rpm.

Result

Primer not OK.

Check solenoid resistance, check for pinched fuel hoses or blockage of injection nipple at bypass cover with Nipple Cleaning Tool No. 326623.

Step 8. Check Recirculation System

With engine idling and motor at full "trim out," remove drain hose from bypass cover. Fuel should pulse from each hose. Check for pinched hoses and proper hose routing.

Result

Recirculation system not OK.

Check bypass cover nipples for blockage with Nipple Cleaning Tool. Check hose routing. See Hose Routing Diagram. Replace manifold nipples.

Figure 9–37   Fuel system trouble isolation chart continued. *(Courtesy of Outboard Marine Corporation)*

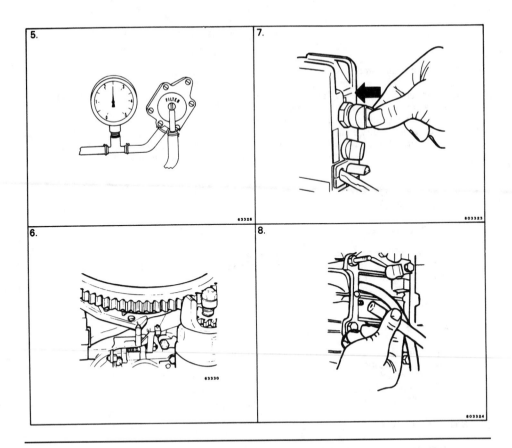

Figure 9–37   Fuel system trouble isolation chart continued. *(Courtesy of Outboard Marine Corporation)*

valve that will pass this test will not offer too great a restriction and cause problems in the fuel distribution line. *Never leave this valve out—it is a safety valve.*

9.26   Safety When Fueling

One of the most important subjects is saved for last. Think about this question for a minute. "As an explosive, which is more powerful, 30 sticks of dynamite or one pint of gasoline?"

As you might have already guessed, the answer is that *one pint of gasoline has as much explosive power as 30 sticks of dynamite.* You'll be carrying several gallons in the boat as fuel for your motor—so handle it, store it, and treat it as if it were a case of dynamite with a short fuse attached.

There are several things you *must* do before you put gas in the boat. Learn them! Do them! The U.S. Coast Guard has issued the following safety procedures that you *must* follow when fueling:

- First, always try to fuel the boat in good light. It's too easy to spill gas when it's dark.
- Don't smoke and don't allow anyone in or near the boat to smoke while you're filling the tank with gas. (It's really not a good way to quit smoking!)
- Make sure all electrical equipment on the boat is shut off and don't operate any switches while you're fueling.
- Keep gas fumes out of the boat by closing all doors and hatches. Gas fumes (vapor) are much more treacherous than dynamite.
- After fueling, open up the boat to air it out. Walk around inside of the boat and give it the sniff test, especially in low places. Gasoline vapor is heavier than air and sneakier than nitroglycerine. Wait a few minutes before you start the engine.
- If you use a portable tank fill it away from the boat or vehicle, placed on the ground (slab), and with the pump nozzle *in contact with the tank* so static electricity will be discharged from the tank to the ground.
- Keep the metal nozzle of the gas pump hose in contact with the plastic or metal tank *at all times* to prevent a static spark.
- Don't spill gas—pay attention to what you're doing. If some gas does spill—don't let it turn to vapor—wipe it up immediately. Get rid of the rag you wiped it up with.
- When you're through, put the gas tank cap back on and tighten it.

9.27   Speed Control System

The Speed Control System is a safety system. What the operator does at the helm or steering handle must happen at the powerhead. If not, loss of control could be experienced.

The steering handle incorporates a twist-grip throttle, which turns a shaft or pulls cables to advance the armature plate. In turn, the armature plate cam moves the throttle primary pickup or the cam follower, which opens the carburetor throttle plate. Upon returning the twist grip to idle position, a spring on each carburetor shaft(s) returns the throttle blade(s) to the closed position. This is a totally closed position. Air to operate the powerhead goes through holes or a cut-out in the throttle blade. There are synchronization adjustments with the armature plate that have to be made.

Remote control of the powerhead speed is controlled by pushing the control lever through 32 degrees of shift engagement. After this movement the speed cable and linkage move the armature plate and then the carburetor shaft(s). This is a long cable, and the turns put into the cable should be very gentle. When undue pressure is applied there is a restriction of cable movement. When properly installed there is little effort for movement, with smooth operation. These cables are in specific lengths for given applications, and are coated to keep moisture away. Very often we find that these cables can seize up during the storage season simply because of lack of movement. It is suggested that the controls be worked from time to time during the storage period so as to keep the cable free. When trailering the boat or when the boat is in storage, place the outboard in a port turn position. This will bring the cable end into the tilt tube, protecting the cable end from road grime and weather (Figure 9–38)

9.27.1   Troubleshooting the Speed Control System

The speed control should be easy to operate in all operational positions of the outboard—from trim out/in through port and starboard turns. Smooth absolute control of powerhead RPM is a necessity. If there is any binding, check for lubrication requirement, routing of cable, cable mounting, and cable adjustment at the powerhead. Also check the position of the friction adjustment at the control. If the shifting cable is out of adjustment the speed control may be affected. These adjustments are simple and can be made by the operator using the instructions in the service manual.

Some steering-handle throttle controls provide for increasing resistance in the handle to set a trolling speed. If this has been set, increased turning effort is required. Loosen the adjustment knob before service checks are made on this type of speed control.

Remember, what happens at the control must happen at the powerhead, or operator control will be lost. On some of the later models there is a neutral RPM limiter. This prevents high RPM in neutral, which could damage the powerhead.

9.28   Fuel Requirements

All Mercury and Mariner outboards can use any major brand of unleaded gasoline with a minimum pump-posted octane rating of 87. Outboards may use gasoline containing up to 10 percent ethanol, but the addition of a Quicksilver water-separating fuel filter is recommended. Mid-grade automotive gasoline advertised to contain fuel injector cleaning agents is recommended for adding internal engine cleanliness. Mix to the proper ratio with TC-W3 oil, for two-stroke outboards using carburetors (Figure 9–40).

Johnson/Evinrude outboards may use any regular unleaded, regular leaded, or premium unleaded gasoline having the recommended octane rating of 87 and not extended with alcohol. Any of the above gasolines are acceptable fuel with up to the following percentage alcohol by volume: 10 percent Ethanol or 5 percent Methanol with 5 percent of solvents.

Do not use any gasoline having more than 10 percent Ethanol or 5 percent Methanol even if it contains cosolvents or corrosion inhibitor, regardless of octane rating. If your outboard is not oil-injected, mix to the proper ratio with TC-W3 oil, for two-stroke outboards (Figures 9–40 and 9–41).

9.29   Outboard Fuel Systems and Gasohol

In high humidity climates, storage of gasoline (Gasohol) in fuel tanks (carburetors) must be limited to a few days. Under those conditions, either drain the fuel tank and idle the engine until carburetor(s) run dry, or run the fuel tank nearly empty and refill with alcohol-free gasoline. Having changed to alcohol-free gasoline, you must run the outboard long enough to purge the fuel lines, filter, carburetor(s), or fuel injection system from Gasohol before stopping the engine.

Alcohol containing fuels will absorb moisture from the air. At first, this moisture will remain in solution, but once the water content of the fuel has built up to about one-half of one percent, it will separate out (phase separation), bringing the alcohol with it. This al-

Outboard Speed Control System and Persons Safety

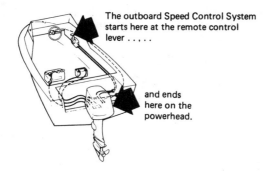

The outboard Speed Control System starts here at the remote control lever . . . . .

and ends here on the powerhead.

What's Most Important?

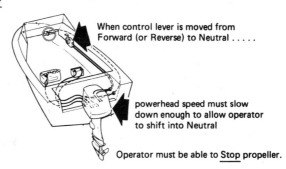

When control lever is moved from Forward (or Reverse) to Neutral . . . . .

powerhead speed must slow down enough to allow operator to shift into Neutral

Operator must be able to <u>Stop</u> propeller.

What Could Happen?

If Operator can't slow down the motor Or shift into Neutral gear (stop propeller), Operator could panic and lose control of boat.

How Can Loss of Speed Control be Minimized?

● Read, Understand, and <u>Follow</u> manufacturers <u>Instructions</u>.

● Follow warnings marked " ⚠ " closely.

when Rigging or after Servicing

● Assemble Parts Carefully
● Make Adjustment Carefully
● Test Your Work. Don't Guess. Make Sure Motor Changes Speed Smoothly, Quickly.
● Make Sure Full Throttle Can be Obtained so Operator Won't Overload Parts.

Figure 9–38 Outboard speed control system and persons safety. *(Courtesy of Outboard Marine Corporation)*

Outboard Fuel, Electrical System and Persons Safety

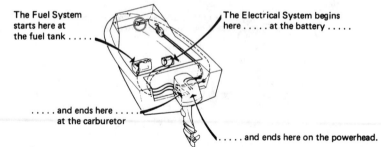

The Fuel System starts here at the fuel tank . . . . .

The Electrical System begins here . . . . . at the battery . . . . .

. . . . . and ends here . . . . . at the carburetor

. . . . . and ends here on the powerhead.

What's Important?

● Fuel Leakage should be minimized.

● Stray electric sparks must not happen.

What Could Happen?

Gasoline can ignite and burn easily. Gasoline vapors can ignite and explode.

● When not boating, fuel leaking in car trunk or van, or a place where portable tank is stored (basement, cottage) could be ignited by any open flame or spark (furnace pilot light, etc.).

● When boating, fuel leaking under the motor cover could be ignited by a damaged or deteriorated electrical part or loose wire connection making stray sparks.

How Can Fire and Explosion Be Minimized?

● Read, Understand, and Follow manufacturers Instructions.

● Follow warnings marked " ⚠ " closely.

● Do not substitute fuel or electrical system parts with other parts which may look the same. Some electrical parts, like starter motors, are of special design to prevent stray sparks outside their cases.

● Replace wires, sleeves, boots which are cracked or torn or look in poor condition.

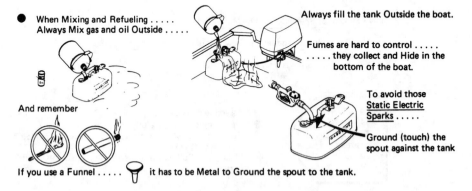

● When Mixing and Refueling . . . . . Always Mix gas and oil Outside . . . . .

And remember

If you use a Funnel . . . . . it has to be Metal to Ground the spout to the tank.

Always fill the tank Outside the boat.

Fumes are hard to control . . . . they collect and Hide in the bottom of the boat.

To avoid those Static Electric Sparks . . . . .

Ground (touch) the spout against the tank

Figure 9–39   Outboard fuel, electrical system, and persons safety (continued). *(Courtesy of Outboard Marine Corporation)*

● <u>After Repair</u> on <u>any</u> part of the fuel system, pressure test engine portion of fuel system as shown: (See Chapter ⌐ for testing the fuel tank portion.)

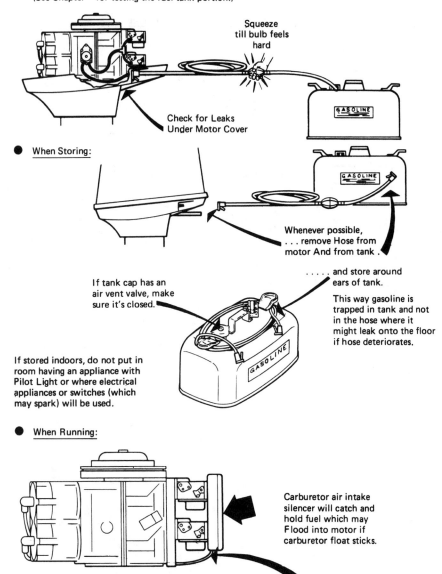

Squeeze
till bulb feels
hard

Check for Leaks
Under Motor Cover

● When Storing:

Whenever possible,
. . . remove Hose from
motor And from tank .

. . . . . and store around
ears of tank.

If tank cap has an
air vent valve, make
sure it's closed.

This way gasoline is
trapped in tank and not
in the hose where it
might leak onto the floor
if hose deteriorates.

If stored indoors, do not put in
room having an appliance with
Pilot Light or where electrical
appliances or switches (which
may spark) will be used.

● When Running:

Carburetor air intake
silencer will catch and
hold fuel which may
Flood into motor if
carburetor float sticks.

So make sure Silencer and all its Gaskets are on motor <u>And</u> Drain Hose is in place.

Air silencer mounting Screws are special lock screws. Use only the special Screws.

Figure 9–39   Outboard fuel, electrical system, and persons safety (continued). *(Courtesy of Outboard Marine Corporation)*

Mix Ratio	Ounces (Canadian) of Oil to be Added to Gasoline for Desired Mix Ratio				
	1 Gallon	2 Gallons	3 Gallons	4 Gallons	5 Gallons
16:1	10	20	30	40	50
20:1	8	16	24	32	40
24:1	7	13 $1/2$	20	27	33 $1/2$
32:1	5	10	15	20	25
40:1	4	8	12	16	20
50:1	3 $1/2$	6 $1/2$	9 $1/2$	12 $1/2$	16
100:1	1 $2/3$	3 $1/4$	4 $3/4$	6 $1/3$	8
			Imperial Gallons		

Mix Ratio	Pints or Ounces (U.S. Measure) of Oil to be Added to Gasoline for Desired Mix Ratio											
	1 Gallon		2 Gallons		3 Gallons		4 Gallons		5 Gallons		6 Gallons	
	Pt.	Oz.	Pt.	Oz.	Pt.	Oz.	Pt.	Oz.	Pt.	Oz.	Pt.	Oz.
16:1	$1/2$	8	1	16	1 $1/2$	24	2	32	2 $1/2$	40	3	48
20:1	$3/8$	6	$13/16$	13	1 $3/16$	19	1 $5/8$	26	22	32	2 $3/8$	38
24:1	$1/3$	5	$11/16$	11	1	16	1 $5/16$	21	1 $11/16$	27	2	32
32:1	$1/4$	4	$1/2$	8	$3/4$	12	1	16	1 $1/4$	20	1 $1/2$	24
40:1	$3/16$	3	$3/8$	6	$5/8$	10	$13/16$	13	1	16	1 $3/16$	19
50:1	$1/6$	3	$5/16$	5	$1/2$	8	$11/16$	11	$13/16$	13	1	16
100:1	$1/16$	1	$3/16$	3	$1/4$	4	$5/16$	5	$3/8$	6	$1/2$	8
					(United States Gallons)							

Figure 9–40 Fuel mixing table.

Figure 9–41 Quicksilver TC-W3 Premium Plus. All the moving parts in your engine never really touch each other. They glide on a thin film of oil. *(Courtesy of Mercury Marine)*

cohol-water mixture settles to the bottom of the fuel tank and if this mixture gets into the engine, the engine can be seriously damaged internally, as it may wash the protective film of oil off the bore of any cylinder it enters. Before the engine can be restarted, it is necessary to remove the separated alcohol and water layer, flush out the fuel system with clean fuel, and remove and dry the spark plugs.

The use of gasoline/alcohol blends may cause deterioration/corrosion of the following: portable metal fuel tanks, fixed metal or fiberglass fuel tanks, fuel lines, fuel filters, and carburetor float bowls. Use of blends may also create injector problems. The clogging of a fuel filter or carburetor jets may be evidence of alcohol reaction on fuel system components. Technicians should inspect the fuel systems for any of the above listed corrosion problems when servicing a fuel system.

Older outboards require a frequent inspection of the fuel system to detect and correct deterioration of elastomers and plastic parts, such as hoses, seals, and gaskets. Deterioration may be caused by the alcohol and acids in today's gasoline.

Fuel stored more than 15 days may have lost some of the desired properties. This is dependent upon temperature and storage conditions.

Always keep the tank vent closed on portable fuel tanks when not in use, to prevent air exchange and water absorption.

KNOW THESE PRINCIPLES OF OPERATION

- Four basic purposes of the carburetor.
- Carburetor circuitry.
- Recirculation of puddled fuel.
- Fuel pump operation and testing.
- VRO oil injection system.
- Ficht fuel injection.
- Production V-6 electronic fuel injection.
- Portable fuel tank testing.
- Fixed fuel tank system.
- Safety in refueling.
- Speed control and persons safety.
- Fuel mix ratio.

REVIEW QUESTIONS

1. What moves the fuel within the carburetor?
2. The primary and secondary ports (orifices) are in which circuit?
3. Why is air bled into a carburetor circuit?
4. What happens to air flow in the venturi at wide-open throttle?
5. Air-fuel ratio is a measurement of _____.
6. Octane rating is the measurement of _____.
7. Atomization is the process of _____.
8. What are the four basic purposes of the carburetor?

9. When the powerhead is idling through the primary port, the secondary port is functioning as an _____.
10. What causes fuel to flow from the high speed nozzle?
11. Carburetor jet size should be changed, if continuous operation is above ___ feet.
12. A low float setting will cause a _____ _____.
13. What causes the fuel pump diaphragm to pulse?
14. An atmospheric leak into the cylinder crankcase operating the fuel pump causes

    _____ _____ _____.
15. List the parts of the recirculation system.
16. How is the recirculation system checked on an OMC V-6 model?
17. What is the purpose of the primer system?
18. The primer system delivers fuel to the _____.
19. When the choke is fully applied, fuel is delivered through which ports of the carburetor?
20. The portable fuel tank is not repaired because it may rupture at the weakened repair area creating a _____ _____.
21. How do you determine if there is a restriction in the fixed fuel tank system?
22. Describe the purpose of the anti-siphon valve. (Write on back of answer sheet.)
23. For a 235-horsepower outboard, what size fuel line should be used?
24. Give a specification for an outboard fuel filter.
25. What component triggers the Ficht fuel injector?
26. In the Ficht Fuel Injection System, oil and fuel are both injected into the combustion chamber.
    a. true
    b. false
27. In the Ficht Fuel Injection System, the low pressure fuel pump brings fuel from the gas tank.
    a. true
    b. false
28. The Ficht Fuel Injection ECU is designed to fire the spark with very high voltage, one time per power stroke.
    a. true
    b. false
29. The Ficht Fuel Injection System uses an engine-pulse-driven pump to move oil into the compartment in the powerhead.
    a. true
    b. false
30. Ficht Fuel Injection System pressures can be as high as 250 PSI.
    a. true
    b. false
31. The Ficht Fuel Injection system uses a vapor separator between the high and low pressure fuel pumps.
    a. true
    b. false

32. For the in-line, new type bleed system (Mercury) restrictors are used in the transfer port bleed passage.
    a. true
    b. false
33. Reed stop setting is measured between the reed block and reed stop.
    a. true
    b. false
34. The VRO pump is an electrical pump, which pumps oil and fuel.
    a. true
    b. false
35. When adding fuel to the portable fuel tank, the nozzle must be kept in contact with the tank.
    a. true
    b. false
36. When put into storage, the portable fuel tank vent is left open.
    a. true
    b. false
37. 5 US gallons of gasoline: How much TC-W3 oil is used to make a 50-1 ratio?
    a. 11 oz.
    b. 8 oz.
    c. 13 oz.

Mercury Production V-6 EFI Fuel System

38. The water-separating filter contains a sensor
    a. which monitors water level.
    b. which activates a red light at the module.
    c. which will activate a beep.
    d. All of the above are correct.
39. The vapor separator is considered to be a fuel reservoir which continuously blends and circulates fuel.
    a. true
    b. false
40. Where does the unburned excess fuel in the bleed system flow to?
    a. water separator
    b. fuel rail
    c. vapor separator
41. Excess fuel from the electric fuel pump is circulated
    a. back to the water separator.
    b. back through the fuel rail.
    c. back to the vapor separator.
    d. Both b and c are correct.

42. What component controls the pulse width of the injector?
    a.   diaphragm fuel pump pressure
    b.   Electronic Control Unit
    c.   Both a and b are correct.
43. The fuel pressure regulator controls fuel pressure in the
    a.   vapor separator.
    b.   water separator.
    c.   fuel rail (and out line).
44. The induction manifold is used as a plenum chamber and contains the fuel rail.
    a.   true
    b.   false

Mercury Production V-6 EFI Electrical System

45. The ECU constantly adjusts the pulse width _____ or _____ to compensate for operating conditions.
46. When does the air temperature sensor resistance decrease?
    a.   as air temperature lowers
    b.   over 4000 RPM
    c.   as air temperature increases
47. The air temperature sensor transmits manifold absolute air temperature
    a.   through full RPM.
    b.   through 2500 RPM.
    c.   to the ECU.
    d.   Both a and c are correct.
48. The ECU uses the MAP sensor to determine
    a.   fuel flow.
    b.   ignition timing above 2000 RPM.
    c.   changes in manifold pressures.
    d.   Both a and c are correct.
49. The TPI transmits information to the ECU during
    a.   slow speed operation.
    b.   mid-range operation.
    c.   slow speed and mid-range operation.
50. Ignition detonation control is activated by combustion noises which excite the piezoelectric circuit located inside the detonation sensor.
    a.   true
    b.   false
51. When will the water sensing system function?
    a.   When voltage is present at the sensing probe.
    b.   When water reaches the top of the probe.
    c.   Both a and b are correct.

52. When are injector pulse widths tripled?
    a. when cylinder head is at normal temperature
    b. when cylinder head temperature is below 90 degrees
    c. when key switch is in start position
    d. Both b and c are correct.

Mercury Oil Injection, with Variable Ratio Oil Pump

53. What alerts the operator to a low oil condition?
    a. engine RPM slows down
    b. warning horn will beep intermittently
    c. warning horn will sound continuously
54. The oil pump is a gear driven pump.
    a. true
    b. false
55. What keeps fuel out of the oil line?
    a. ECU
    b. 5 PSI check valve
    c. 2 PSI check valve
56. Oil and gasoline flow together into the diaphragm fuel pump.
    a. true
    b. false

CHAPTER 10

Tune-Up

Objectives

After studying this chapter, you will know

- What a tune-up consists of.
- What *sync* and *link* is all about.
- Test wheel use.
- Dynamometer testing.
- Effects of altitude on outboard performance.
- Effects of marine growth.
- Combustion—Normal and Abnormal.
- Spark plug analysis.
- Outboard troubleshooting.

10.1   The Need for a Tune-Up

The outboard motor periodically needs a tune-up, or maybe it has a miss and needs to be repaired. A set procedure needs to be followed to determine the cause of the miss. If you are doing the repair work for someone else, talk with the customer and get his comments on the performance of the outboard. Did the miss just start? Has it been hard starting? What type of fuel is being used and at what oil mix? What type of oil is used? Has it been running in the operating range? How long has it been since it was tuned up? The answers to these questions will give you some idea of what the problem may be in the outboard.

What should be accomplished if a tune-up is in order to repair the miss? Talk to five different people and you will get five different answers. What a good tune-up covers will be discussed in this chapter.

10.2   What a Tune-Up Consists of

It is important to know how to identify model and serial number of an outboard, for the purpose of using the correct specifications, factory service bulletins, and installing the correct parts. So, you will need to find the number. Some have the serial number on the transom brackets.

Start by checking for any service bulletins that may apply. Visually inspect the engine for leaks and missing parts. Determine if the powerhead is in acceptable mechanical condition by making a compression test. Compression should not vary more than 15 PSI (100 kPa) between cylinders. With equal compression, is the engine hard to start and does it operate poorly? If the compression test shows a variation greater than 15 PSI (100 kPa), there is the possibility of a damaged piston, a damaged or leaking head gasket, scored cylinder walls, and stuck or broken piston rings. Also, a carbon buildup may need to be removed from the piston head(s) and the combustion chamber(s). This can be done using a Power Tune (Mercury) additive, which is sprayed through the carburetor inlet when the engine is running. If the aluminum cylinder heads are removed, check them for warpage. If they are warped, the head(s) should be resurfaced on a flat surface. Inspect the spark plugs for signs of performance problems. Also inspect the wiring leads and connectors in the ignition system; check ignition coil(s), ECU, sensors, trigger coils, modules, and do a test for alternator output. Check for spark on each cylinder. Will the spark jump the prescribed gap? If not, sensors, charge coils, driver coils, stop switch, etc., will have to be checked. Inspect the carburetor(s) for gum and varnish left over from stale fuel. The carburetor bowl will quickly give an indication of untreated gasoline being left in the carburetor during storage. The fuel pump should be inspected and the filter cleaned or replaced. Check the lines for good condition. The timing synchronization/linkage procedure should be gone through to insure that the timing and carburetion are working together as a team (Figure 10–1).

10.3   Synchronization of Timing/Carburetor

The throttle control movement changes ignition timing and the amount of throttle opening (Figure 10–2). Setting the timing to advance with the carburetor opening must be carefully synchronized. (This is commonly known as **sync** and **link**.) For a powerhead to run satisfactorily, the correct volume of air and fuel must be delivered to the cylinder. Then a spark must occur at the spark plug at just the right time. This must take place throughout all operating speeds if the best performance and economy are to be realized. Longevity of the powerhead can be affected if timing or fuel adjustments are not properly set. (See "Combustion—Normal and Abnormal," Figure 10–12)

 Note

Reliable troubleshooting *cannot* be done if linkages and settings are out of adjustment. These are critical adjustments and must be made according to service manual instructions!

Listed below are procedures for *specific outboards* from OMC and Mercury. In general, the sequence of adjustments for OMC 50 through V-6 are the following (Figures 10–1 and 10–2):

1. Determine top dead center for cylinder 1 and align the pointer. This brings the piston to the top of the compression stroke and then the timing pointer is aligned for top dead center.
2. Synchronize the throttle valve. With the roller cam backed off, the throttle valve linkage is released to insure that the throttle valve(s) completely close.
3. Adjust the throttle pickup point. This adjustment will cause the carburetor(s) to start to open at a predetermined point.
4. Synchronize the choke valves, and adjust the choke solenoid. This adjustment is made to insure that the multi-carburetor chokes completely close.
5. Adjust wide-open throttle setting. Clearance is set at wide-open throttle to insure that there will be no binding of the throttle shaft and blade.
6. Synchronize throttle and timer base linkage. Adjust the linkage for minimum lost motion. Set idle timing and index throttle cam yoke with center of cam roller. [Place outboard in test tank with test wheel (propeller) installed.]

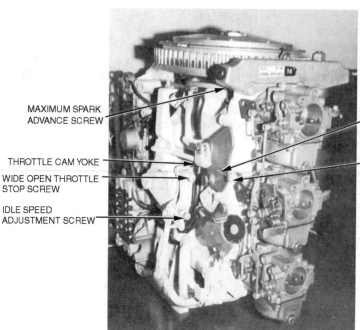

MAXIMUM SPARK
ADVANCE SCREW

FIRST EMBOSSED MARK
ON THROTTLE CAM

THROTTLE CAM YOKE

WIDE OPEN THROTTLE
STOP SCREW

CAM FOLLOWER
ROLLER

IDLE SPEED
ADJUSTMENT SCREW

Figure 10–1   Location of adjustments for synchronization—70/75 HP *(Courtesy of OMC).*

7. Adjust maximum spark advance. Sets maximum timing advance at a specified powerhead RPM.
8. Set idle RPM. Adjust powerhead idle RPM in forward gear to specification.
9. Adjust throttle cable to remove looseness and preload the cable.

In general the sequence of adjustments for Mercury 50 HP (above 4576236) are as follows (see Figure 10–3):

1. With the outboard in a test tank, preset the carburetors. This sets the idle mixture for initial start-up.
2. Run the outboard in gear at 7 to 9 degrees advance and set the primary pickup arm on the carburetor cluster. This determines when the carburetors will start to open in regard to powerhead timing.
3. Set the timing. Run the outboard in forward gear, increasing RPM until maximum timing specification is reached and a stop is set.
4. Adjust secondary pickup. This permits further carburetor opening beyond maximum timing stop.
5. Adjust wide-open throttle. Clearance is set at wide-open throttle to ensure that there will be no binding of the throttle shaft and blade.

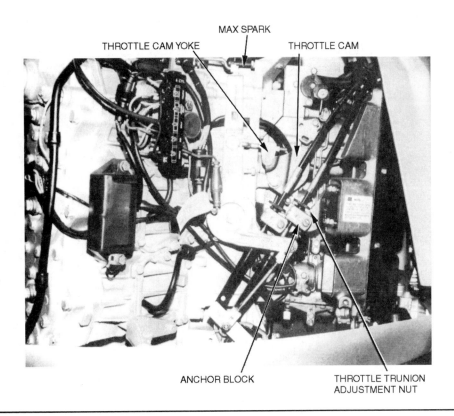

THROTTLE CAM YOKE   MAX SPARK   THROTTLE CAM

ANCHOR BLOCK   THROTTLE TRUNION
ADJUSTMENT NUT

Figure 10–2   OMC sea drive linkages for synchronization.

6.   Make carburetor adjustments. This sets carburetor idle mixture and powerhead idle RPM.
7.   Adjust throttle cable. Cable barrel is adjusted to remove looseness and preload the cable.

These two sets of sync and link (timing/carburetor) adjustments are *particular to certain horsepower outboards*. All outboards have a sequence of adjustments that are required to bring the synchronization of timing and carburetion into a working team for performance. If the sequence given in the service manual is not followed, timing (engine bogs down, runs rough) and/or carburetion will be off, and major damage will occur (mainly at maximum timing advance) to the powerhead. This damage will be caused by preignition/detonation which will be evident on the piston head and ring land areas (Figure 10–4). Piston skirts can also be damaged by overheating and breakdown of the lubricant. Incorrect propeller pitch (too high), which allows the powerhead to run below recommended maximum RPM range, can contribute to piston damage as well. These adjustments and the proper pitch propeller will allow the powerhead to run within the operating range at wide-open throttle.

THROTTLE STOP SCREW   THROTTLE PICKUP SCREW

MAXIMUM IGNITION
ADVANCE SCREW   IDLE SPEED SCREW   DISTRIBUTOR

Figure 10–3   Timing and synchronizing screws. *(Courtesy of Outboard Marine Corporation)*

Figure 10–4   Piston damaged
by preignition/detonation.

10.4   **Propeller Considerations**

The propeller is the connection between the water and the boat. Without it nothing happens. Check the condition of the propeller. Small nicks should be dressed out with a file. The propeller may need to be changed depending on the type of boating the customer is doing. The correct propeller will insure that the tune-up work accomplished will be satisfactory.

Next, tank-test the outboard. If the outboard is an OMC product, a test wheel (propeller) should be installed to determine if the operating RPM is obtained. This will load the powerhead and prove the tune-up work (see Figure 10–5).

A dynamometer run can be made on medium through V-6 outboards to prove the powerhead performance (see Figure 10–6).

A tank test with a standard propeller installed is of very little use, because you cannot load the powerhead properly. While the tank test or dynamometer run is being accomplished, check the cooling system and powerhead temperature. This may indicate that the

Figure 10–5   Test wheel used to determine operating RPM. *(Courtesy of Outboard Marine Corporation)*

Marine Dynamometer

Solve The Tough Problems

So a good customer has handed you his problem. Fix that engine, or maybe he just wants it ready for spring. Maybe it starts and idles, stirs up your "tank" OK, but he says it lacks "punch" on the lake.

Or how do you catch the "Gremlin" that only shows up when your customer is trying to get a skier up to speed? Or find an intermittent miss that only shows up when he's 20 minutes out? DYNO testing helps by simulating on-water reality right in your shop.

DYNO Lets You

- Run up engines from idle to full throttle under measurable load with shop test equipment for analysis.

- Locate high-speed and mid-range problems at working temperatures and loads.

- Duplicate various "bog" situations.

- Time engines at recommended RPMs.

- Properly set I/O shifters.

- POWER TUNE to max engine performance.

- LOAD TEST to confirm repair or service.

What Is The DYNO?

Dynamometer is a "closed" hydraulic system capable of fully "loading" marine outboard and sterndrive engines for diagnostic and performance tuning and testing. Precise adjustment of "load" and measurement of output are possible throughout the engine's entire RPM range.

Easy To Use

The International DYNO® Dynamometer consists of two major components and associated support accessories. The major components are: (1) The DYNO control unit that produces, controls and measures the load, engine speed and temperature, and (2) the DYNO power-head pump that attaches to the engine's prop shaft to "load" the engine. Accessories include prop shaft connection adapters and cooling and exhaust systems.

Typical Test Run

1. Remove prop from engine.

2. Select correct prop shaft adapter.

3 Slide DYNO pump onto prop shaft.

4. Connect tachometer and water flushette.

5. Connect ignition analyzer or timing light if appropriate.

6. Start and warm up engine.

7. Adjust throttle and load valve to achieve appropriate RPM and measure the output.

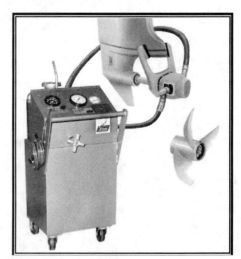

Figure 10–6   Marine dynamometer **(continued).** *(Courtesy of International Dyno Corporation)*

DYNO set-up is a one man,
10 minute shop job.

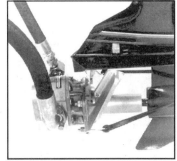

DYNO power-head pump slips
on prop shaft to load engine .

INTERNATIONAL DYNO® Marine
Dynamometer. One man can hook up the
DYNO and in 10 minutes have that engine
under test at full throttle, full load "reality". And it all happens where you can use
your test equipment and where special
tools and parts are handy.

And when you think you've solved your
customer's problem it can be tested and
verified easily before your customer is out
on the water.

EXHAUST SYSTEM reduces noise,
directs exhaust away from operator.

EXHAUST SYSTEM includes exhaust
cooling, muffler and back pressure.

Figure 10–6   Continued. *(Courtesy of International Dyno Corporation)*

water pump needs to be serviced. It is a good idea to replace the water pump impeller each spring.

After the tank test or dynamometer run is completed, the lower unit should be drained and refilled with the correct gear oil. Any water in the oil or oil that is milky-looking indicates that a seal (s) in the lower unit is (are) leaking. If that is the case, the lower unit should be pressure-tested to determine where the leak is, and then resealed.

If the outboard is going into storage, the powerhead should be protected from corrosion during extended storage periods. To do so, inject rust-preventive oil through the carburetor air horn while the powerhead is running. This bathes the cylinders, bearings, crankshaft, etc., with a storage seal (Quicksilver) oil. (Do not use a hardware product.) Also, add a fuel stabilizer in the fuel system and run the outboard for at least five minutes in water.

 **Warning**

Do not store the outboard with Gasohol-type gasoline in the fuel system!

10.5   **Effects of Weather**

Weather conditions have an effect on horsepower output of the outboard. The power produced is dependent upon the density of air that the powerhead consumes. Air density is dependent upon ambient temperature, barometric pressure, and humidity (Figure 10–7).

Summer conditions of high temperature, high humidity, and low barometric pressure reduce powerhead performance. When cool dry weather returns, so will performance. This brings up the question of propping the outboard. Was it done on a hot muggy day or in the spring when air was dry and cool? If the boat was rigged in the spring, the outboard was equipped with a propeller which allowed the powerhead to run in operating range at full throttle. In the summer months the power falls off because of hot humid weather. In effect, the propeller has become too large, reducing boat speed and powerhead RPM at wide-open throttle. This is a subtle loss of power and boat speed. This RPM loss can be regained by changing the propeller to bring the powerhead RPM back up into the upper part of the recommended operating RPM. Horsepower cannot be regained until cool dry weather conditions return.

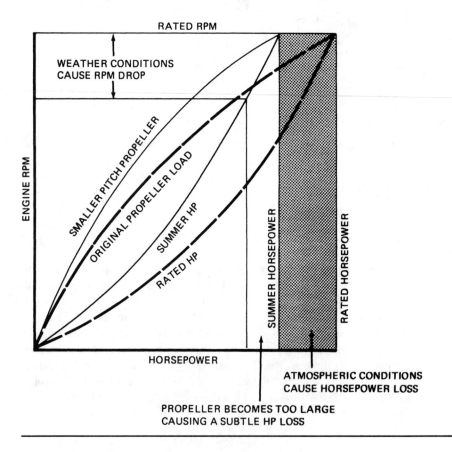

Figure 10–7   Effects of weather on horsepower.

10.6   Effects of Altitude

Altitude also affects powerhead performance. As altitude increases the air becomes less dense (less pressure differential) and the powerhead does not breathe as well as at sea level. Engine power is reduced by about three percent for every 1,000 feet of altitude above sea level. If the outboard runs primarily in high-altitude areas, then the carburetor jetting should be changed. The manufacturers offer different sized jets, beginning at approximately 2500 feet elevation. If you plan on a boating vacation in the higher elevations, you should consider a reduction in propeller pitch and also a jet change that will allow the powerhead to operate within the upper part of the operating range in that altitude. Once down from the higher elevations, the propeller and jets should be changed again. For more information on propellers and the reasons for changing them, see Chapter 13, *Propeller Performance*.

All fuel injection systems, such as OMC's Ficht Fuel Injection system, will automatically adjust to altitude and temperature. This is because of the sensors supplying information to the ECU.

10.7   Hull Conditions

Hull condition affects boat operation and miles per hour. Any condition, such as a hook or rocker, or surface roughness in the critical center aft portion, will have a negative effect on speed. The bottom of the boat can be checked with a long straight edge.

Marine fouling reduces the ability of the boat to slip through the water. This results in fewer miles per hour and requires more work from the outboard. A definite loss of performance and lost operating economy occur. Fouling by marine growth adds up to drag (see Figures 10–8, 10–9 and 10–10).

When marine growth advances on the bottom of the boat to the point of creating a performance problem, the growth needs to be removed. The best time to remove the growth is immediately after taking the boat from the water. It can be removed easily by using household bleach. Apply the bleach to the marine growth on the bottom of the boat, lower unit, etc. by using a spray bottle or garden sprayer. Let the bleach do its work for about ten minutes. Then rinse the affected area using a sharp water spray from a water hose and attachment. Modern finishes should be tested to insure that there will be no adverse reaction to the finish before general cleaning is attempted.

Marine fouling can also develop on the lower unit if left in the water. This growth can form quite quickly and can even cause powerhead overheating. If the growth is severe enough, the water passages may become clogged or reduce water intake. Marine growth on the lower unit should be removed before outboard operation is begun.

This chapter indicates that several systems need to be serviced during the tune-up. The end result of the tune-up should be that the powerhead is set to specifications and will run in the upper part of the operating range. The work or inspection on the hull should result in maximum miles per hour and operating economy.

Figure 10–11 describes some typical problems for high-speed boating. Remember that high-performance boating requires the driver to give a high degree of attention to driving.

Figure 10–8   Boat bottom deterioration.

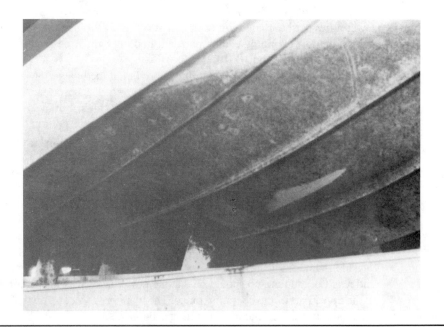

Figure 10–9   Marine growth on boat bottom.

MARINE GROWTH

Figure 10–10   Accumulation of marine growth on lower unit and boat bottom.

Somewhere out there ahead of the boat there are always different wind currents, wakes from other boats, and debris floating low in the water. When you are moving fast, almost any disturbance in the water can cause the boat to roll or yaw, requiring the proper corrective action. The faster you go, the quicker things will happen. Only you can decide when you are going fast enough.

10.8   High Performance

High-performance boating is an exciting, exhilarating sport. The sport boater must, however, be considerate of others who may use the same waters. High-performance boating requires a lot of water. Pursue your sport where you have plenty of room and where your noise and wake won't annoy others. Make sure your pursuit of pleasure does not create a hazard or annoyance to nearby homeowners, fishermen, swimmers, water skiers, sailors, or other powerboaters. Make common sense and courtesy a regular part of your boating routine. Good boating! Figures 10–12 through 10–16 describe additional tune-up information for outboards.

Problem	Possible Cause	Correction
Lack of speed	1. Power low.	1. Check and tune engine — ensure throttles open fully. Ignition Timing. See Service Manual and check.
	2. Wrong type of propeller for application.	2. Select correct type of propeller (SST II, SST R, SST R_x).
	3. Motor RPM low/high.	3. Select lower/higher pitch propeller.
	4. Unable to trim out due to ventilation.	4. a. Add additional cup to propeller. b. Check for speedometer pickup location. If less than 15″ from gearcase center line, a trail of air bubbles may cause ventilation. c. Lower motor on transom.
	5. Too much drag. (Motor mounted too low.)	5. Raise motor on transom.
Difficulty in planing boat	1. Wrong type of propeller for application.	1. Select correct type of propeller for application (SST II, SST R).
	2. Propeller ventilates as boat tries to get on plane.	2. a. Trim in further until on plane or lower angle adjusting rod if not in lowest position. Refer to **Minimum Trim Angle Position** in Chapter 12. b. Add additional cup to propeller. c. Lower outboard on transom. d. Install wedges to increase transom angle. (Increases available negative trim.)
	3. Motor bogs and lacks power coming "out of hole" (getting on plane).	3. a. Check and tune motor. b. Change to lower pitch propeller. c. Shift weight forward in boat or reduce weight located aft in boat. d. Install wedges to increase transom angle.
Boat porpoises	1. Out-of-balance rig.	1. Shift weight forward.
	2. Motor trimmed too far out.	2. Trim in until porpoising stops.
	3. Cannot trim in far enough.	3. Lower angle adjustment rod if not in lowest position.
Boat "fishtails" or "chin walks" at high speed	1. Loose steering.	1. Check steering system integrity. Eliminate as much steering play as possible.
	2. Loose motor mounting fasteners.	2. Inspect fasteners for damage, then tighten securely.
	3. Movement in mounts.	3. Inspect motor mounts for damage and make sure mount fasteners are tightened securely.

Figure 10–11 Problem chart for high-speed boating. *(Courtesy of Outboard Marine Corporation)*

Combustion—Normal and Abnormal

The power in an internal combustion engine is developed by expanding gases resulting from the burning of the air/fuel charge. If you have the proper air/fuel mixture, timing is correct, and the anti-knock (octane rating) quality of the fuel meets the engine requirements, the burning process should occur evenly and steadily. This is normal combustion...a cycle which will be repeated many thousands of times every minute the engine is in operation.

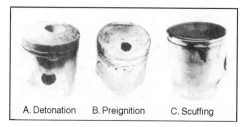

A. Detonation   B. Preignition   C. Scuffing

There are, however, various forms of abnormal combustion which are not only troublesome, but can be expensive with occasional failures to pistons, spark plugs and other engine parts. Abnormal combustion has always been a problem to contend with in all types of engines . . . two or four-cycle. Engine manufacturers are aware of these problems and caution against indiscriminate use of fuels, oils, oil mixes and poor maintenance practices.

The spark plug is rarely the cause, but, rather, the victim of abnormal combustion. Because the spark plug is positioned in the combustion chamber, it is readily exposed to the damaging effects of preignition and detonation. The responsibility for piston damage is often unfairly placed on the spark plug simply because it may show evidence of damage similar to the piston. There are many times, however, that piston failure does occur with no damage to the spark plug.

In any case, damaged engine parts, after thorough examination, can almost always be attributed to some form of abnormal combustion, revealing that preignition or detonation is the real cause. Because preignition and detonation are so closely related, it is difficult to determine where one ends and the other begins.

Detonation occurs when the anti-knock value of the fuel does not meet the engine requirements. A portion of the air/fuel charge begins to burn spontaneously from increased heat and pressures just after ignition. The two flame fronts meet and the resulting "explosion" (as illustrated) applies extreme hammering pressures on the piston and other engine parts. Piston damage occurs from the pounding pressures of severe detonation, and the increased heat factor can cause preignition.

Preignition is just what the term implies. As illustrated, it is the ignition of the air/fuel charge prior to the timed spark. Any hot spot within the combustion chamber can cause preignition.

Detonation and preignition are usually the result of one or a combination of the following conditions.

DETONATION:

1. Ignition timing advanced too far.
2. Fuel octane rating too low.
3. Lean mixtures. Poor carburetion and/or leaks in manifolds-crankcase and/or intake.
4. Compression ratio increase due to combustion chamber deposits or engine modification.
5. Excessive intake manifold temperatures.
6. Lugging the engine.

The piston, see Figure A above, has been damaged by detonation. Because of the hammering

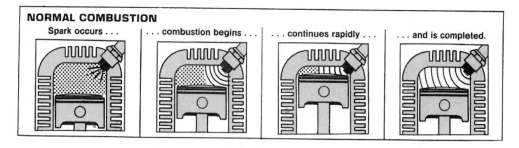

NORMAL COMBUSTION

| Spark occurs . . . | . . . combustion begins . . . | . . . continues rapidly . . . | . . . and is completed. |

Figure 10–12   Normal and abnormal combustion (continued). *(Courtesy of Champion Spark Plug Co.)*

DETONATION

Spark occurs . . .   . . . combustion begins . . .   . . . continues . . .   . . . detonation.

pressures of detonation, piston damage usually appears as fractures on or through the crown, or in the skirt and wrist pin area.

PREIGNITION:

1. Combustion chamber deposits which become incandescent.
2. Hot spots in the combustion chamber due to poor heat dissipation.
3. Scuffing—due to inadequate lubrication or improper clearance on engine parts.
4. Detonation or conditions leading to it.
5. Head gasket protrusion into the combustion chamber. (Thin edges on valves in 4 cycles).
6. Cross firing (electrical induction between spark plug wires).
7. Spark plug heat range too high for engine operating situation.

The piston in Figure 3.14 and in Figure B above was damaged by preignition. This is evidenced by the burned and melted effect produced from the high temperatures of preignition. Damage may appear on and through the crown, through the ring lands, or both.

Another DANGER sign that preignition has been encountered, is evidenced by aluminum throw-off

from the piston onto the spark plug as illustrated. (See Figure 10–16)

The piston in the Figure C above is the victim of a scuffing condition which, almost without fail, will appear on the piston when severe detonation or preignition is encountered. It can, however, occur singly because of a loss of lubricant, improper piston to bore clearance or abnormally high combustion temperatures. It, too, can cause preignition.

Effect of high temperature on the spark plug is usually indicated by a clean white insulator core nose, and/or excessive electrode erosion.

In almost all cases, abnormal combustion can be completely eliminated by: (1) a regular maintenance schedule which includes a proper tune-up with particular attention to spark timing; (2) using spark plugs of the correct heat range for the engine and the type of running situation; (3) using a suitable grade of lubricant in the crankcase for 4 cycles and a proper 2-cycle oil in 2-cycle mixtures; (4) selecting a proper octane-rated fuel and maintaining correct mixtures in 2-cycle engines.

Abnormal combustion can be troublesome and expensive . . . but it can be averted with proper care and attention given to the engine, the fuel, and engine adjustments.

PREIGNITION

Ignited by hot deposit . . .   . . . regular ignition spark . .   . . ignites remaining fuel . .   . . . flame fronts collide.

Figure 10–12   Normal and abnormal combustion continued. *(Courtesy of Champion Spark Plug Co.)*

SURFACE GAP SPARK PLUGS

These plugs are primarily designed for use with two-cycle engines equipped with capacitor discharge (CD) ignition. Most common applications are in motorcycles, outboards and snowmobiles.

CD ignition refers to an electronic system using a special capacitor and coil permitting a very rapid buildup of firing voltage technically referred to as "rise time." Figure 1 shows typical rise time comparisons between CD, transistorized, and conventional systems. Series gap plug types are used in some applications to better match ignition system and engine characteristics to spark plug firing requirements.

This fast rise time permits the surface gap spark plug to fire in spite of fouling deposits that may be present on the insulator firing end. The slower rise time of conventional systems often allows the voltage to "bleed off" to ground rather than form an igniting spark.

When used in the applications for which they are designed, surface gap spark plugs offer a number of advantages:

1. Colder operating—avoids preignition.
2. Less affected by fouling deposits.
3. Longer life—improved performance.

Surface gap plugs may be cleaned exercising care not to subject the firing end to excessive abrasive "blasts." Make certain that all abrasive materials are removed from threads, gasket and firing end before reinstallation. Due to design, no gap adjustment can be made.

Good surface gap plug performance, like any other plug design, depends upon spark energy and the presence of a combustible mixture at the firing gap. If a spark is present for a shorter time than required to initiate combustion, misfire will occur. The length of time a spark continues to cross the gap is referred to as "arc duration" by those who evaluate ignition system characteristics.

A surface gap spark plug is a plug which, by design and construction, delivers a spark that travels from the center electrode to the shell across the surface of the ceramic insulator. The inside diameter of the shell at the firing end is machined to a fixed distance from the center electrode. This forms the surface gap space.

The spark travels from the center electrode above the exposed surface of the insulator to the shell. In essence, the entire end of the shell is the ground electrode. Normally, the spark should rotate to a different location across the insulator surface.

FIRING GAP

The designs of these plug types are such that heat range values associated with standard plug designs do not apply. All surface gap plugs are "cold" types and virtually preignition free.

FIGURE 2

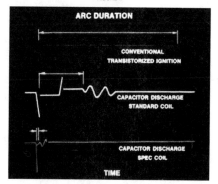

Figure 2 compares the arc duration of typical ignition systems now in use. The obvious difference in firing patterns and spark time illustrates why CD systems are required when surface gap plugs are used. A rapid buildup of voltage and a quick spark release are necessary if these plugs are to perform properly. With one exception (SAAB), they are not now suitable for general automotive application considering automotive heat range requirements and ignition systems.

FIGURE 1

TYPICAL VOLTAGE RISE TIME

Figure 10–13   Surface gap spark plugs. *(Courtesy of Champion Spark Plug Co.)*

SURFACE GAP SPARK PLUG FIRING END ANALYSIS

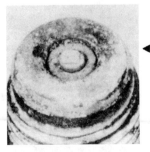

NORMAL: Light tan or gray colored deposits indicate good engine and ignition system condition. Electrode wear indicates normal spark rotation.

WORN OUT: Plugs with excessive electrode wear may cause misfire during acceleration or hard starting. Replace with the recommended Champion plug.

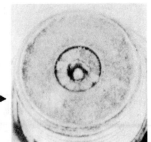

COLD FOULED: Wet fuel/oil deposits can be caused by "drowning" with raw fuel mix during cranking, rich carburetion or improper fuel/oil ratios. Weak ignition can also contribute to this condition. Clean or replace.

LOW TEMPERATURE FOULING: Soft, sooty deposits indicate incomplete combustion. Probable causes: rich carburetion; weak ignition; retarded timing or low compression. Continuous low-speed operation or, with oil injection systems, gunning throttle at idle. Clean or replace.

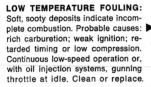

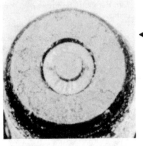

CARBON TRACKING: Electrically conductive deposits on the firing end provide a low-resistance path for the voltage. Carbon tracks are formed and misfire could occur. Plugs should be serviced or replaced.

CHANNELING. Is sometimes incorrectly diagnosed as cracking. Believed to be caused by extreme spark heat. When deposits cover the shallow channels, the rate of insulator erosion is aggravated, the spark is masked and misfire may occur. If so, plugs should be replaced.

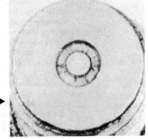

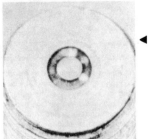

CONCENTRATED ARC: Multi-color appearance is a normal condition. Caused by electrical energy consistently following the same firing path. Arc path will change with deposit conductivity and gap erosion.

ALUMINUM THROW-OFF: Danger. Preignition has occurred. Not a plug problem . . . check engine to determine cause and extent of damage. Replace plugs.

Figure 10–14   Surface gap spark plug firing end analysis. *(Courtesy of Champion Spark Plug Co.)*

Heat Range

Heat range is especially important to a marine engine, because of the great differences in how an engine is used. Racing at high speed requires a much different type of plug in terms of heat range than does prolonged trolling. Then, too, there is that middle path calling for plugs that lie somewhere in between these two extremes.

Heat range refers to a plug's ability to transfer heat from the firing tip of the insulator to the cooling system of an engine. This rate of heat transfer is controlled by the distance the heat must travel to reach the cooling medium. A "cold" plug has a shorter insulator nose and gets the heat quickly to the engine's cooling system, and is recommended for continuous high speed operation to avoid overheating. A "hot" plug has a longer insulator nose and gets heat away more slowly from the firing end. It runs "hotter" to burn off combustion deposits that tend to foul the plug during prolonged idle or low-speed operation, such as trolling.

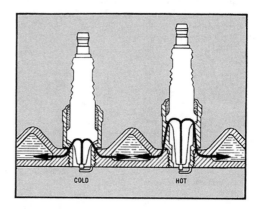

Figure 10–15   Heat range. *(Courtesy of Champion Spark Plug Co.)*

Spark plug analysis

The spark plug is an indicator of an engine's performance. The condition of the plug is often a graphic description of engine problems, and learning to "read" a marine plug may offer the opportunity to prevent future mechanical trouble. Here are some typical conditions and what they may signal.

NORMAL

Correct heat range. Insulator is light tan to gray color. Few deposits present. Electrodes are not burned.

Change plugs at regular intervals, using Champion type of the recommended heat range.

ALUMINUM THROW-OFF

Metallic "gob" of gray pot metal adhering to electrodes and plug core. Rare.

Caused by pre-ignition source within cylinder melting aluminum alloy off piston. Do not install new plugs until source of pre-ignition is determined and piston examined.

WET FOULING

Damp or wet, black carbon coating over entire firing end. Forms sludge in several cases.

Wrong spark plug heat range (too cold)
Prolonged trolling operation
Low-speed carburetor adjustment is too rich
Improper ratio of fuel-to-oil mixture
Induction manifold bleed-off return passage obstructed
Worn or defective breaker points resulting in lack of voltage

CORE BRIDGING

Electrodes not badly burned. Bottom side electrode usually coated with ash-like deposits. Insulator nose "peppered" with tiny beads or small chunks fused to firing end. Sometimes have the appearance of glass-like bubbles.

For cause of core bridging, see "Gap Bridging."

GAP BRIDGING

Spark gap shorted out by combustion particles, wedged or fused between electrodes.

Both core bridging and gap bridging are caused by excessive combustion chamber deposits striking and adhering to the spark plug's firing end. They originate from the piston and cylinder head surfaces. These deposits are formed by one or a combination of the following:

Excessive carbon in cylinder
Use of non-recommended oils
Immediate high-speed operation after prolonged and excessive trolling
Improper ratio of fuel mix

OVERHEATING

Electrodes badly eroded. Premature gap wear. Insulator has gray or white "blistered" appearance.

Incorrect spark plug heat range (too hot)
Ignition timing overadvanced
"Sticky" piston rings
Prop not suited to boat load or motor (lugging engine)
Worn or defective water pump
Restricted water intake
Scale, salt, or mud in water jackets causing restriction in cooling system

NOTE:

Many outboard manufacturers market combustion chamber purging solvents. These can be run in an engine to clean out troublesome deposits. In extreme cases, cylinders and pistons may have to be cleaned manually, as new plugs may give only temporary relief.

Figure 10–16   Spark plug analysis. *(Courtesy of Champion Spark Plug Co.)*

NOTE: When trouble-shooting power heads of 4-cycle outboard motors, follow procedures outlined in Chapter 3.

START ALL MAJOR DIAGNOSES WITH A COMPRESSION TEST, AND TEST WHEEL RPM CHECK.

DO NOT RUN MOTOR OUT OF WATER.

MOTOR REACTION	CHECK POINTS
1. Manual starter rope pulls out, but pawls do not engage.	A. Friction spring bent or burred. B. Excess grease on pawls or spring. C. Pawls bent or burred.
2. Starter rope does not return.	A. Recoil spring broken or binding. B. Starter housing bent. C. Loose or missing parts.
3. Clattering manual starter.	A. Friction spring bent or burred. B. Starter housing bent. C. Excess grease on pawls or spring. D. Dry starter spindle.
4. Electric starter inoperative.	A. Loose or corroded connections or gound. B. Starting circuit safety switch open, or out of adjustment. C. Improper capacity or weak battery or corroded battery terminals. D. Faulty starter solenoid. E. Moisture in electric starter motor. F. Broken or worn brushes in starter motor. G. Faulty fields. H. Faulty armature. I. Broken wire in harness or connector J. Faulty starter key, push button or safety switch. K. Worn or frayed insulation.
5. Electric starter does not engage but solenoid clicks.	A. Loose or corroded connections or ground. B. Weak battery. C. Faulty starter solenoid. D. Broken wire in electric harness. E. Loose or stripped post on starter motor. F. See steps in number 4.
6. Hard to start or won't start.	A. Empty gas tank. B. Gas tank air vent not open. C. Fuel lines kinked or severely pinched. D. Water or dirt in fuel system E. Clogged fuel filter or screens. F. Motor not being choked to start. G. Engine not primed -- pump primer system. H. Carburetor adjustments too lean (not allowing enough fuel to start engine). I. Timing and synchronizing out of adjustment. J. Manual choke linkage bent -- auto choke out of adjustment. K. Spark plugs improperly gapped, dirty or broken. L. Fuel tank primer inoperative (pressurized system). M. Ignition points improperly gapped, burned or dirty or triggering (CD) system inoperative. N. Loose, broken wire or frayed insulation in electrical system. O. Reed valves not seating or stuck shut. P. Weak coil or condenser.

Figure 10–17   Outboard troubleshooting (continued). *(Courtesy of National Marine Manufacturers Association)*

MOTOR REACTION	CHECK POINTS

	Q. Faulty gaskets.
	R. Cracked distributor cap or rotor or shorted rotor.
	S. Loose fuel connector.
	T. Amplifier (CD) inoperative.
	U. Poor engine or ignition ground.
	V. Faulty ignition or safety switch.
7. Low speed miss or motor won't idle smoothly and slowly enough.	A. Improper fuel/oil mixture.
	B. Timing and synchronizing out of adjustment.
	C. Carburetor idle adjustment (mixture lean or rich).
	D. Ignition points improper (gap, worn or fouled) or triggering (CD) system inoperative.
	E. Weak coil or condenser.
	F. Loose or broken ignition wires.
	G. Loose or worn magneto plate.
	H. **Spark plugs (improper gap or dirty incorrect plug).**
	I. Head gasket, reed plate gasket (blown or leaking).
	J. Reed valve standing open or stuck shut.
	K. Plugged crankcase bleeder, check valves, or lines.
	L. Leaking crankcase halves.
	M. Leaking crankcase seals (top or bottom).
	N. Exhaust gases returning thru intake manifold.
	O. Poor distributor ground.
	P. Cracked or shorted distributor cap or rotor.
	Q. Fuel pump diaphragm punctured.
	R. Accessory tachometer shorted or not compatible with ignition system.
	S. Faulty ignition or safety switch.
8. High speed miss or intermittent spark.	A. Spark plugs improperly gapped or dirty.
	B. Loose, leaking or broken ignition wires.
	C. Breaker points (improper gap or dirty; worn cam or cam follower) or triggering (CD) system faulty.
	D. Weak coil or condenser.
	E. Water in fuel.
	F. Leaking head gasket or exhaust cover gasket.
	G. **Incorrect spark plug.**
	H. Engine improperly timed.
	I. Carbon or fouled combustion chambers.
	J. Magneto, distributor or CD triggering system poorly grounded.
	K. Distributor oiler wick bad.
	L. Accessory tachometer shorted or not compatible with ignition system.
	M. Faulty ignition or safety switch.
9. Coughs, spits, slows.	A. Idle or high speed needles set too lean.
	B. Carburetor not synchronized.
	C. Leaking gaskets in induction system.
	D. Obstructed fuel passages.
	E. Float level set too low.
	F. Improperly seated or broken reeds.
	G. Fuel pump pressure line ruptured.
	H. Fuel pump (punctured diaphragm), check valves stuck open or closed, fuel lines leak.
	I. Poor fuel tank pressure (pressurized system).
	J. Worn or leaking fuel connector.

Figure 10–17 Outboard troubleshooting **(continued)**. *(Courtesy of National Marine Manufacturers Association)*

General Guide to Electronic Ignition Troubleshooting

Motor Reaction

Low speed miss   High speed miss
Won't start   Limited RPM

Ignition Systems

Make visual inspection and lubricate all system wiring connectors with isopropyl alcohol or dielectric silicon compound before testing. Test complete system!

Battery CD Ignition System—Mercury 3, 4, and 6 cylinder.

1. Load test and check battery voltage.

2. Check operation of ignition switch circuit—On/Off.

3. Test voltage input at switch box.

4. Check primary input voltage to coil.

5. Test coils for high voltage output.

6. Check spark plug leads and spark plug condition.

ADI Ignition—Mercury (6–25 HP)

Keep switch boxes grounded to engine at all times!

1. Check switch box stop circuit.

2. Check coil primary voltage at low and high speeds.

3. Test stator voltage at low and high speeds.

4. Check coils for secondary voltage output.

5. Check spark plug leads and spark plug condition.

CD II Ignition—OMC (40–50 HP)

1. Check stator and charge coil assembly while cranking.

2. Check sensor coil output while cranking the engine.

3. Check output of power coil.

4. Check power pack for charge coil and output voltage.

5. Check stop circuit which grounds power pack output.

6. Test ignition coils primary/secondary windings and for power output.

7. Check spark plug leads and spark plug condition.

Figure 10–18   General guide to electronic ignition troubleshooting.

KNOW THESE PRINCIPLES OF OPERATION

- Basics of a tune-up.
- Purpose of *sync* and *link*.
- Why a test wheel is used.
- Dynamometer testing.
- Effects of altitude on performance.
- How to clean marine growth.
- Indications of normal and abnormal combustion.

REVIEW QUESTIONS

1. An acceptable maximum compression variation between cylinders would be
   a. 10 percent
   b. 15 percent
   c. 20 percent
2. Why should cylinder heads be removed during tune-up?
   a. to inspect for piston damage
   b. to remove carbon
   c. Both a and b are correct, as damaged pistons may not show up during a compression test.
3. The throttle control changes ignition timing and the amount of throttle opening, during low RPM operation (below 1000 RPM).
   a. true
   b. false
4. For OMC 50 through V-6 timing/synchronization procedure, the maximum timing is set before the throttle pickup point adjustment is made.
   a. true
   b. false
5. For Mercury 50 HP timing/synchronization adjustments, the secondary pickup is adjusted after the maximum timing is set.
   a. true
   b. false
6. If the sequence given in the service manual for timing/synchronization adjustments is not followed, timing and/or carburetion will be off, and major damage will occur to the powerhead.
   a. true
   b. false
7. Incorrect propeller pitch will allow the powerhead to run out of the recommended _____ _____.
8. An OMC test wheel is used to determine if the powerhead will operate to the designed _____ .

9. If the outboard is going into storage, the outboard should be
   a. run in water until the carburetors are dry.
   b. fogged with rust-preventive oil through the carburetor(s).
   c. Both a and b are correct.
10. The power produced by the outboard is dependent upon the _____ of air the powerhead consumes.
11. Summer conditions of high temperature, high humidity, and low barometric pressure reduce powerhead performance.
    a. true
    b. false
12. If powerhead RPM is lost during the high-humidity summer months, propeller pitch should be changed to regain lost RPM due to the effects of the weather. However, MPH (miles per hour) probably will not be regained.
    a. true
    b. false
13. If boating is to be consistently done in altitudes above 2,500 feet, the float setting in the carburetor should be lowered.
    a. true
    b. false
14. A loss of performance and operating economy may be blamed on
    a. a hook not designed in the boat hull.
    b. a rocker.
    c. marine fouling, plus a and b.
15. A tune-up results in an outboard set to specifications and performing in the upper one quarter of the operating range.
    a. true
    b. false
16. The power in an internal combustion engine is developed by _____ _____ resulting from the burning of the air-fuel charge.
17. With proper air-fuel mixture, timing/synchronization is correct and the anti-knock (octane rating) quality of the fuel meets the powerhead requirements. Therefore the burning process should occur
    a. with abnormal combustion.
    b. with preignition.
    c. evenly and steadily.
18. The two flame fronts meet in the combustion chamber and the resulting explosion applies extreme hammering pressures on the piston and other engine parts. This is the definition of
    a. detonation.
    b. preignition.
    c. normal combustion.

19. The ignition of the air-fuel charge prior to the timed spark. This is the definition of
   a. detonation.
   b. preignition.
   c. normal ignition.
20. Usually the result of one or a combination of the following conditions: ignition timing is too far advanced, lean mixture, compression ratio, increase because of carbon buildup, lugging the powerhead.
   a. Increased RPM above the operating range
   b. Normal combustion
   c. Detonation
21. Preignition is caused by incandescent combustion chamber deposits, hot spots, scuffing, spark plug heat range too high, and conditions leading to detonation.
   a. true
   b. false
22. A surface gap spark plug is a plug which, by design and construction, delivers a spark that travels from the center electrode to the shell across the surface of the ceramic insulator.
   a. true
   b. false
23. Normally the spark of a surface gap spark plug should rotate to a different location across the insulator surface.
   a. true
   b. false
24. All surface gap spark plugs are _____ types, and are virtually preignition free.
   a. "hot" (in heat range)
   b. "cold" (in heat range)
   c foul proof
25. Surface gap spark plugs may be used for greater performance in standard ignition systems.
   a. true
   b. false
26. Normal color for a surface gap spark plug is light tan or grey-colored deposits.
   a. true
   b. false
27. Wet fuel-oil deposits can be caused by "drowning" with raw fuel mix during cranking, rich carburetion, or improper fuel-oil ratios. Weak ignition can also contribute to this condition. This is the definition of a _____ spark plug.
   a. low temperature fouling
   b. carbon tracking
   c. cold-fouled

28. When the spark plugs are removed and there is evidence of aluminum throw-off on the plugs, this indicates that _____ has occurred.
    a. detonation
    b. preignition
    c. cold running
29. Heat range refers to the plug's ability to transfer heat from the firing tip of the _____ to the _____ of the powerhead.
30. A "hot" plug has a _____ insulator nose and gets heat away more slowly from the firing end.
    a. longer
    b. shorter
    c. colder
31. What type of lubricant is used in electrical connectors?
    a. 2-4-C marine lubricant
    b. vaseline
    c. dielectric grease/isopropyl alcohol
32. What can you use to clean marine fouling on the lower unit and boat bottom?
    a. dishwashing soap
    b. corrosion guard
    c. bleach

11

Midsection/Lower Unit

Objectives

After studying this chapter, you will know

- What the mid-section and lower unit consist of.
- How the outboard is mounted.
- How water cools the exhaust housing.
- Causes for uneven drive shaft spline wear.
- What drives the water pump impeller.
- What three gears are always rotating.
- Service and pressure testing of the lower unit.
- Outboard jet operation.
- Outboard shift system.

11.1   Exhaust Housing

In the outboard's mid-section, the transom brackets (stern brackets) mount to the transom and, in turn, are mounted to the steering bracket. This provides a means of turning the outboard for steering purposes. A friction device may be located in this bracket to apply a slight drag against turning. This will permit a release of the steering handle during trolling.

The exhaust housing mounts to the steering bracket through rubber mounts. In this way, all powerhead and drive vibrations are dampened out. This prevents vibrations from being transmitted into the steering, transom brackets, and on into the boat. In essence, the exhaust housing floats in the rubber motor mounts. The exhaust housing is equipped with an exhaust relief hole that relieves exhaust pressure upon initial start-up. This relief hole also serves as an exhaust water discharge during operation. The main exhaust is carried completely through the housing and is released below the anti-ventilation plate or through the lower unit and out through the exhaust passage of the propeller when underway. There may be a contoured inner exhaust housing to increase flow of exhaust and exhaust scavenging throughout the power range. This will increase performance (Figures 11–1 and 11–2). The inner housing may be surrounded by water, cooling the housing and effectively quieting the exhaust. The water will drain out of the housing when the outboard is stopped or tilted out of the water.

Another feature found on the housing is the spray flange deflector plate. It is found near the level of the flange between the lower unit and the driveshaft housing. It extends from about the center line of the driveshaft on either side of the unit around the front, usually an inch or more in width. This spray flange stops the spray which is deflected up behind the leading edge of the lower unit (Figure 11–2). Without the spray flange, water could go up high enough to wet the operator at the tiller handle in the rear seat. The spray flange knocks this spray down into a horizontal direction.

Because the exhaust housing supports the lower unit, it is susceptible to all the forces transmitted by the lower unit. This includes the drive forces and forces created by impact upon the lower unit when struck by a submerged object. The exhaust housing can be sprung, if the impact is heavy enough. This may be very subtle but nevertheless there will be evidence of the impact when the lower unit is removed after some hours of operation. When the lower unit is removed for water pump service or resealing, you may notice uneven spline wear on the driveshaft splines and crankshaft splines (Figure 11–3).

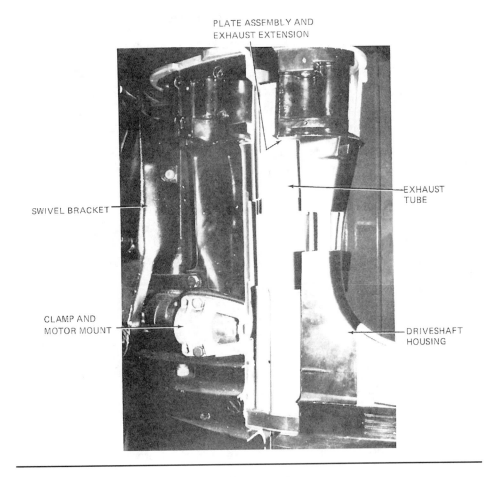

Figure 11-1   Cutaway of Mercury midsection.

If the splines are worn *unevenly*, this is evidence of a damaged exhaust housing or lower unit housing, or that both housings are bent. The exhaust housing can be checked using special tools to measure housing ends, to see that they are parallel. If the housing is bent, then a replacement is necessary. If the lower unit housing is bent, generally it can be seen and will have to be replaced.

11.2   Adapter

The adapter (lower motor cover) is sandwiched between the powerhead and the exhaust housing. It provides a means of mounting the fuel line connector, steering handle, and throttle assembly on small outboards (Figures 11-4 and 11-5).

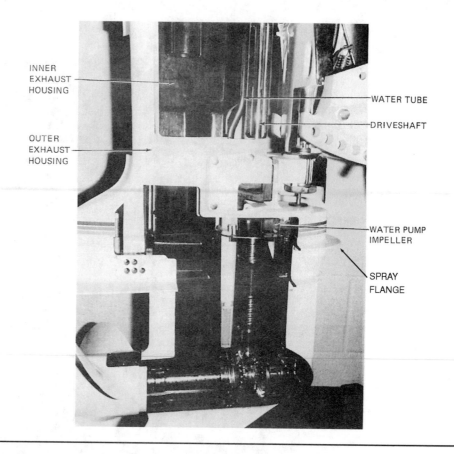

Figure 11–2   Cutaway of Outboard Marine Corporation exhaust housing.

The steering handle (tiller) is generally able to fold up for storage and transportation. Certain models may have an ignition stop button and lanyard located towards the end of the steering handle. The larger outboards have a mounted fuel line connector, electrical harness, and remote control cables passing through the adapter. The upper motor cover attaches to the adapter and provides silencing and safety.

11.3   Operating Principles of the Lower Unit

The lower unit consists of the driveshaft, water pump, pinion gear, bearings, forward and reverse gears, propeller shaft, a shift mechanism, and a housing. The lower unit housing is bolted to the exhaust housing, which places the driveshaft and water tube(s) through the center of the exhaust housing (Figures 11–6 and 11–7). The water tube carries water from the water pump to the powerhead. The driveshaft splines insert into the crankshaft ac-

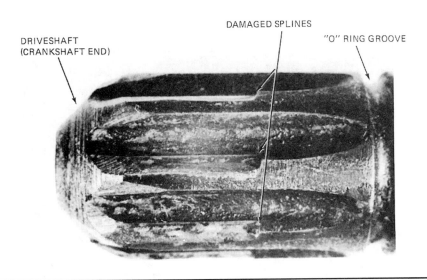

Figure 11–3   Wear-damaged driveshaft splines.

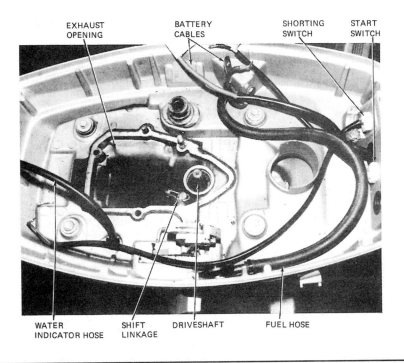

Figure 11–4   Outboard Marine Corporation lower motor cover (adapter).

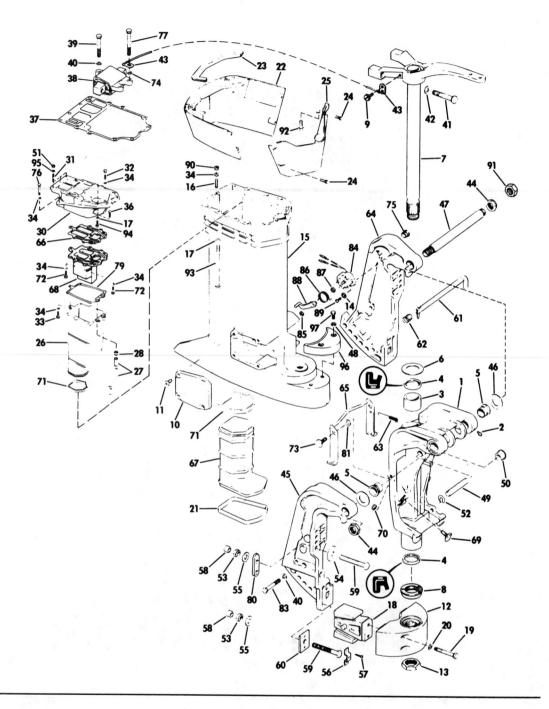

Figure 11–5   Midsection (continued). *(Courtesy of Outboard Marine Corporation)*

OP0924C

Ref. No.	P/N	Name of Part	Qty.	Ref. No.	P/N	Name of Part	Qty.
1	389335	SWIVEL BRACKET	1	53	321861	NUT, Stern bracket screw	4
2	313607	. LUBRICATION FITTING	4	54	307238	WASHER, Nut	2
3	320962	. BUSHING	1	55	320248	WASHER	4
4	320937	. SEAL	2	56	318273	RETAINER	2
5	321035	BUSHING, Swivel bracket	2	57	319886	SCREW	4
6	320866	THRUST WASHER, Pivot shaft	1	58	318572	CAP, Mounting screws	4
7	392518	STEERING ARM ASSY.	1	59	313327	SCREW, Brkt. to transom, 4"	4
8	327696	. KEEPER, Pivot shaft	1	59	326614	SCREW, Brkt. to transom, 4-1/2"	
9	510193	SCREW, Ground lead	1			STD	4
10	327801	COVER, Mount bracket lower	2	59	321577	SCREW, Bracket to transom, 6"	4
11	322000	SCREW, Cover	8	60	318272	PLATE	2
12	327051	MOUNT BRACKET	1	61	387278	THRUST ROD ASSY.	1
13	320746	NUT, Mount bracket	1	62	321053	SPRING, Thrust rod	1
14	313396	WASHER, Sending unit screw	2	63	323142	SPRING, Trail lock	1
15	392144	EXHAUST HOUSING, 2.6 litre	1	64	392203	STERN BRKT. ASSY., Port	1
16	314956	. STUD, 2.6 litre	1	65	390442	LOCKING LEVER, Rod	1
17	306470	WASHER, Exhaust housing	8	66	323292	GASKET, Adapter to megaphone, 2.6	
18	391774	RUBBER MOUNT, Lower Hi Perf.				litre	1
		green	2	67	327796	EXHAUST HOUSING, Inner lower	1
19	328078	SCREW, Mount	4	68	329789	MEGAPHONE, 2.6 litre	1
20	306314	LOCKWASHER, Mount	4	69	321499	THRUST PAD	2
21	320961	SEAL, Gearcase to inner housing	1	70	315077	LOCKNUT	2
22	390131	COVER, Exhaust housing, rear	1	71	320936	SEAL, Exhaust housing	1
23	318258	. SEAL, Exhaust housing cover	1	72	306418	SCREW, Megaphone, front	6
24	321527	SCREW, Front to rear cover	4	73	321389	SHOULDER SCREW	2
25	323800	COVER, Exhaust housing, front	1	74	307161	LOCKWASHER, Mount	1
26	329788	EXH. HSG., Inner upper, 2.6 litre	1	75	321243	BUSHING, Stern bracket	1
27	321016	WATER TUBE	1	76	302030	SCREW, Hsg. to adapter, 2.6 litre	2
28	305145	GROMMET, Water tube	1	77	317222	SCREW	2
30	392891	ADAPTER, Exh. hsg., 2.6 litre	1	79	323222	GASKET, Inner exh. hsg. to plate	1
31	301185	. STUD, 2.6 litre	3	80	321735	SUPPORT, Stern bracket	4
32	303459	SCREW, Adapter to inner housing	3	81	552993	WASHER, Screw to bracket	2
33	306778	SCREW, Megaphone rear	1	83	552968	SCREW, Manifold to stern bracket	
34	302290	LOCKWASHER	AR			1-7/16" long	4
36	306409	SCREW, Adapter to housing, outer	2	83	318612	SCREW, Stern bracket to manifold	
37	323214	GASKET, Adapter to powerhead	1			2-3/4" long	2
38	387098	RUBBER MOUNT ASSY., Upper	1	84	582153	SENDING UNIT ASSY.	1
39	301431	SCREW, Upper mount	1	85	511527	. RETAINER	1
40	307708	LOCKWASHER, Mount screw	9	86	511528	. SPRING	1
41	316406	SCREW, Upper mount	2	87	320689	. NUT	1
42	313169	LOCKWASHER, Mount screw	2	88	511530	. ARM	1
43	392520	LEAD ASSY., Ground	1	89	319682	SCREW, Sending unit to bracket	2
44	316828	NUT, Tilt tube	2	90	306338	NUT, Stud, exh. hsg., 2.6 litre	1
45	392204	STERN BRACKET, Starboard	1	*91	322376	WIPER NUT, Tilt tube to cable	1
46	327699	THRUST WASHER, Stern bracket	2	92	321527	SCREW, Lower cover, front	2
47	321052	TILT TUBE	1	93	908931	SCREW, Exhaust housing	6
48	307555	LOCKWASHER	2	94	302479	SCREW, Adapter to powerhead	2
49	320792	PIN, Tilt cylinder	1	95	309192	WASHER, Manifold to adapter, 2.6	
50	322481	BUSHING, Tilt cylinder	2			litre	3
51	306422	NUT, Manifold to adapter, 2.6 litre	3	96	392123	ANODE, Exhaust housing	1
52	321054	SPRING CLIP	2	97	328049	SCREW, Anode	2

* Recommended to prevent corrosion in salt water usage.

Figure 11–5   **Midsection continued.** *(Courtesy of Outboard Marine Corporation)*

cepting the torque of the crankshaft. Power flow in the manual lower unit goes through the driveshaft into the pinion gear, which constantly turns the forward and reverse gears in opposite directions. The clutch dog is part of the manual shift mechanism and is splined to the

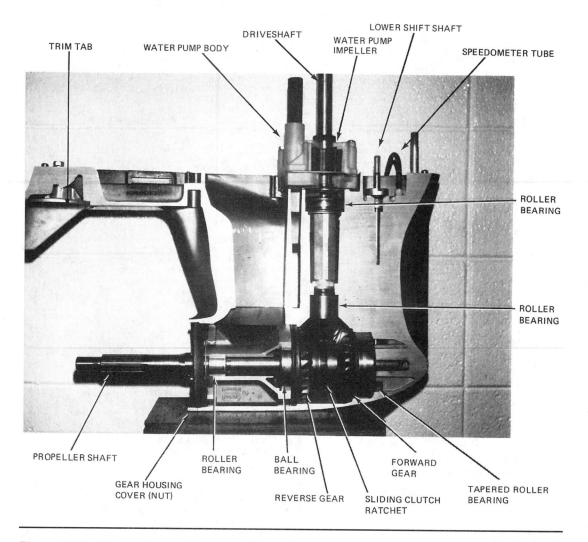

Figure 11–6 Mercury lower unit.

propeller shaft. The clutch dog is held in the central position (neutral) between the forward and reverse gears. When the shift shaft (rod) is moved, the shift cam (shifter) moves the follower (shift shaft), which, in turn, moves the clutch dog into mesh with the selected gear. Power is then transmitted from the gear through the clutch dog into the propeller shaft, and finally on to the propeller (Figures 11–8, 11–9 and 11–10).

To determine the gear ratio for a given lower unit, you first need to know the number of teeth on the pinion (driving) gear and the forward (driven) gear. Let's say that the pinion gear has 14 teeth. The forward gear has 26 teeth. The torque increase is equivalent to the

WATER TUBE
INSIDE MID-SECTION   WATER PUMP

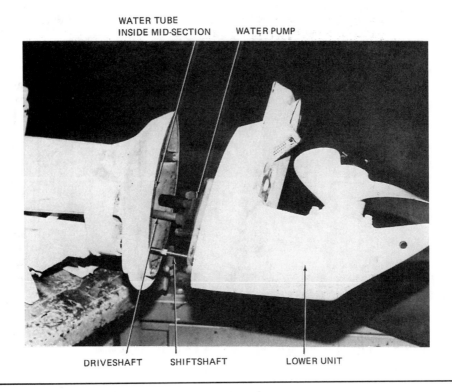

DRIVESHAFT   SHIFTSHAFT   LOWER UNIT

Figure 11–7   Removing lower unit exposing driveshaft, shift shaft, and water pump.

forward gear tooth number (26 divided by 14 = 1.85 to 1). Some of the sales brochures list the pinion gear and forward gear teeth numbers instead of the actual ratio. As an example one manufacturer lists a 14:26 gear ratio for a 225 HP and an 11.23 for a 15 HP model.

11.4   Lower Unit Housing Design

The lower unit drive train receives torque from the powerhead and transfers it to the propeller, delivering a thrust to propel the boat. The gear case housing must be strong enough to accept the pressures, but also have a design that will allow the water to move smoothly around the housing. The torpedo-type design offers little resistance to the flow of water. Water flowing by the unit should not disturb the water the propeller will be meeting. This is critical at high speeds so cavitation and ventilation will not be a problem.

Another consideration is how far down into the water the unit should be placed. This all depends on application: work, high-performance, or recreational boating. The lower unit needs to extend down far enough to allow the propeller to work and not draw air from the surface; however, it must not be too far down to create an unacceptable drag in the water. If it is too far down into the water, speed and efficiency will be lost.

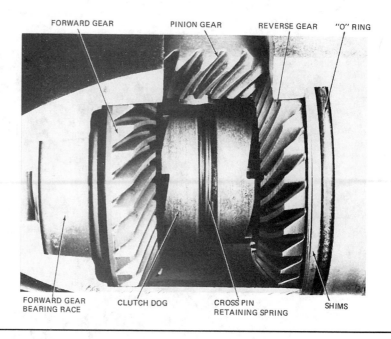

Figure 11–8   Neutral gear. *(Courtesy of Mercury Marine)*

Figure 11–9   Clutch dog moved into forward gear. *(Courtesy of Mercury Marine)*

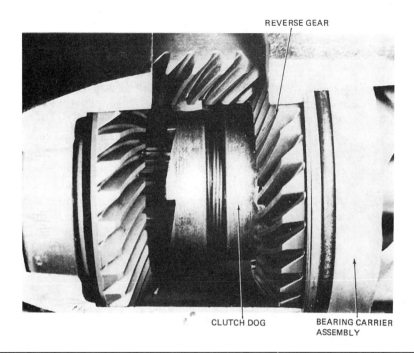

Figure 11–10   Clutch dog moved into reverse gear. *(Courtesy of Mercury Marine)*

There are normally three sizes of transom heights available—15-, 20-, and 25-inch. To accommodate these transoms, manufacturers produce a short shaft (driveshaft), a long shaft, and an extra long shaft model, enabling a dealer to install the proper length outboard on the transom. This, in turn, will cause the anti-ventilation plate to be positioned at or slightly above the bottom of the boat. With proper trim adjustment, the anti-ventilation plate should be running parallel to and just below the water surface. With the use of a high rake and cup propeller, the anti-ventilation plate can run above the water surface. The portion of the lower unit which is in the water acts as a rudder. It is aided by directional thrust from the propeller and is very efficient at low speeds, making maneuvering easy.

11.4.1   Troubleshooting the Lower Unit

The lower unit is basically trouble-free until there is an impact, water enters the gear case, the operator shifts incorrectly, or the oil is not changed regularly and corrosion gets into the unit. Because the lower unit is normally underwater, extra care and constant maintenance must be taken to prevent these problems.

Shifting the unit in and out of gear needs to be quick and positive to prevent rounding over of the clutch dog and/or ratchet teeth. Slow engagement will damage the parts. This problem is evident when the lower unit jumps out of gear. Operators of the outboard need to be instructed on how to properly shift the unit.

Service manuals give a schedule indicating when to change the lower unit gear oil. This schedule should be followed exactly. When oil is drained, it gives you a chance to see if the lower unit is taking on water and therefore needs to be resealed. A milky-looking oil indicates that water has entered the lower unit (Figure 11–11).

Replacement of gear oil is not the answer. The lower unit must be pressure/vacuum-tested to locate leaks and repaired, otherwise bearings and gears become pitted and rusted (Figures 11–12, 11–13, and 11.14).

When the boat is moored, the lower unit should be tilted out of the water. If the unit is left in the water, corrosion accelerates and marine growth accumulates on the unit and will probably restrict the inlet water passage. It is fairly common to have electrolysis present in marina waters, caused by a poor marina electrical system. This, along with the dissimilar

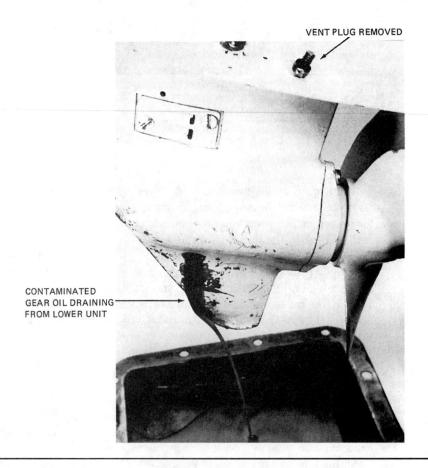

VENT PLUG REMOVED

CONTAMINATED GEAR OIL DRAINING FROM LOWER UNIT

Figure 11–11   Draining lower unit gear oil.

Figure 11–12   Pressure-testing the lower unit.

Figure 11–13   Thrust washers damaged by water (rust marks).

Figure 11–14   Pinion and driveshaft bearings damaged by water (rust-pitted).

metals of the lower unit and the salt or brackish water, really accelerate corrosion on the lower unit. Corrosion damage can become very expensive to repair.

 Note

When the outboard is taken from saltwater it *should be flushed for at least five minutes, running with freshwater.*

This will help control the corrosion inside the water passages of the outboard. The outboard should also be washed externally each time it is removed from saltwater.

Occasionally you see external damage to the lower unit. This may be caused by hitting a submerged object or from being struck while trailered. Care should be exercised to prevent this type of damage. Casting replacement becomes very expensive (Figure 11–15).

11.5   The Shift System

This is a safety system. What the operator does at the helm or control lever must happen at the propeller. This system makes it possible to shift the lower unit and therefore the propeller into forward, neutral, and reverse. Each time there is a shift, there is sudden resistance or shock to the drive train, caused by the bite of the propeller. A rubber propeller hub may be used to cushion and absorb this shock. Of course, this type of hub also helps when a submerged object is struck.

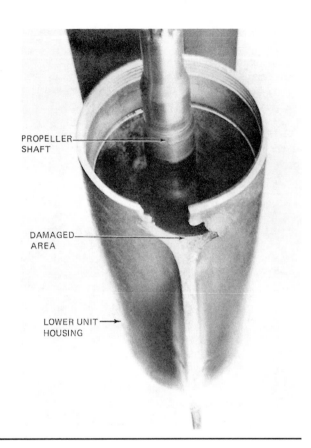

PROPELLER
SHAFT

DAMAGED
AREA

LOWER UNIT →
HOUSING

Figure 11–15   External
damage to the lower unit.

There have been many types of shifting systems in lower units over the years. The shift systems are classified as mechanical, hydroelectric, hydro-mechanical or electromatic. All electrical start units will use a neutral start switch to prevent starting in gear (see Chapter 6, *Starter System*). This text will explain only the mechanical shift system.

11.5.1   Mechanical Shift System

This system is manually operated by the use of a cable from the helm control or shift lever on the lower motor cover. The cable or shift lever moves an upper shift shaft, which, in turn, rotates the lower shift shaft (Figure 11–16, #11), shift cam #35, moves the cam follower #44, and then the clutch #47 into mesh with the gear selected, either #42 or #51. The pinion gear (#9), forward, and reverse gears are rotating whenever the powerhead is running. The clutch must be *shifted quickly* and with *positive force* to prevent rounding over the corners of the clutch. There is no mechanism to ease the engagement of the clutch in either forward or reverse gears (Figure 11–16).

1 - Gear Housing
2 - Drain Screw
3 - Screw
4 - Washer
5 - Dowel Pin (Hollow)
6 - Dowel Pin (Solid)
7 - Pipe Plug
8 - Roller Bearing
9 - Pinion Gear
10 - Nut
11 - Shift Shaft
12 - Retaining Ring
13 - Bushing
14 - O-ring
15 - Oil Seal
16 - Drive Shaft
17 - Shim Set
18 - Tapered Roller Bearing
19 - Nut
20 - Water Pump Base
21 - O-ring
22 - Oil Seal
23 - Oil Seal
24 - Face Plate
25 - Impeller
26 - Key
27 - Water Pump
28 - Gasket (Lower)
29 - Gasket (Upper)
30 - Seal
31 - Screw
32 - Washer
33 - Insulator
34 - Insulator
35 - Shift Cam
36 - Spacer
37 - Coupler
38 - Plate
39 - Seal
40 - Shim Set
41 - Tapered Roller Bearing
42 - Forward Gear
43 - Roller Bearing
44 - Cam Follower
45 - Guide
46 - Spring
47 - Clutch
48 - Cross Pin
49 - Spring
50 - Propeller Shaft
51 - Reverse Gear
52 - Ball Bearing
53 - Bearing Carrier
54 - O-ring
55 - Roller Bearing
56 - Oil Seal
57 - Oil Seal
58 - Screw
62 - Thrust Hub
63 - Propeller Nut
64 - Tab Washer

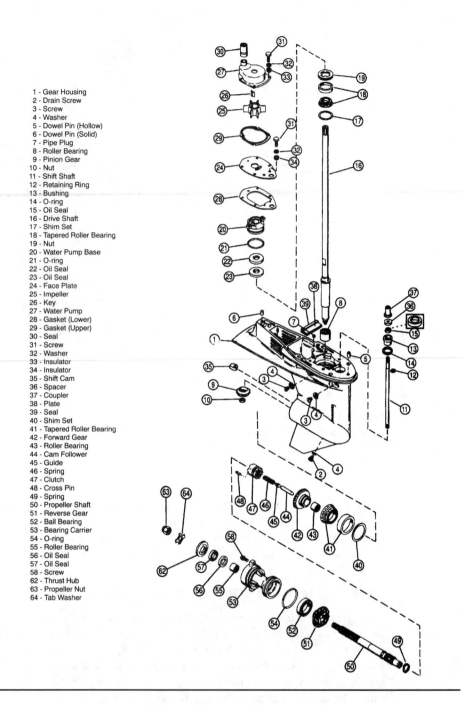

Figure 11–16 Gear case components for 50/55/60 HP (3 stroke engine with 3 carbs).
(Courtesy of Brunswick Corporation)

 **Warning**

Shifting *should not* be attempted when the throttle is beyond the shift position. Older models do not have a control to prevent shifting at higher RPM. Newer models incorporate a starter lock out that prevents starting in gear and at the higher RPM settings (beyond the start position).

Probably the most common problem with the manual shift is damage done to the clutch (lugs) and their adjoining gears. Slow, easy shifting takes a toll, rounding over the lugs. To avoid this, make quick positive shifts at idle speed each time. If shifting has become hard, possibly the shift linkage is bent or it may be seized up. Try lubrication and then check the shift shaft adjustment. If there is a jumping out of gear, then the adjustment is not allowing full engagement of the clutch into the gear. The clutch lugs and their mating gear surfaces may be rounded over. (Move shift and throttle levers occasionally while the boat is in storage.)

When the manual shift is made from the helm, there can be increased resistance to the shift caused by cables binding or dragging excessively within the cable housing. There is a marine lubricant that can be applied to the control and ends of these cables. Cable adjustments and mountings in the lower motor cover should be checked. If cable action does not have that good smooth feeling, check the routing and turns of the cable. The speed control cable adjustment may also affect the shift cable operation. These cables must work smoothly in all trim/tilt and turn positions.

If the lower unit is put into gear, you expect to move by propeller thrust. The propeller is the final drive unit. If there is damage to the propeller rubber hub, it may slip, causing reduced or no forward movement. This hub can be checked out using an adapter and a torque wrench.

At the back and on the inside of many lower units (bullet), there are hydrostatic rings. These are not threads for the cover nut, but are used to help control the exhaust gas flow leaving the lower unit. If there is a large chip out of these rings (housing broken out), then exhaust gas can flow over the outside of the propeller hub (Figures 11–15 and 11–17). This allows exhaust gas bubbles into the propeller blades causing a ventilation condition, similar to the propeller breaking the surface of the water, and causing powerhead speed-up. Thrust is reduced when these bubbles are in the water as the blade enters. To make an effective repair to these hydrostatic rings, the housing has to be replaced or the housing may possibly be welded, rebuilding the final part of the broken housing.

11.5.2   Gear Case Lubrication

Within the gear case there are needle bearings, ball bearings, and gears that need a special type of gear oil to survive in the marine environment. Probably the highest stress load for the lubricant to carry is the sliding contact between gears. The shear load is heavy and the gear oil is formulated to take the load. Also it must be compatible with water, in case of water leakage into the gear case. When water has entered into the oil, the gear oil looks milky. If this color is noticed during routine service. the oil should be changed, at the very

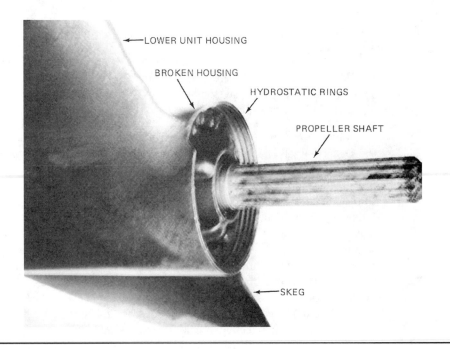

LOWER UNIT HOUSING

BROKEN HOUSING

HYDROSTATIC RINGS

PROPELLER SHAFT

SKEG

Figure 11–17   Lower unit with broken hydrostatic rings.

least. The correct repair is to reseal the gear case. The gear case oil should be checked each 50 hours of operation; drained and refilled each 100 hours of operation when operated in fresh water; and drained and refilled each 50 hours of operation when operated in saltwater. Check the service manual for service procedures (Figures 11–11, 11–18 and 11–19).

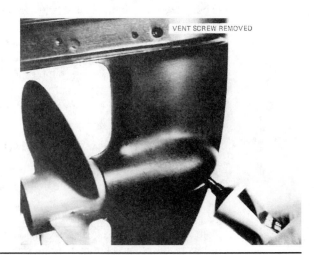

VENT SCREW REMOVED

Figure 11–18   Filling lower unit with gear oil.

Figure 11–19 Types of gear oil.

11.6   Jet Propulsion

There is an outboard jet available that enables a boat to operate safely in shallow water areas that are inaccessible to conventional propellers. The jet propulsion unit attaches directly to the exhaust housing in the same manner as the conventional lower unit. The powerhead drives a "jet" pump that is attached to the lower end of the drive shaft. The pump draws water through a grilled inlet from beneath the jet unit and expels it aft through a nozzle. This thrusts the boat forward. This action is similar to an airplane jet engine ejecting air. There is no gear box to leak or cause shifting problems, and no propeller to be damaged (Figure 11–20).

The jet unit is mounted with the leading edge of the water intake even with the bottom of the boat (see Figure 11–21).

11.6.1   Steering

Without a propeller churning away underwater, there is no torque resistance and turns are equally easy in either direction. Steering is accomplished by directing the nozzle from side to side. Reverse is accomplished by lowering a reverse gate (deflector), which deflects the jet stream forward under the boat. In the neutral position, the reverse gate directs equal thrust of water forward and aft.

The jet unit is capable of planing a light shallow draft hull with a relatively flat wide bottom, in six inches of water. It can idle through water that is only one foot deep. If the boat runs aground, the jet will act like a dredge and suck up sand and gravel from the river bed. This will shorten the life of the pump. No attempt should be made to power the boat out of a grounding. Instead, shut off the motor quickly and drag the boat off.

There is a trade-off using the jet drive. Horsepower is reduced by as much as 50 percent on the smaller jet drives. Thirty percent loss can be expected on the larger horsepower out-

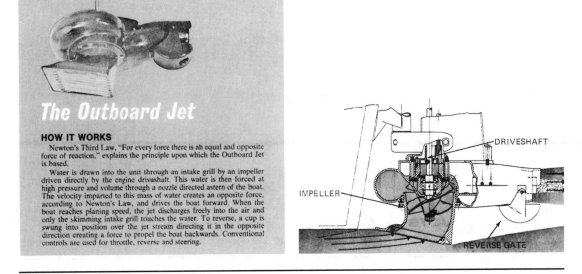

The Outboard Jet

HOW IT WORKS

Newton's Third Law, "For every force there is an equal and opposite force of reaction," explains the principle upon which the Outboard Jet is based.

Water is drawn into the unit through an intake grill by an impeller driven directly by the engine driveshaft. This water is then forced at high pressure and volume through a nozzle directed astern of the boat. The velocity imparted to this mass of water creates an opposite force, according to Newton's Law, and drives the boat forward. When the boat reaches planing speed, the jet discharges freely into the air and only the skimming intake grill touches the water. To reverse, a cup is swung into position over the jet stream directing it in the opposite direction creating a force to propel the boat backwards. Conventional controls are used for throttle, reverse and steering.

Figure 11–20   The outboard jet. *(Courtesy of Specialty Manufacturing Co.)*

Figure 11–21   Outboard jet installed.

boards. As with all outboards, you will get maximum mileage by not running at full throttle. Two-thirds throttle greatly improves economy. This trade-off is worth it, if you must go into the shallow waters of a lake or river.

Maintenance on some jets is minimal. Jets need to be lubricated with marine wheel bearing grease after ten hours of continuous operation, or after weekly use. When you add grease, look at the old grease being flushed out through the vent hole. If it contains an abnormal amount of water or is beginning to turn dirty gray, it is time to check the condition of the seals and bearings.

As sand and gravel pass through the pump, performance will drop off. Performance can be regained by filing the impeller vanes back to their original sharpness. When this is done, a shim pack may have to be readjusted to maintain the 1/32-inch clearance required.

The original factory water pump for the outboard cooling system is still used. It is mounted directly over the jet pump. It is recommended that the water pump be serviced yearly. It is advisable to use the chrome pump if it is available. At the same time renew the driveshaft spline lubricant with moly lube before installation.

11.7   Maintaining and Repairing the Lower Unit

Once a season the lower unit needs to be dropped and lubrication on the driveshaft splines renewed. An extreme pressure moly lube is applied directly to the splines. Seals or an O-ring are used around the driveshaft to retain the lubrication and keep the water and exhaust from the splined joint area. This is mandatory service to keep the splines from rusting together. Exhaust pressure and water are both present at the joint seal. If seal failure occurs, water washes the lubricant from the splines and rusting will occur. The rust can be so severe that the two shafts will be rusted together. This will first be known when there is an attempt to drop the lower unit and it will not separate from the powerhead. Now what can be done? First, a journeyman outboard technician needs to do the work. If hammering and prying with long pry bars are decided upon, damage to very expensive housings may occur. These methods may prove effective if the rust at the spline joint is not too bad. If rust has seized the joint, the drive shaft will have to be cut with a saw or a cutting torch. Some drive shafts use stainless steel and others use a good grade of steel. You will know once the cutting process begins. The exhaust housing will need to have a hole made in it, to enable the mechanic to get to the driveshaft to cut it. If this hole is put in with a standard hole saw, an automotive brass core plug can be installed to cover the hole for cosmetic purposes. The best service is to not let it happen. Service the spline joint each season along with a water pump impeller replacement.

When water is found in the lower unit oil, let the milky-looking oil drain out. The unit should then be pressure tested with air to 16–18 PSI (110–124 kPa) and then submerged under water or sprayed with soap suds (Figure 11–12). Bubbles will come from the point of leakage. Also the unit should be subjected to 3–5 inches (76–127mm) of mercury and see if vacuum holds. If it holds, then pump vacuum out to 15 inches (381mm) of mercury and check. This will stress the seals in the opposite direction. The pressure/vacuum tests should hold for a few minutes.

To make repairs, the shafts and housings are cleaned up and examined, and special attention to the area of leakage can be exercised. It is common for grit in the water to wear the shafts at the seal contact points or corrosion to erode away some of the housing. An examination will determine if the mating parts can be reused with a new seal. Also, examine the gears and bearings. Parts inside the lower unit can become rusted and pitted (Figures 11–22 and 11–23).

After being repaired, the reassembled lower unit should be pressure/vacuum tested again to prove that a total seal was actually obtained.

In repairing the lower unit, several shims are used in different locations to position gears for proper mesh. Shims are also used to preload some of the bearings. The manufacturers have made available shimming gauges to determine the pinion gear's height (position) and

Figure 11–22 Water-damaged, rusted, and corroded shaft, gears, and bearings.

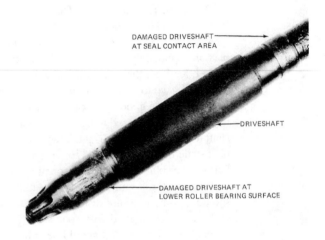

DAMAGED DRIVESHAFT AT SEAL CONTACT AREA

DRIVESHAFT

DAMAGED DRIVESHAFT AT LOWER ROLLER BEARING SURFACE

Figure 11–23 Water-damaged bearing area and rust-pitted seal contact area.

preloading of the bearings. The clearance between pinion and forward or reverse gear is the *backlash*. This is necessary for heat expansion of the gears and to make correct tooth contact (mesh) between pinion, forward and reverse gears.

The lower unit should be pressure-tested after any seals/gaskets are replaced (Figure 11–12). Consult the service manual for the necessary tools, specifications, and procedures.

All bolts that attach the lower unit to the exhaust housing should be coated with a nonhardening sealing compound, such as OMC Gasket Sealing Compound or Mercury Anti-Corrosion grease. This will help control galvanic corrosion between the two dissimilar metals.

Troubleshooting the gear case and remote controls is shown along with the outboard shift system and persons safety in Figures 11–24 through 11–28.

Troubleshooting - All Models

PROBLEM: GEARSHIFT JUMPS OUT OF GEAR

STEP	PROCEDURE	RESULT
1 Check detent balls and spring, clutch dog, gears, and shifter lever.	Remove and disassemble gearcase. Inspect components for wear, breakage or bent conditions.	Replace a bent shifter lever, damaged shifter dog and gears.

PROBLEM: GEARSHIFT WON'T SHIFT

STEP	PROCEDURE	RESULT
1 9.9 - 40 Check pivot pin.	Inspect to see that pivot pin is in place thru shifter lever.	Pivot pin OK – Go to step 2. Pivot pin missing – Replace it.
2 Check shift lever or control cable adjustment.	Follow procedure in Section 6, Adjustments.	Lever OK – Go to step 3. Lever out of adjustment – Adjust.
3 Check shift rod connection.	Inspect for loose or missing screws, disconnected shift rods.	Connector screws OK, shift rods in place – Go to step 4. Missing screw, shift rods disconnected – Connect shift rods and replace screw. Check for stripped threads in connector.
4 Check gearcase components and drive shaft.	Remove and disassemble gearcase. Inspect all components.	Replace worn, corroded, broken or bent components. Replace grommets and seals.

PROBLEM: GEARCASE SEIZED

STEP	PROCEDURE	RESULT
1 Check gearcase for lubricant.	If lubricant is present, drain and disassemble gearcase. Inspect all components for breakage and corrosion.	Replace broken or corroded components. Check for bent gearcase.

PROBLEM: MOTOR DOESN'T HOLD IN TILT POSITION

STEP	PROCEDURE	RESULT
1 Check tilt/run mechanism.	Inspect for broken springs, loose nut, linkage and missing components.	Replace components as necessary.

Figure 11–24   Troubleshooting for all models. *(Courtesy of Outboard Marine Corporation)*

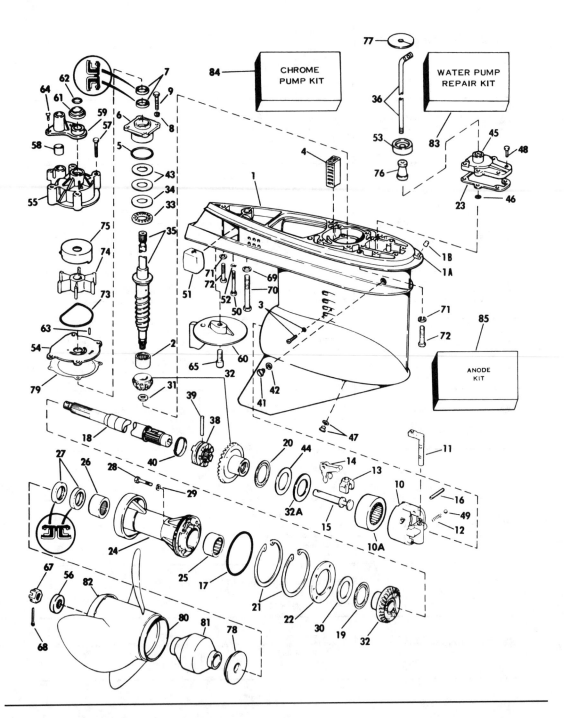

Figure 11–25 Gearcase (continued).

Ref. No.	P/N	Name of Part	Qty.	Ref.No.	P/N	Name of Part	Qty.
*	393459	GEARCASE ASSY., Complete, 2.6 litre	1	45	388618	. COVER & SEAL ASSY.	1
1	393461	. GEARCASE & BEARING ASSY., 2.6 litre	1	46	318372	. . O-RING, Shift rod	1
1A	326266	. . FITTING, Speedometer	1	47	318544	. SCREW, Drain & fill	1
1B	305749	. . PLUG, Fitting	1	*	311598	. . O-RING	1
2	387817	. . BEARING, Pinion	1	48	318139	. SCREW, Cover	6
3	324852	. SET SCREW, Bearing	1	49	160084	. BALL, Detent	1
*	307853	. . . O-RING	1	50	328096	. SCREW, Anode	1
4	321123	. . SCREEN, Water intake	1	51	393023	. ANODE	1
5	314728	. . O-RING, Driveshaft bearing	1	52	303886	. WASHER, Anode	1
6	387245	. BEARING HOUSING & SEAL	1	†53	321008	GROMMET, Exhaust hsg. to shift rod	1
7	321453	. . OIL RETAINER, Driveshaft	2	†54	327782	PLATE, Impeller housing	1
8	301250	. WASHER	4	†55	393508	IMPELLER HOUSING, Gearcase	1
9	316534	. SCREW, Housing to gearcase	4	56	320570	SPACER, Propeller nut	1
10	389455	. HOUSING & BEARING ASSY.	1	†57	323626	SCREW, Impeller housing	4
10A	389039	. . BEARING	1	†58	314008	GROMMET, Water tube	1
11	323946	. DETENT, Shifter	1	†59	321159	COVER, Impeller housing	1
12	312870	. SPRING, Detent	1	60	327810	TRIM TAB, Gearcase	1
13	322900	. CRADLE, Shifter	1	†61	320943	GROMMET, Impeller housing	1
14	322938	. SHIFTER	1	†62	310585	O-RING, Driveshaft to crankcase	1
15	322941	. SHAFT	1	†63	323950	KEY, Driveshaft to impeller	1
16	328825	. PIN, Shift lever	1	†64	306643	SCREW, Cover to impeller	2
17	321119	. O-RING, Propeller shaft	1	65	313715	SCREW, Trim tab	1
18	391472	. PROPELLER SHAFT ASSY.	1	67	320569	NUT, Propeller	1
19	382408	. THRUST BRG. ASSY., Rev. gear	1	68	309955	COTTER PIN, Propeller nut	1
20	389042	. THRUST BRG. ASSY., Fwd. gear	1	69	306314	LOCKWASHER, Screw, long	1
21	321171	. RETAINING RING	2	70	313697	SCREW, Gearcase to housing, long	1
22	321174	. RETAINER PLATE, Brg. hsg.	1	71	307708	LOCKWASHER, Screw, short	5
23	324670	. GASKET, Shift rod cover	1	72	320741	SCREW, Gearcase to housing, short	5
24	393300	. BEARING HSG. SEAL ASSY.	1	†73	321120	SEAL, Impeller housing plate	1
25	382407	. . NEEDLE BEARING ASSY.	1	†74	391538	IMPELLER ASSY.	1
26	387247	. . NEEDLE BEARING ASSY.	1	†75	327781	LINER, Impeller	1
27	320862	. . SEAL	2	76	321122	SPACER, Shift rod	1
28	316563	. SCREW, Bearing hsg.	4	77	322302	SEAL, Shift rod	1
29	317178	. O-RING	1	78	321173	THRUST BUSHING	1
30	314731	. THRUST WASHER, Reverse gear	1	†79	324701	GASKET, Impeller housing plate	1
31	314730	. NUT, Pinion to driveshaft	1	80	387158	PROPELLER, 15-1/2 x 13	1
32	392166	. GEARSET, 2.6 litre (inc. fwd. rev & pinion gr)	1	80	387159	PROPELLER, 15-1/2 x 15	1
32A	327753	. . SHIM, Fwd. gear (.015)	1	80	387160	PROPELLER, 15 x 17	1
32A	327754	. . SHIM, Fwd. gear (.017)	1	80	391200	PROPELLER, 15 x 17	1
32A	327755	. . SHIM, Fwd. gear (.019)	1	80	387161	PROPELLER, 14-1/2 x 19	1
32A	327756	. . SHIM, Fwd. gear (.021)	1	80	391201	PROPELLER, 14-1/2 x 19	1
32A	327757	. . SHIM, Fwd. gear (.023)	1	80	391202	PROPELLER, 14-1/4 x 21	1
32A	327758	. . SHIM, Fwd. gear (.025)	1	80	387163	PROPELLER, 14-1/4 x 23	1
32A	327759	. . SHIM, Fwd. gear (.027)	1	80	389925	PROPELLER, 15 x 16 SST	1
32A	327760	. . SHIM, Fwd. gear (.029)	1	80	389924	PROPELLER, 14-1/2 x 19 SST	1
33	387656	. THRUST BEARING, Pinion	1	80	389923	PROPELLER, 14-1/4 x 21 SST	1
34	327656	. THRUST WASHER, Pinion	1	80	389019	PROPELLER, 14-1/4 x 23 SST	1
35	391380	. DRIVESHAFT ASSY.	1	80	387995	PROPELLER, 14-1/2 x 24 SSTR	1
36	324407	. SHIFT ROD	1	80	387996	PROPELLER, 14-1/2 x 26 SSTR	1
38	910198	. SHIFTER, Clutch dog	1	80	387997	PROPELLER, 14-1/2 x 28 SSTR	1
39	313448	. PIN, Clutch dog	1	80	390750	PROPELLER, 14 x 25 SST	1
40	324369	. SPRING, Clutch dog pin	1	80	390831	PROPELLER, 15-1/2 x 14 SST	1
41	307551	. PLUG, Fill & drain	1	80	391863	PROPELLER, 14-1/4 x 27 SSTRX	1
42	311598	. . WASHER, Nylon	1	80	391864	PROPELLER, 14-1/4 x 29 SSTRX	1
43	314742	. SHIM, .002	AR	80	391290	PROPELLER, 15 x 17 SST II	1
43	323362	. SHIM, .003	AR	81	387157	. BUSHING ASSY., Prop., alum.	1
43	323361	. SHIM, .004	AR	82	321295	. CONVERGING RING, Std. prop.	1
43	314745	. SHIM, .005	AR	83	393082	WATER PUMP REPAIR KIT	1
44	327439	. THRUST WASHER, Forward gear, 2.6 litre	1	83	392750	WATER PUMP REPAIR KIT, Includes Impeller Housing	1
				84	392749	CHROME PUMP KIT	1
				85	392160	ANODE KIT, Includes 50,51	1

* Not Shown
† Contents of Water Pump Repair Kit

Figure 11–25   Gearcase (continued).

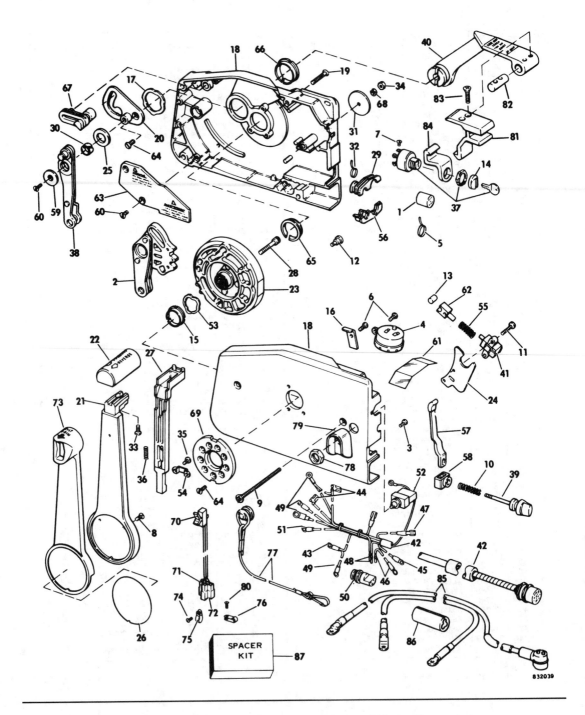

Figure 11–26   Remote control (continued). *(Courtesy of Outboard Marine Corporation)*

OP0896C

Ref. No.	P/N	Name of Part	Qty.	Ref.No.	P/N	Name of Part	Qty.
*	394084	REMOTE CONTROL ASSY.	1	45	204038	. . RING TERMINAL	1
1	324831	. CAP, 3-way connector	1	46	390760	. . LEAD ASSY.	1
2	390610	. SHIFT LEVER ASSY.	1	47	204034	. . TERMINAL	1
3	307019	. SCREW, Ground	1	48	581656	. . SOCKET	2
4	389263	. WARNING HORN	1	49	327166	. . TERMINAL	4
**5	320107	. TIE STRAP	8	50	511990	. . PLUG	1
6	324064	. SCREW, Horn & plate	2	51	581959	. . LEAD ASSY., Ground	1
7	306559	. SCREW, Ignition switch	6	52	392286	. . EMERGENCY IGNITION	
8	205042	. SCREW, Cover, handle	1			CUT-OFF SWITCH ASSY.	1
**9	326593	. SCREW, Single control, portside .	2	53	323765	. WAVE WASHER, Shift & throttle	
10	325790	. SPRING, Friction adjusting screw	1			hub	1
11	324524	. SCREW, Mounting bracket switch	2	54	323771	. PLATE, Slide control lever	1
**12	323776	. CLEVIS PIN, Shift & throttle		55	318625	. SPRING, Detent	1
		lever	2	56	323763	. GROMMET, Lower	1
13	321888	. ROLLER, Detent, shift & throttle		57	389267	. FRICTION ADJUSTMENT	
		plate	1			LEVER	1
**14	322176	. PUSH KNOB, Key	1	58	325191	. NUT, Friction lever	1
15	323764	. BUSHING, Cam, hub cover	1	59	318228	. WASHER, Fast idle lever	1
16	323769	. RETAINER PLATE	1	60	318277	. SCREW	3
17	325780	. SPRING WASHER, Cam to hsg. ..	1	61	328168	. INSULATOR, Horn to cover	1
18	394253	. HOUSING & COVER ASSY.	1	62	321889	. SHOE, Detent, shift & throttle	
19	308527	. . SCREW, Access cover to housing	3			plate	1
20	393219	. CAM, Shift lockout	1	63	390179	. SEPARATOR, Levers	1
21	327164	. HANDLE, Remote control	1	64	323760	. SCREW, Index plate	4
22	323774	. KNOB, Handle	1	65	323749	. FLANGED BUSHING, Shift &	
23	389266	. SHIFT & THROTTLE PLATE				throttle plate	1
		ASSY.	1	66	325779	. FLANGED BUSHING, Fast idle	
24	323766	. MOUNTING PLATE, Neutral start				lever	1
		switch	1	67	325781	. LEVER, Shift lockout	1
25	321966	. SPACER, Throttle lever to cam	1	**68	306396	. LOCK WASHER, Control to boat .	2
26	327490	. APPLIQUE	1	69	323759	. INDEX PLATE, Neutral lock	1
27	323770	. SLIDE, Neutral lock	1	*	391858	. COVER & SWITCH ASSY.	1
28	323775	. SCREW, Handle	1	70	582078	. . SWITCH & LEAD ASSY.	1
29	323762	. GROMMET, Upper half	1	71	313113	. . . CONNECTOR	1
30	323757	. BUSHING, Throttle lever	1	72	510541	. . . TERMINAL	3
**31	318321	. WASHER, Remote control to boat	2	73	327165	. . COVER, Handle	1
32	318322	. TIE STRAPS, Grommets	1	74	205042	. . SCREW, Clamp	1
33	318324	. SCREW, Knob to control handle .	1	75	322389	. . CLAMP	1
**34	318349	. NUT, Mounting screw	2	**76	316245	. . CLAMP	1
35	323772	. SCREW, Plate	2	77	393079	. CAP & LANYARD	1
36	323773	. SPRING, Slide return	1	78	328169	. NUT	1
37	393301	. SWITCH & KEY ASSEMBLY,		79	328170	. GUARD	1
		Ignition	1	**80	305009	. SCREW	1
*	322693	. . IGNITION KEY BLANK	AR	*	394079	. LATCH KIT, Fast idle lever	1
38	389261	. THROTTLE LEVER ASSY.	1	81	329358	. . LATCH, Fast idle lever	1
39	391036	. KNOB & FRICTION ADJ.		82	329359	. . FILLER ROD, Fast idle lever	1
		SCREW ASSY.	1	83	329361	. . SCREW, Latch	2
40	329357	. LEVER, Fast idle	1	84	329362	. . LATCH, Remote control box	1
41	389265	. SWITCH ASSY., Neutral start	1	85	391767	BATTERY CABLE ASSY.	1
42	393129	. INST. CABLE ASSY.	1	86	327120	GROMMET	1
43	204036	. . RING TERMINAL	1	87	390795	SPACER KIT, Remote control	
44	309437	. . SLIDE ON TERMINAL	2			mounting	1

* Not Shown
** Also Available In Installation Hardware Kit Part No. 390009

Figure 11–26   Remote control continued.

Commander 2000 Features

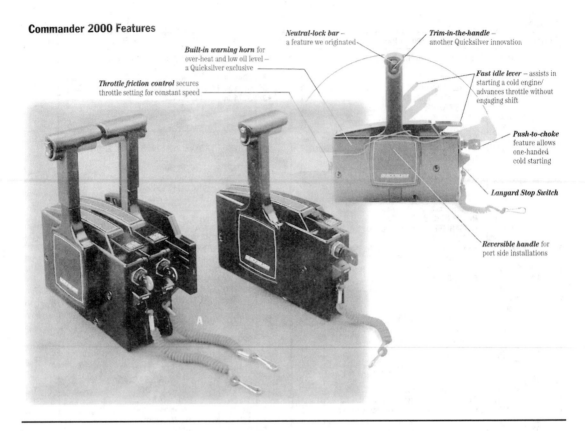

Built-in warning horn for
over-heat and low oil level –
a Quicksilver exclusive

Throttle friction control secures
throttle setting for constant speed

Neutral-lock bar –
a feature we originated

Trim-in-the-handle –
another Quicksilver innovation

Fast idle lever – assists in
starting a cold engine/
advances throttle without
engaging shift

Push-to-choke
feature allows
one-handed
cold starting

Lanyard Stop Switch

Reversible handle for
port side installations

Figure 11–27   Helm controls. *(Courtesy of Mercury Marine)*

Outboard Shift System and Persons Safety

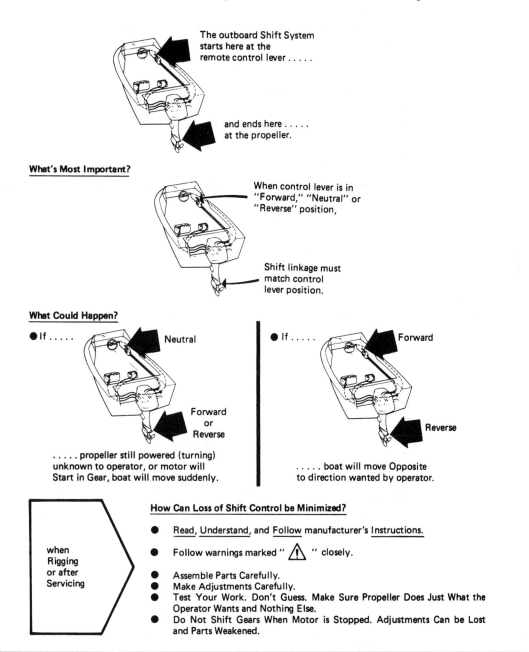

The outboard Shift System starts here at the remote control lever . . . . .

and ends here . . . . . at the propeller.

What's Most Important?

When control lever is in "Forward," "Neutral" or "Reverse" position,

Shift linkage must match control lever position.

What Could Happen?

● If . . . . .   Neutral

Forward or Reverse

. . . . . propeller still powered (turning) unknown to operator, or motor will Start in Gear, boat will move suddenly.

● If . . . . .   Forward

Reverse

. . . . . boat will move Opposite to direction wanted by operator.

when Rigging or after Servicing

How Can Loss of Shift Control be Minimized?

● Read, Understand, and Follow manufacturer's Instructions.

● Follow warnings marked " ⚠ " closely.

● Assemble Parts Carefully.
● Make Adjustments Carefully.
● Test Your Work. Don't Guess. Make Sure Propeller Does Just What the Operator Wants and Nothing Else.
● Do Not Shift Gears When Motor is Stopped. Adjustments Can be Lost and Parts Weakened.

Figure 11–28   Outboard shift system and persons safety. *(Courtesy of Outboard Marine Corporation)*

KNOW THESE PRINCIPLES OF OPERATION

- How the outboard is mounted and steered.
- How circulating water is used to cool the exhaust housing.
- Power flow through the lower unit.
- How to shift into gear.
- Principles upon which the outboard jet is based.
- Indications of water leaking into the lower unit.
- Pressure-testing the lower unit.
- Indications that water has damaged lower unit parts.

REVIEW QUESTIONS

1. The exhaust housing mounts to the steering bracket through rubber mounts to dampen vibrations into the boat.
   a. true
   b. false

2. The inner exhaust housing on larger outboards may be surrounded by water when underway.
   a. true
   b. false

3. If the driveshaft splines are worn excessively, this indicates (a) _____.
   a. bent exhaust housing
   b. bent driveshaft
   c. normal wear

4. The water tube carries water from the water inlet to the powerhead.
   a. true
   b. false

5. In reverse the clutch dog is moved forward.
   a. true
   b. false

6. With proper trim adjustment, the anti-ventilation plate should be running parallel to and just below the water surface.
   a. true
   b. false

7. Slow lazy shifting will damage the _____ .
   a. clutch dog
   b. pinion gear
   c. driveshaft splines

8. When the fishing trip is completed, the outboard should be removed from the salt or brackish water and flushed for at least _____ of running time.
   a. 15 minutes
   b. 10 minutes
   c. 5 minutes

9. The rubber propeller hub is used to dampen vibrations.
    a. true
    b. false
10. The pinion, forward, and reverse gears are continually rotating when the power-head is in operation in any gear.
    a. true
    b. false
11. Hydrostatic rings help control the flow of _____ .
    a. water
    b. exhaust gases
    c. propeller thrust
12. In the mechanical shift system, the speed control cable adjustment may affect the shift cable operation.
    a. true
    b. false
13. Water in the lower unit is indicated by a milky color of the gear oil. This indicates a seal is leaking and the unit needs to be pressure/vacuum-tested to determine leak location.
    a. true
    b. false
14. Gear case oil needs to be checked each 30 days of operation and changed every _____ days of operation in freshwater.
    a. 50
    b. 60
    c. 70
15. Once a season, the lower unit needs to be removed and the ___.
    a. lubrication replenished on the driveshaft splines
    b. water pump impeller replaced
    c. Both a and b are correct, plus coat the bolts or studs on the lower unit with gasket sealing compound
16. Shims are used in the lower unit to position gears and to preload bearings.
    a. true
    b. false
17. The use of an outboard jet will increase horsepower.
    a. true
    b. false
18. Steering when using the jet is accomplished by directing the nozzle from side to side.
    a. true
    b. false
19. The jet reverse gate is used to_____.
    a. direct thrust downward
    b. direct thrust under the boat
    c. increase horsepower

20. When the control lever is in "forward," neutral" or "reverse," the shift linkage at the outboard must match the control lever position.
    a. true
    b. false
21. What is the pinion gear mounted to?
    a. clutch
    b. forward gear
    c. drive shaft
22. The clutch is located between the _____.
    a. propeller shaft and forward gear
    b. driveshaft and reverse gear
    c. forward and reverse gears
23. The water pump impeller is turned by the _____.
    a. drive shaft
    b. propeller shaft
    c. shift shaft
24. When checking for a lower unit water leak, the lower unit is _____.
    a. vacuum tested
    b. pressure tested
    c. Both a and b are correct.

CHAPTER 12
Trim and Tilt System

Objectives

After studying this chapter, you will know

- Purpose of the power trim.
- Electrical circuitry of the trim and tilt systems.
- Hydraulic Up and Down circuits.
- Manual tilt system.
- Trail out and reverse lock system.
- Bounce system.
- Trim/Tilt Out/Up.
- What happens when a submerged object is hit.
- Troubleshooting systems and common problems.

12.1   Power Trim (Mercury)

This system consists of an electrical circuit, electric/hydraulic pump with valve body, an oil reservoir incorporated within, hydraulic hoses, and a trim cylinder (Figures 12–1 and 12–2).

The pump is mounted inside the boat, and the trim cylinder is mounted to the outboard swivel bracket and transom brackets. The torque of the electric motor is converted into hydraulic pressure at the pump. Hydraulic pressure is routed through the pump body into the hoses and to the trim cylinder. This forces a movement of the trim cylinder, piston, and ram, which converts the hydraulic pressure back to a physical force to trim or tilt the out-

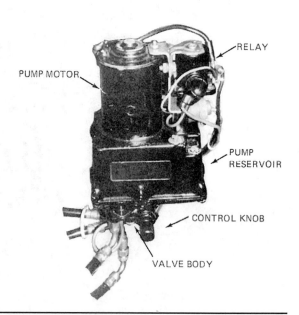

Figure 12–1 Hydraulic trim pump (as used by Mercury).

Figure 12–2 Hydraulic trim cylinder (as used by Mercury).

board. For practical purposes, assume that the motor oil used in the system cannot be compressed.

There are two types of electrical circuits: a dual solenoid system and a single solenoid system. The single solenoid circuit consists of the trim pump motor, trim solenoid, a limit switch, and the power trim control, and is used on the "Midi" models (Figures 12-3 and 12–4).

Electricity from the battery is routed to a single solenoid terminal B and then to the power trim control at the helm. When the Up button is pushed, current is sent through wire I and limit switch H to the solenoid, creating a magnetic field, pulling the solenoid switch closed. This routes direct battery current to the electric pump motor through terminal D and wire J. The motor turns the pump and oil is forced into the up side of the trim cylinders. This trims the outboard up (out). If the Up button is held long enough, the trim limit switch will open the circuit. This stops the outboard swivel bracket within the confines of the transom bracket supporting flanges for good side thrust support.

To trim in (down) the unit, less current is used as there is less hydraulic pressure involved, so no solenoid is used (see Figure 12–4). The weight of the outboard and propeller thrust aids in trimming in the unit. When the Trim In button is pushed, current flowing through the control trim switch and wire F will activate the electric hydraulic pump in the reverse direction. The pump forces oil into the down side of the trim cylinders, trimming in (down) the outboard.

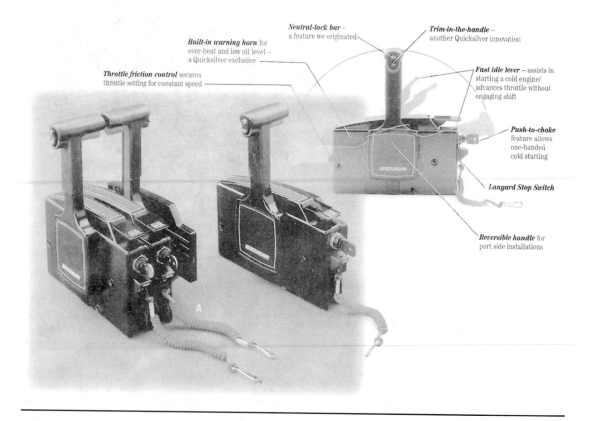

Neutral-lock bar –
a feature we originated

Trim-in-the-handle –
another Quicksilver innovation

Built-in warning horn for
over-heat and low oil level –
a Quicksilver exclusive

Throttle friction control secures
throttle setting for constant speed

Fast idle lever – assists in
starting a cold engine/
advances throttle without
engaging shift

Push-to-choke
feature allows
one-handed
cold starting

Lanyard Stop Switch

Reversible handle for
port side installations

Figure 12–3   The Mercury Marine Commander 2000 helm. *(Courtesy of Mercury Marine)*

To place the outboard in the trailering position, the Up/Out switches are pushed, which activate the up solenoid through wire K. This sends direct battery current to the electric motor and pump through wire J, forcing oil into the up side of the trim cylinders. This circuit bypasses the trim limit switch and the outboard will raise all the way to the trailering position. If the operator chooses, the trailering circuit can be used to raise the outboard for shallow water operation and docking. In this position, make sure that the water inlet is under water.

When the dual solenoid electrical system is used, there is a solenoid placed in the up circuit and also in the down (trim in) circuit (see Figure 12–5). This is because there is a greater pressure against the lower unit on the models where it is used. Therefore, more current is required to bring the unit to the down (trim in) position. By adding a down solenoid, the down switch is relieved of the heavy current flow and gives longevity to the switch. With the trim In button pushed, electricity flows through wire L, closing the down solenoid. Battery current then flows to the down solenoid terminal O to wire M, operating the motor in the reverse direction.

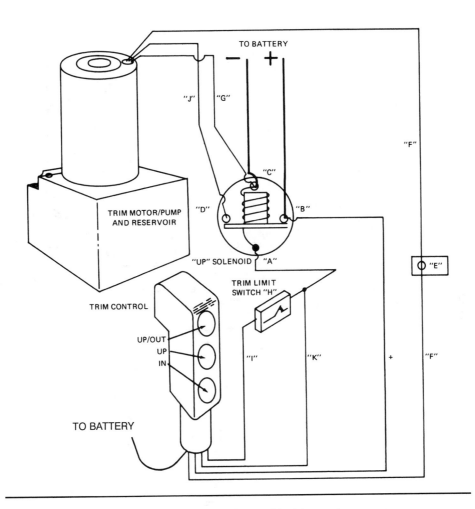

Figure 12–4   Single solenoid trim circuit (as used by Mercury).

The simplified circuit illustrations which support the theory of electrical and hydraulic operations are intended for instructional purposes only.

12.1.2   Troubleshooting the Power Trim

In troubleshooting, there are two systems to consider: electrical and hydraulic. Let us consider the electrical system first. The heart of the system is the battery. It needs to be fully charged and have at least 465 MCA. If there are problems with the battery, the pump motor cannot operate satisfactorily.

There are five nonoperational conditions that could affect control of the trim operation of the single solenoid system:

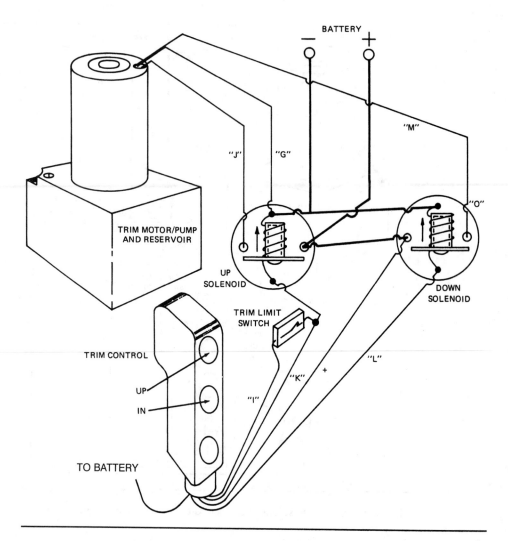

Figure 12–5 Dual solenoid trim circuit (as used by Mercury).

1. Trim in, Trim out, and Trailer up
2. Trim out and Trailer Up
3. Trim out
4. Trailer up
5. Trim in/dn

When there is a customer complaint, generally one or all of these circuits will not work.

Troubleshooting involves verifying the customer's complaint, and determining exactly which circuits are operating and which circuits are not operating. First, determine if it is a single or dual solenoid system. Once this is accomplished you will be able to choose the appropriate troubleshooting chart in the service manual. These charts will guide you over each individual circuit, checking voltage at the solenoid, trim control, motor, and the battery.

To repair the trim circuits that do not operate, you must locate and repair the loose connections, repair broken wires, and clean up the corrosion that may exist in the connections. If the wires and connections prove out, then it will be necessary to replace the solenoid trim switch or repair the pump motor. There are five inoperative electrical conditions in trouble isolation procedures given for a power trim troubleshooting exercise. Work between the schematic and the inoperative conditions that follow.

12.1.3   All Circuits Are Inoperative (Single Solenoid System)

Prove the battery state of charge and load-test the battery. Then check to see if voltage is at the solenoid battery terminal B. If voltage is present, prove the pump motor by placing a remote starter switch across the solenoid battery terminal B and terminal E. The trim motor should operate. If no voltage is present at the battery terminal B, there may be a bad lead between the battery and the solenoid battery terminal. If the trim motor still does not operate, there may be a bad ground G between the pump motor and the solenoid terminal C. Also check out the motor by jumping directly from the battery. If the trim motor operates, check for electricity at the trim control. There may be an open in the battery (+) wire leading to the control, or the trim buttons may not be passing the current to the solenoid or motor. There are connectors in the circuit, so check that they are fully pushed together and free of corrosion (see Figure 12–4).

12.1.4   Trim Out and Trailer Up circuits when Trim In Works

When the trim in works, we already know that the battery is good, that there is electricity to the trim control, and that the pump motor armature and brushes are good for the reverse rotation. To be able to operate in the Up mode, electricity must be at the control Up button and must be sent through the trim limit switch H and then on to the solenoid. To systematically check this circuit, test for battery voltage at the UP button. If voltage is present, the next check is at the connector and then at terminal A of the solenoid. If there is no voltage at the Up button, there is a problem in the control Up/Out trim button (switch).

If voltage is present at the A terminal of the solenoid, then the solenoid needs to be checked. If the solenoid clicks, check for voltage at the D terminal. This proves the contact within the solenoid. If the pump still does not operate, check the wire between the pump motor and the solenoid D terminal. If the wire is good, then there are problems with the field winding connection within the motor. If there is no voltage at the solenoid terminal A, then there is an opening between the trim limit switch and the trim Up button (see Figure 12–4).

12.1.5   Trim Out When Trim In and Trailer Up Work

When the outboard will trim in and will also lift to the trailering position, we know that the control buttons, solenoid, pump motor, and battery are all working. A look at the schematic will reveal that the trim limit switch H could be the problem. It can be checked by using an ohmmeter. The outboard should tilt so that the swivel bracket stops with no less than 1/2-inch minimum engagement within the clamp and swivel bracket flanges. This means that there should be continuity through the switch until this position is reached. At this point the switch should open, breaking the circuit, and stopping the trim action. If the switch and adjustment of the switch check out according to the service manual, then the wire between the Out button and the limit switch is open. It is normal to have a lot of grease around the switch. It may have been placed there to help reduce corrosion.

If the trim limit switch contact lever is stuck in the closed position, the trim limit switch may not open, and the outboard will trim out beyond safe operating limits (Figure 12–4).

12.1.6   Trailer Up When the Trim Out and In Work

To place the outboard in the trailering position, the limit switch is bypassed. Electricity is routed from the Up button to the Up/Out button. Check for electricity at the Up/Out button when the Up button is depressed. If there is battery voltage at this point, then wire K is open. If there is no voltage at the Up/Out button (with the Up button depressed), there is a problem with the Up trim switch (Figure 12–4.)

12.1.7   Trim In/Down When the Trim Out and Trailer Up Work

If the outboard will not come down or trim in, then check to see if there is voltage at both of the In button terminals when depressed. Voltage on only one terminal indicates a faulty switch. If voltage is present at both terminals, then there is an open wire F. Check out the wire from the pump motor to the In button.

In the dual solenoid system, there will be additional checks involving the down solenoid operation (Figure 12–5).

 Note

Such electrical testing of the trim system is not difficult. When the service manual is followed, it will guide the mechanic through a systematic approach to repair the problem. The theoretical troubleshooting given here is intended as an in-depth study, providing a systematic approach to repair of a faulty trim system.

12.1.8   Hydraulic Up Circuit

When the hydraulic trim system is activated by the Up button at the control, the pump motor shaft and gear A rotate. The pump draws some oil from the reservoir through the oil inlet B, unseating check valve J.

The shuttle valve E is moved to the right by oil pressure. The shuttle valve makes contact with the pilot check valve F, opening the down circuit, which allows for the oil to return from the down chamber of the trim cylinder. This returning oil supplies most of the oil to operate the up circuit. Some oil comes from the reservoir.

As oil pressure increases to an approximate 20 PSI, the up circuit pilot check valve D is forced to the open position. This allows oil to flow through the up port K and through the

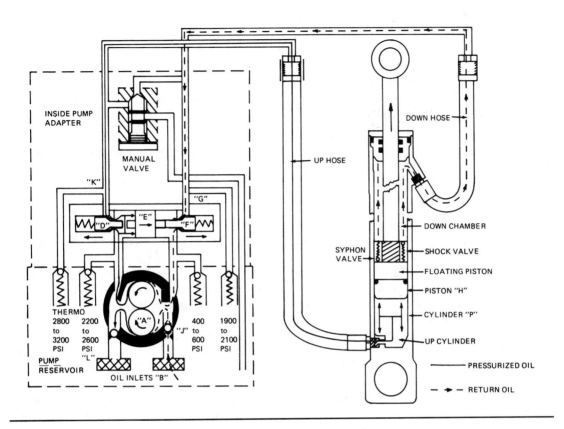

Figure 12–6   Up circuit power trim hydraulic system.

up hose to the trim cylinder, filling the up chamber. This moves the cylinder piston and rod out and up.

When the cylinder piston H is fully extended and the operator continues to hold the Up button depressed, oil pressure will rise. When the oil pressure reaches 2200 to 2600 PSI, the up pressure regulator and relief valve L will be forced from its seat. Oil will then return back to the pump reservoir (Figure 12–6).

The type of oil used in the older systems is SAE 10W-30 or 10W-40 SE automotive oil. In tropical areas, a single grade SAE30 oil may be used. The oil is checked when the outboard is in the full up position. Newer models may use power trim/tilt and or power steering fluid. The oil level is checked as required.

12.1.9   Hydraulic Principles

Power trim hydraulic systems use an oil to transmit pressure, force, and motion. In a closed system, pressure coming from the pump, outboard weight, or an impact, is produced and exerted undiminished throughout the system.

Several principles apply to the operation of the hydraulic power trim system.

1. Liquids in a confined area will *not* compress.
2. When pressure is applied to a closed system, pressure is exerted *equally* in all directions.
3. The power trim hydraulic system is used to increase or decrease force to trim or tilt the outboard.

Pressure is measured in pounds per square inch or in kilopascals. Pressure is the amount of force produced by the electric hydraulic trim pump to push the trim/tilt cylinder's piston out or in to trim or tilt the outboard.

12.1.10   Hydraulic Down Circuit

When the hydraulic system is activated by the Down button at the control, the pump motor shaft and gear A rotate in the reverse direction. The pump draws some oil from the reservoir through the oil inlet B, unseating check valve C.

The shuttle valve E is moved to the left by oil pressure. The shuttle valve makes contact with the pilot check valve D, opening the up circuit, which allows for oil to return from the up side of the trim cylinder. This returning oil supplies most of the oil to operate the down circuit. Some oil comes from the reservoir.

As oil pressure increases to an approximate 20 PSI, the down pilot check valve F is forced into the open position. This allows oil to flow through down port "G" and through the down hose to the trim cylinder, filling the down chamber. This moves the cylinder piston and rod in and down.

When the cylinder piston H is fully retracted and the operator continues to hold the Down button depressed, oil pressure will rise. When the oil pressure reaches 400 to 600 PSI, the down pressure regulator and relief valve I will be forced from its seat. Oil will return back into the pump reservoir (see Figure 12–7).

12.1.11   Manual Tilt System

When the lower unit is to be raised manually, the manual lock release valve knob M must be fully opened by turning in the counterclockwise direction. When the manual valve is in the fully open (out) position, oil is no longer locked into the up and down circuits, but is free to flow from the up side to the down side, or from the down side to the up side, of the trim cylinder. Oil line N compensates for different capacities in the trim cylinder chambers. Excess or needed oil will flow through this line as necessary. The manual valve is normally kept closed (see Figure 12–8).

12.1.12   Trail Out and Reverse Lock System

Upon shifting into reverse gear, there is a reverse thrust that tries to pull the lower unit away from the transom. Also, upon rapid deceleration of the powerhead the lower unit tries to pull away from the transom. For these reasons, the oil in trim system circuits must be held in a static position when the trim system is out of operation. The reverse lock-trail out relief valve O is set at 1900 to 2100 PSI, and the up and down circuit pilot check valves D

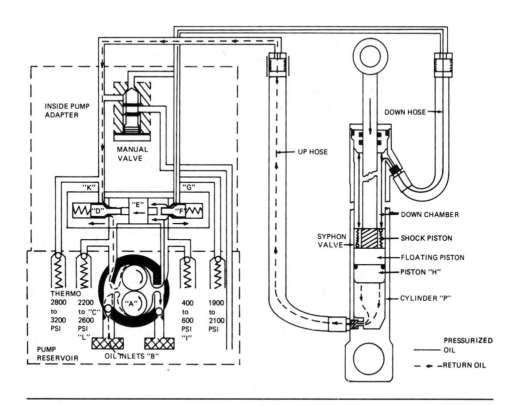

Figure 12–7   Down circuit power trim hydraulic system.

and F are spring-loaded to the closed position. Also, the manual lock release valve is closed (in). This prevents trailing out, as the oil is held in a static position. (The manual lock release valve is shown open to illustrate the manual tilt position. It would be closed when underway, preventing a trailing out condition. See Figure 12–8.)

12.1.13   Manual Valve Open (Leaking) While Trimming

With the manual valve M either open or leaking, there is an approximate equal oil pressure in the up and down chambers of the trim cylinder when trimming. Therefore, there will be little or no movement of the lower unit. Oil pressure could be from 20 to 100 PSI. The lower unit could slowly trail out when underway because oil is moving through the manual valve to the opposite chamber.

12.1.14   Bounce System

While underway in the forward gear, if the lower unit should strike a submerged object with light steady pressure the hydraulics must give. This is because the hydraulic system is

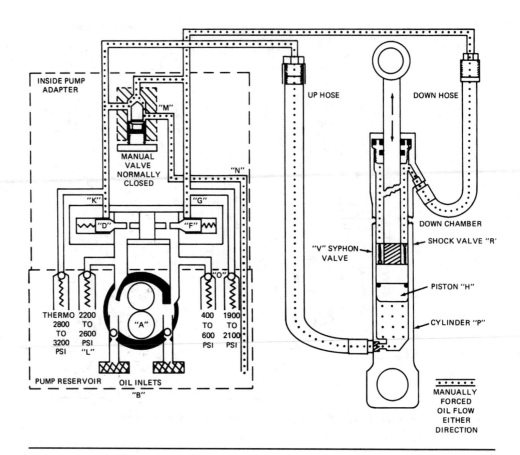

Figure 12–8   Manual tilt system power trim hydraulic system.

locked in a static position by the reverse lock-trail out relief valve O set at 1900 to 2100 PSI. Also, the up and down pilot check valves D and F and the manual lock tilt valve M are closed. Upon a light steady pressure, there can be no oil movement in the trim cylinder P, but the piston rod Q moves outward. This will create vacuum between the shock valve R and the floating piston S. This condition momentarily exists until propeller thrust and lower unit weight return it to the normal running position. Because the down chamber of the trim cylinder becomes pressurized, oil will unseat the reverse lock-trail out relief valve O, spilling oil back into the reservoir. If the lower unit stays trailed out, it might be necessary to activate the down circuit to refill the down chamber in the trim cylinder (Figure 12–9).

12.1.15   Shock Absorber System

Upon sudden heavy impact with a submerged object, trim cylinder oil pressure in the down side chamber will force the opening of the shock relief piston valve R at 1100 to 1300 PSI.

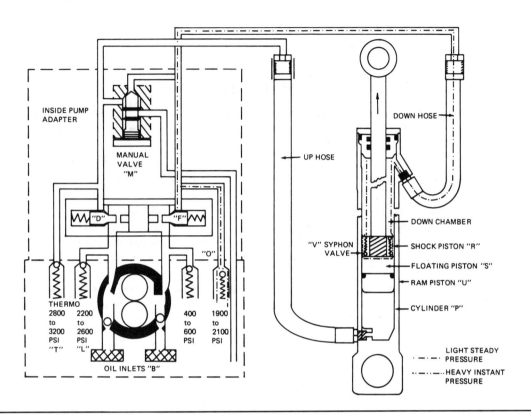

Figure 12–9   Bounce and shock system power trim hydraulic system.

The up chamber oil is locked in the up circuit but the pilot check valve D and the thermal relief valve T, which can open at 2800 to 3200 PSI. As pressure rises, the shock piston relief valve R opens, allowing oil to pass from the down chamber into a vacuum area between the ram piston U and the floating piston S. As the lower unit returns to the normal position, the syphon valve V opens at 50 to 150 PSI, allowing oil between the shock piston and the ram piston to return to the down chamber of the trim cylinder. If the oil temperature and pressure reach a high value, the thermal valve T will open at 2800 to 3200 PSI, spilling additional oil back into the reservoir. After impact, the lower unit is somewhat trailed out. Propeller thrust and weight of the lower unit provide the force necessary to move the drive unit back into the normal trimmed position (Figure 12–9).

12.1.16   Trim and Tilt Angles

In 1984, Mercury increased the trim range for the 75, 90, and 115 HP motors. The power trim delivers a full 20 degrees of trim, from 8 degrees to 28 degrees. This allows handling of heavier loads in the stern and a greater trim out, reducing lower unit drag at high speeds.

The maximum tilt height of 75 degrees allows for easy beaching of the boat and good highway clearance when the boat is trailered.

12.2   Integral Power Trim and Tilt (OMC)

This unit is a self-contained unit mounted between the stern brackets and the swivel bracket. The manifold houses the oil pump, valve body, and two trim cylinders. The pump motor and reservoir sit on top of the manifold. The combination tilt cylinder also acts as a shock absorber upon impact, and is mounted in the center. To trim the outboard the trim cylinders push on the swivel bracket at the trim pads (see Figures 12–10 and 12–11).

When the Up button is pressed, the pump motor is turned on and oil is forced into the tilt cylinder and trim cylinders. The trim cylinders move the outboard through the first 15 degrees of movement by mechanical and hydraulic advantage. This is called the ***trim range***. The center tilt cylinder takes up the load from there, and moves the outboard through 50 degrees of movement into the trailering position. This is called the ***tilt range*** (see Figure 12–12).

When underway, be careful not to tilt the lower unit up too high and expose the water pickup to air. This causes overheating of the powerhead. If the powerhead is accelerated above

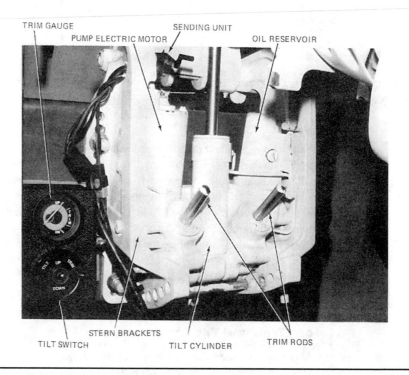

Figure 12–10   Integral power trim and tilt. *(Courtesy of Outboard Marine Corporation)*

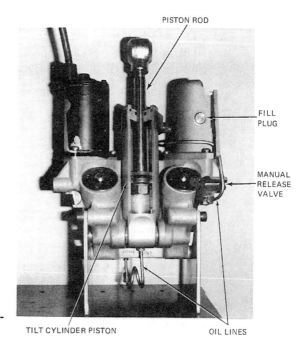

PISTON ROD

FILL
PLUG

MANUAL
RELEASE
VALVE

TILT CYLINDER PISTON

OIL LINES

Figure 12–11 Cutaway of integral power trim and tilt.

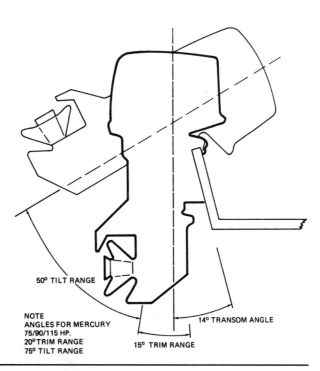

50° TILT RANGE

14° TRANSOM ANGLE

NOTE
ANGLES FOR MERCURY
75/90/115 HP.
20° TRIM RANGE
75° TILT RANGE

15° TRIM RANGE

Figure 12–12 Tilt and trim angles.

1500 RPM when above the 15-degree position, pressure from the forward thrust will automatically open the trim up relief valve, lowering the lower unit to the fully trimmed out position. If the operator attempts to tilt the outboard when underway, it will not be effective. This system is used on 70 HP through V-6, and there are seven hydraulic modes of operation:

1. Trimming or tilting down/in
2. Running in forward
3. Running in reverse
4. Trimming or tilting out/up
5. Shallow low water drive
6. Striking a submerged object
7. Returning to the fully trimmed out position, after striking a submerged object

To understand the power trim, you must be familiar with the oil flow during each mode of operation.

12.2.1   Trim/Tilt Down

When the trim control Down/In button is pressed, the pump motor is turned on. The torque of the motor shaft turns the pump, developing hydraulic pressure within the cylinders. The tilt cylinder moves first until the swivel bracket is resting on the trim rods, which then begin to move. The cylinders convert the hydraulic force into a trimming force or a tilt force to move the lower unit (see Figure 12–13).

Hydraulic pressure is routed to the pump control piston A. It then opens the trim check valve C and the reverse lock check valve B. This permits oil pressure to flow to the down side of the trim and tilt cylinders. Oil on the up side of the cylinders returns through the trim/tilt separation valve H and the open trim check valve C, and returns to the pump. Because of different sized trim/tilt rods and cylinders, a different volume of oil is required above the tilt or trim pistons than is being supplied from beneath the pistons. Therefore, some of the pressurized oil flows through the trim down pump relief valve I into the reservoir. This reduces the total flow of oil into the cylinders. Control of the piston movement equals the flow of returning oil to the pump, oil flow into the cylinders, plus oil returning to the reservoir.

When the control button is held after the trim and tilt cylinders are fully retracted, hydraulic pressure will continue to rise. When pressure reaches 800 PSI, the trim down pump relief valve I will open. This will bypass oil back to the reservoir.

When the control button is held after the trim and tilt cylinders are fully retracted, hydraulic pressure will continue to rise. When pressure reaches 800 PSI, the trim down pump relief valve I will open. This bypasses oil back to the reservoir.

12.2.2   Running Forward

When underway and running forward, the trim hydraulic oil must be held in a static condition (Figure 12–14).

This is accomplished by the closed trim check valve C, the trim/tilt separation valve H, the piston seals, and the expansion valve N.

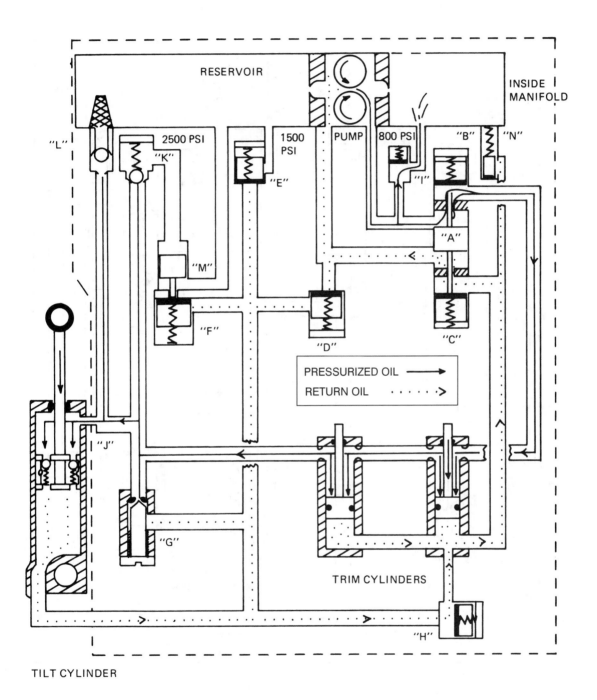

Figure 12–13   Trim/tilt down/in hydraulic circuit. *(Courtesy of Outboard Marine Corporation)*

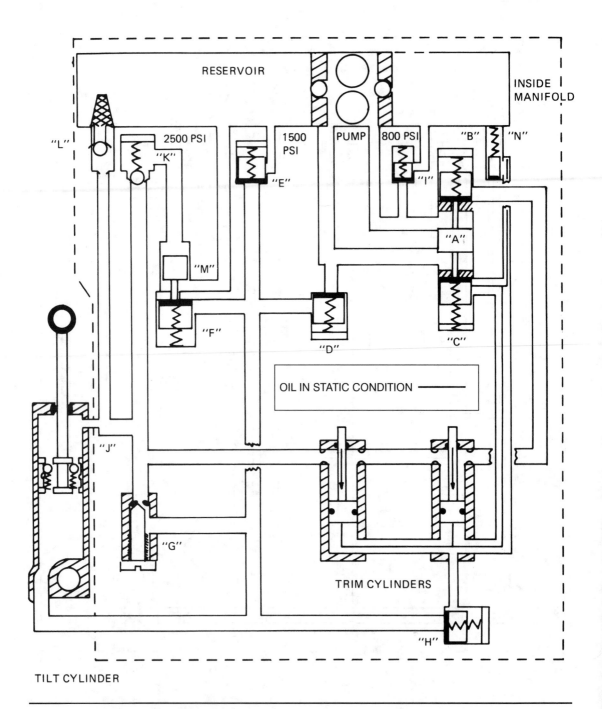

Figure 12–14   Running in forward, hydraulic circuit Outboard Marine Corporation 70 HP through V-6.

12.2.3   Running in Reverse

When the lower unit is shifted into reverse, the propeller thrust will be in the opposite direction (see Figure 12–15). Therefore, the shock absorbing valves J, manual release valve G, reverse lock check valve B, impact sensor valve K (not used on newer models), piston seals, and the filter valve L must hold against the reverse thrust. Oil pressure created by the propeller's reverse thrust pulls on the lower unit and tilt cylinder. This applies pressure to the down side of the trim cylinders. However, these trim pistons cannot move down because the oil in the up side is locked in. The trim/tilt separation valve H, trim check valve C, and the expansion relief valve N are closed. This results in the reverse lock features.

12.2.4   Trim/Tilt Out/Up

Torque of the pump motor shaft is converted into hydraulic pressure at the pump (see Figure 12–16). This pressure is routed to the pump control piston A, forcing it up, opening the reverse lock check valve B as oil pressure opens the trim check valve C. Oil continues to flow to the up chamber of the trim cylinders. As pressure increases, oil opens the tilt check valve D and oil continues to flow to the tilt cylinder. Oil that is in the down chambers of the tilt and trim cylinders is forced by the pistons through the reverse lock check valve B, which is held open by the pump control piston A, so the oil can return to the pump. This return oil is not sufficient to supply the pump with oil, because of the different displacements between up and down chambers within the cylinders. Therefore, additional oil is drawn from the pump reservoir. When the maximum travel of the trim and tilt cylinder pistons is reached, and the operator continues to depress the button, pressure will rise to 1500 PSI. At this point the trim up relief valve E will open, allowing oil to return to the reservoir. The impact letdown valve F, the manual release valve G, and trim/tilt separation valve H are closed.

12.2.5   Tilting or Shallow Water Drive Operation

As the outboard is in the process of being tilted, held in the tilt position or held in the shallow water mode of operation, certain seals and valves are under pressure and must hold. (See Figure 12–17.)

These are the seals on the trim pistons, trim check valve C, seals on the tilt piston, tilt check valve D, impact letdown valve F, shock absorber valves J, trim up relief valve E and manual release valve G.

 **Warning**

When the outboard is placed in the trailering position, these valves and seals must be relieved of hydraulic pressure. If not, damage may be done to the seals as the boat and outboard bounce during travel.

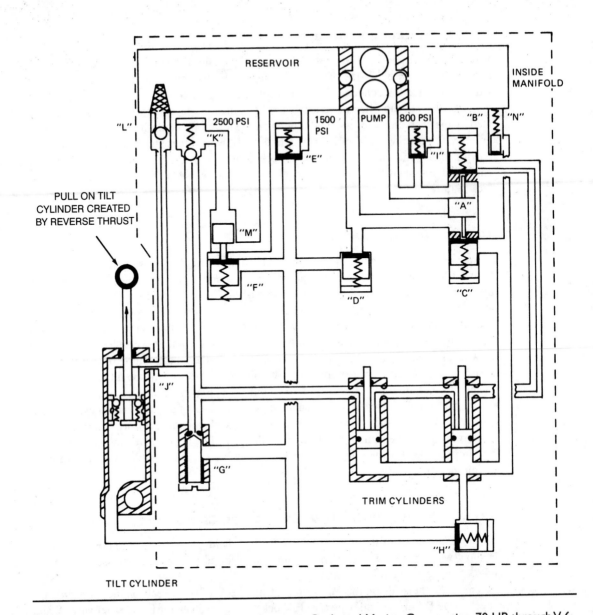

Figure 12–15   Running in reverse, hydraulic circuit Outboard Marine Corporation 70 HP through V-6.

There is a trail lock provided to relieve this type of condition. The trail lock is moved into position when the outboard is in the fully tilted position. Then the trim and tilt unit is run down until the locks are held in position on the stern brackets and the trim rods are retracted completely. This applies the weight and bounce of the outboard directly to the stern brackets, and there will be no damage to the tilt circuit components.

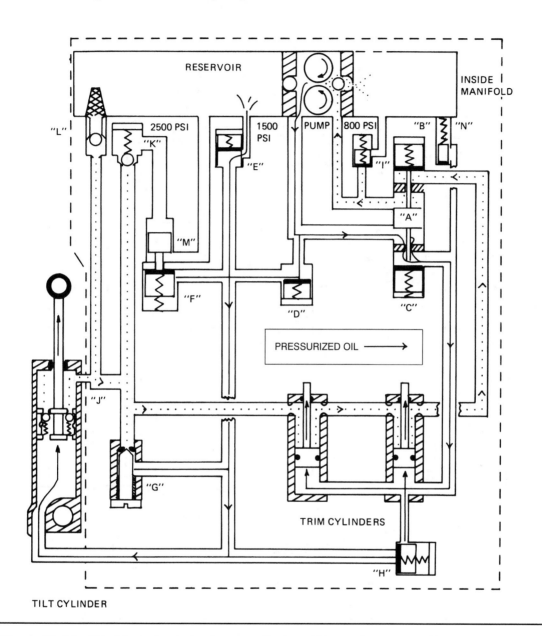

Figure 12–16   Trim/tilt out/up, hydraulic circuit Outboard Marine Corporation 70 HP through V-6.

12.2.6   Striking a Submerged Object

If by chance you strike a submerged object while underway, there must be relief for the sealed hydraulics. The outboard must tilt up instantly (Figure 12–18).

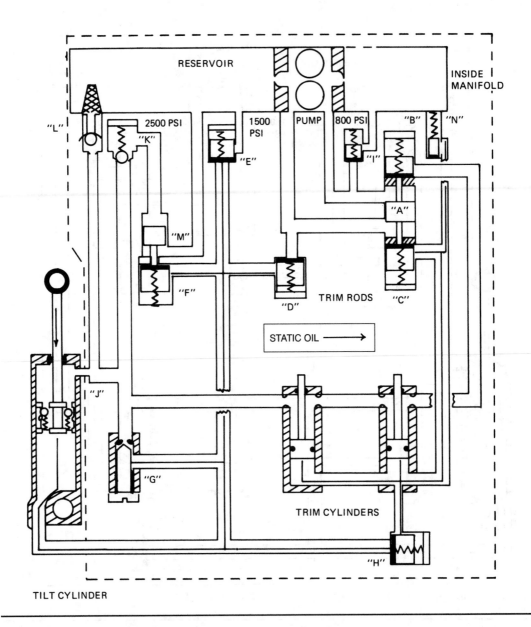

Figure 12–17   Shallow water drive operation, hydraulic circuit Outboard Marine Corporation 70 HP through V-6.

This is accomplished by opening the shock absorber valve J in the tilt cylinder, which allows the oil to flow from the down side to the up side of the tilt cylinder. As pressure reaches 2500 PSI, the impact sensor valve K opens. Oil from this valve moves the letdown

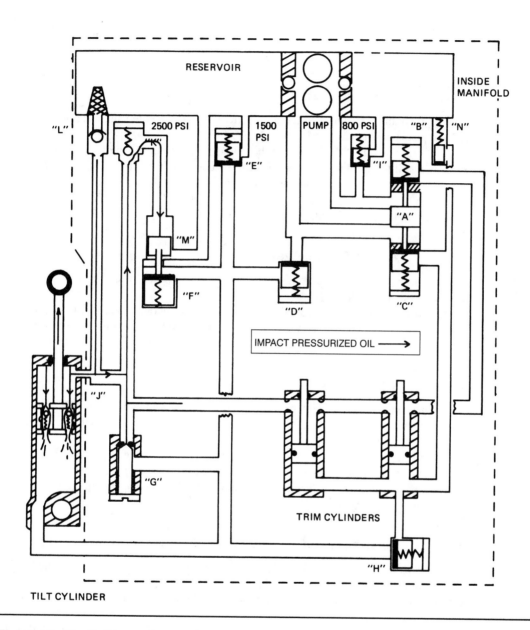

Figure 12–18   Striking a submerged object, hydraulic circuit Outboard Marine Corporation 70 HP through V-6.

control piston M and opens the impact letdown valve F. When the submerged object is passed and the tilt piston stops moving outward, the impact sensor valve K closes. This keeps the letdown piston M pressurized which in turn holds the impact letdown valve F open. Instantly after the impact sensor valve K closes, there is a controlled leak down accomplished by the letdown control piston K.

12.2.7   Return to Water

The weight of the outboard motor and propeller thrust as it reenters the water forces the tilt rod and piston downward (see Figure 12–19).

This forces oil out of the bottom of the cylinder through the open impact letdown valve F, allowing oil to return to the reservoir. At the same time, the filter valve L is opened and oil is replenished to the down side of the tilt cylinder. As the leak down at the letdown control piston M continues, the impact letdown control valve F will close. The lower unit has then returned to the trim position.

12.2.8   Manual Release Valve

There is a manual release valve G provided to permit manual raising or lowering of the outboard (Figure 12–20).

It has a screwdriver access hole through the stern bracket on the starboard side. With the valve opened, oil can pass from the down side to the up side of the tilt cylinder, permitting the operator to manually tilt the outboard.

12.2.9   Thermal Valve

The expansion relief valve N operates only under extremely high pressures. This causes elevated temperatures and the valve opens, allowing oil to return to the reservoir, preventing a lock up in the unit.

There is also a trim gauge provided to indicate the position of the lower unit. It is located in the helm, while the port stern bracket mounts the sending unit.

The simplified circuit illustrations that support the theory of operation are intended for instructional purposes only.

12.2.10   Troubleshooting the Trim/Tilt System

When there are problems noted with the tilt/trim system, it can be easily determined which circuit is the problem. If it's electrical the pump motor does not operate. If it's hydraulic the pump operates but the unit does not move, or there is a leak down or a trailing out/in when under power.

To check the hydraulic system, first determine if the oil pump reservoir is full with the prescribed oil. Check the service manual. Are the trail locks placed in the trim/tilt position? Inspect for visual leaks at the fittings and lines, and make sure the manual release valve is closed. After performing these checks and making any corrections, if the operation is still not satisfactory, two gauges will have to be used to test the circuit pressures—one gauge for the up circuit and one gauge for the down circuit. The manual relief valve is removed and the gauge is installed in the relief valve hole. The system is

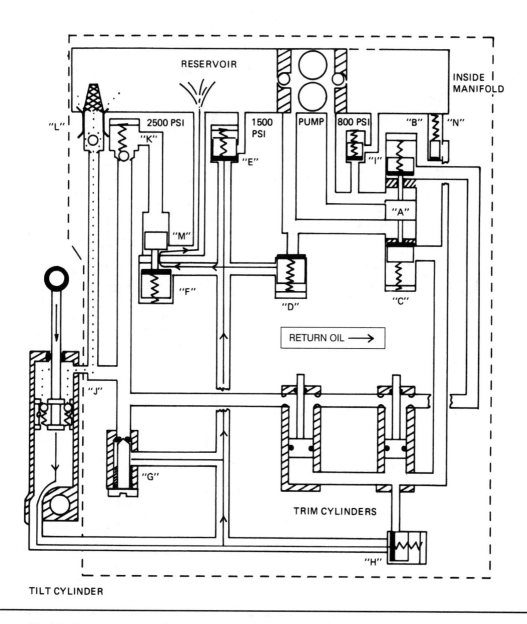

Figure 12–19   Return to water, hydraulic circuit Outboard Marine Corporation 70 HP through V-6.

cycled several times, and the reservoir rechecked for proper oil level. Run the unit up until the tilt cylinder bottoms out and pressure rises to the maximum. The Up button is released, and the pressure should not drop off by more than 100–200 PSI after the motor stops. If the gauge shows a greater pressure drop, then there is leakage in one or more of

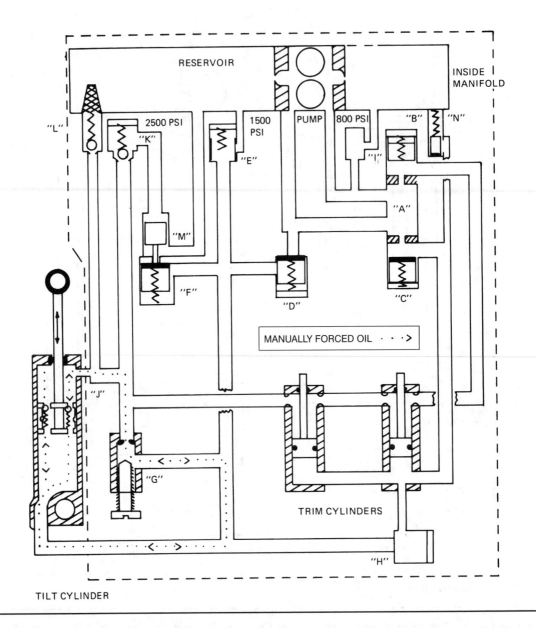

Figure 12–20 Manual release valve open, hydraulic circuit Outboard Marine Corporation 70 HP through V-6.

the following valves (see Figure 12–16), E: trim up relief valve; F: impact letdown valve; C: trim check valve; H: trim/tilt separation valve; D: tilt check valve; and J: shock absorber valves.

12.2.11   Seals on Trim Rod or Tilt Piston Rod

Individual valves can be checked using a special adapter, which pressurizes the valves when they are removed from the manifold.

12.3   Trim/Tilt Electrical System (OMC)

This system consists of a battery, remote control, a control box where all the connections and the relay are installed, a pump motor, and a trim indicator circuit. There are a single relay system and a dual relay system.

The starting battery should have a rating of at least 465 MCA. All the electrical power to start the outboard and operate the trim/tilt system comes from the battery.

In the *single relay system*, the remote control Up/Down switches receive electricity through a fuse in the control box (see Figure 12–21).

When the Up switch is pressed, electricity flows to the up control terminal A, through the coil and to ground D. This activates the relay, closing the contacts within. Twenty-three to 27 amps of current are then permitted to flow directly from the battery to terminal C and to the pump motor, which supplies torque for the hydraulic pump. From the pump motor electricity returns through the ground wire back to the relay ground D and then to the negative side of the battery, completing the circuit. This operates through the first 15 degrees of trim range, and also through the remaining 50 degrees of tilt range to the trailering position (Figure 12–12).

The down circuit is without a relay. When the Down control switch is pressed, electricity flows directly from the switch, through a control box terminal to the pump motor. This turns the pump motor on, in the reverse direction, and activates the down hydraulic circuit. Electricity returns from the motor to relay ground D, then to the battery negative terminal, completing the circuit.

In the *dual relay system*, the same 465 MCA battery is required. There is an up relay and a down relay (see Figure 12–22). Thirteen to 28 amps of current are carried in the down circuit. This relay adds longevity to the Down switch.

In the neutral position (no up/down movement), there is electricity to each relay battery terminal B. Motor terminals C are *normally closed* through the contacts (see Figure 12–23).

Electricity is also routed to the remote control. When the Up switch is pressed, electricity flows to the up relay, control terminal A through the coil to ground D (see Figure 12–24). This activates the relay closing the contacts. Electricity flows directly from the battery terminal B of the relay through the relay contacts to the motor terminal C. Then it continues to the pump motor, driving the pump and activating the hydraulic up circuit. Electricity flows from the pump motor to the down relay motor terminal C across the contacts to the ground terminal D. Then it continues back to the battery negative terminal, completing the circuit. If the operator continues to hold the Up switch depressed, pressures will rise in the hydraulic circuit. At 1500 PSI the trim up relief valve will open, bypassing oil back to the reservoir.

When the Down switch is depressed, electricity flows to the down relay control terminal A through the coil to ground D (Figure 12–25). This activates the relay, closing the

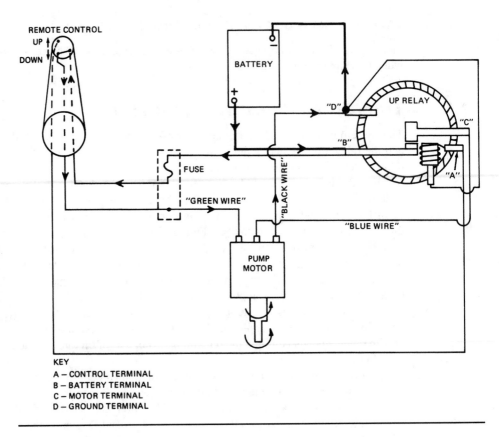

KEY
A – CONTROL TERMINAL
B – BATTERY TERMINAL
C – MOTOR TERMINAL
D – GROUND TERMINAL

Figure 12–21   Single delay down circuit operation. *(Courtesy of Outboard Marine Corporation)*

contacts, and permits electricity from battery terminal B to flow across the contacts to the motor terminal C and to the pump motor, reversing the rotation and activating the down hydraulic circuit. Electricity flows from the pump motor to the up relay motor terminal C through the contacts to the up relay ground terminal D, and on to the battery, completing the circuit. If the operator keeps the Down switch depressed, hydraulic pressures will rise. At 800 PSI the trim down pump relief valve will open in the hydraulic circuit, bypassing oil back to the reservoir.

12.3.1   Troubleshooting the Trim/Tilt Electrical System (OMC)

First, perform a battery capacity test to prove the battery's ability to operate the tilt/trim system properly. The battery should be rated at a minimum of 465 MCA and be fully charged.

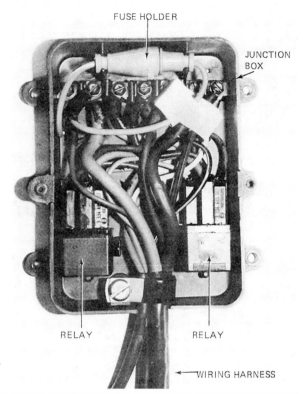

FUSE HOLDER

JUNCTION
BOX

RELAY   RELAY

WIRING HARNESS

Figure 12–22   Junction box of
power trim/tilt.

To test the control circuit, test for voltage at both relay battery terminals B (see Figure 12–24). Also, test for battery voltage in the trim control at the helm. To test the single and dual relay system, press the Up button. Voltage should be present at the up relay control terminal A and the relay should click. With the button still depressed, check for voltage at the motor terminal C of the up relay. The next check is at the pump motor connector. Battery voltage should be indicated on the voltmeter at each position. If battery voltage is not indicated at a given point, check the wire leading to that check point. There is an open or high resistance which is preventing the flow of electricity to that test point. Clean and repair the wire or connections. If voltage is indicated at the relay and the relay clicked, but no voltage is indicated at the relay motor terminal C, replace the relay. Next check for voltage at the motor connector. If voltage is indicated at the pump motor connector but the pump does not operate, what happened to the ground side of the circuit? Check the down relay to make sure it is in the normally closed position, having continuity between the motor terminal C and the ground terminal D. If it is not, replace the relay.

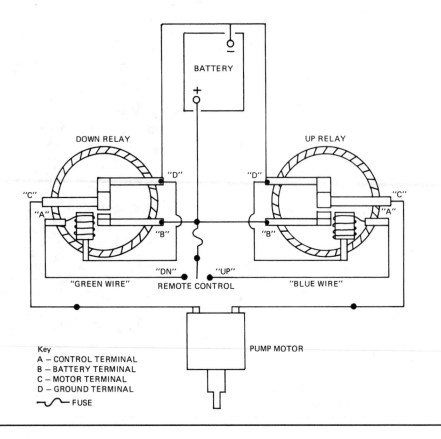

Figure 12–23   Dual relay neutral circuit operation.

The down circuit dual relay testing is basically the same (see Figure 12–25). Check for voltage coming to the down relay control terminal A. Does the relay click? With the button still depressed, check for voltage at the motor terminal C of the down relay. Next, check the pump motor connector. Battery voltage should be indicated on the voltmeter at each position. If voltage is not indicated at a given point, inspect the wire leading to that check point. There is an open or high resistance which is preventing the flow of electricity to the test point. Clean and repair the wire or connections. If voltage is indicated at the relay and the relay clicked, but no voltage is indicated at the relay motor terminal C, replace the relay. If voltage is indicated at the pump motor connector but it does not operate, check the ground circuit. Check the up relay to see if it is in the normally closed position between the motor terminal C and the ground terminal D.

If the circuits check out but the pump motor still does not operate, place a mechanic's remote starter switch between the relay battery terminal B and the green wire at the motor. Jump the blue motor wire to ground. Press the remote starter button, and the pump motor

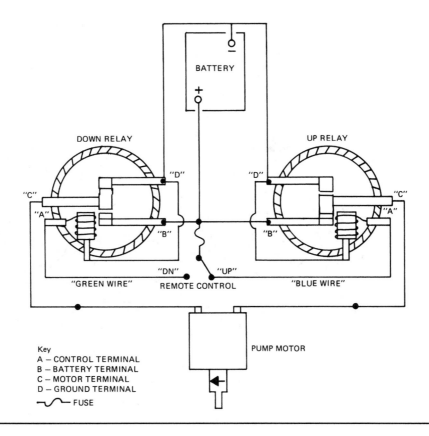

Figure 12–24 Dual relay up circuit operation

should operate. Then move the remote starter switch lead to the blue motor wire. Jump the green wire to ground. The pump should run in the opposite direction. If either test failed, repair the brushes and leads in the motor. The pump armature can be tested at your local dealer for shorts and grounds. Brushes and armature are available as replacement parts. You may have to replace the motor.

The down circuit of the single relay system is without a down relay (see Figure 12–21). To prove the circuit, test for electricity at the connector, then move to the control box terminal green wire. Next check the connector near the pump motor. Battery voltage should be present at all of these check points. If not, there is an open in the circuit. Inspect the wire and connections and make the necessary repairs. If there is voltage at the pump motor and it does not run, inspect the ground side of the circuit for loose, corroded connections, and disconnected or broken wires. If nothing is found that will prevent electrical flow, the next step is to install a mechanic's remote starter switch from battery terminal B to the green wire at the motor. Jump the black wire of the motor

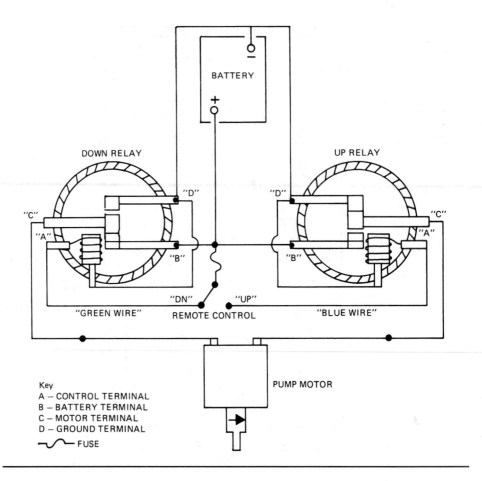

Figure 12–25   Dual relay down circuit operation.

to the negative terminal of the battery. Push the remote starter button and the motor should run. If it does not, repair or replace the pump motor. This testing includes a systematic approach of checking for battery voltage at accessible check points along the circuit. In this way, you will not overlook potential problems. Very often there is more than one repair necessary in a circuit to return that circuit to normal.

Another test that can be used to determine electrical and hydraulic condition of the trim/tilt system is performed using an ammeter. The ammeter is placed in series at the positive battery cable just before the relay. When the circuit is activated, an amperage reading will be registered. This reading can be compared to the normal reading in the service manual. This test gives an indication of the condition of seals, valves, and the pump motor. (See Figures 12–26, 12–27 and 12–28 for additional test information.)

COMMON PROBLEMS

a. *Tilt Leakdown* (Make sure that the trim does not leak.)

  1. Trim up relief valve.

  2. Manual release.

  3. Tilt cylinder valve or seals.

  4. Tilt check valve.

  5. Impact letdown valve.

  6. Oil lines.

b. *Trim and Tilt Both Leak.*

  1. Trim cylinders—sleeve O-rings or piston seals.

  2. Trim check valve.

  3. Expansion relief valve or O-rings.

c. *Reverse Lock Does Not Hold.*

  1. Filter valve seat.

  2. Impact sensor valve.

  3. Manual release valve.

  4. Tilt cylinder valves or seals.

  5. Reverse lock check valve.

  6. Oil lines.

d. *Trim Pulls in When Reverse Lock Does Not Hold.*

  1. Trim/tilt separation valve.

  2. Trim check valve.

e. *Slow Up or Down* (See "Troubleshooting with Ammeter" for maximum time.)

  1. Pump malfunctions.

  2. Motor malfunctions.

  3. Trim down pump relief valve (down only).

  4. Mechanical binding.

  5. Expansion relief valve O-ring.

f. *Will Not Trim or Tilt in Either Direction.*

  Motor runs fast—low current draw.

  1. Pump coupling broken.

  2. Low oil—or no oil in pump cavity.

  Electric motor does not run—high current or no current.

  1. Motor or pump bound up.

  2. Electrical problems—See "Troubleshooting the Electrical System."

g. *Will Not Trim Out Under Load Or Will Not Tilt.*

  1. Trim up relief valve.

  2. Manual release.

  3. Trim cylinder—sleeve O-rings, piston seals.

  4. Expansion relief valve O-ring.

h. *Will Not Trim Or Tilt Down.*

  1. Filter valve—seat.

  2. Manual release.

  3. Impact sensor valve.

  4. Trim down pump relief valve.

  5. Pump control piston not opening trim check valve.

  6. Trim check valve plugged.

i. *Unit Locked in Full Tilt Position.*

  1. Expansion relief valve.

Figure 12–26   Common problems of trim and tilt system. *(Courtesy of Outboard Marine Corporation)*

KNOW THESE PRINCIPLES OF OPERATION

- Basic principles of the power trim system.
- Operation of the electric and hydraulic systems.
- How the outboard is manually tilted.
- When trail out and reverse lock out occur.
- Reasons for the bounce system.
- How electrical/hydraulic circuits work together to trim and tilt.
- How impact shock is absorbed.
- Troubleshooting systems and common problems.

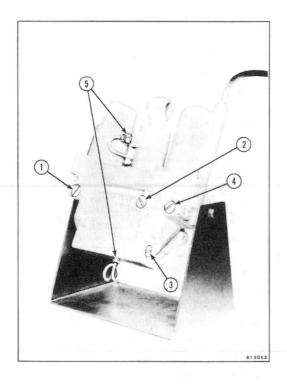

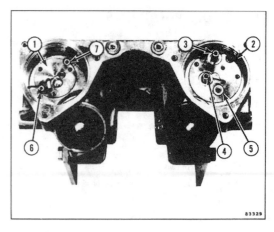

1. Motor and pump cavity
2. Reservoir cavity
3. Trim up relief valve
4. Impact sensor valve
5. Filter valve
6. Trim down pump relief valve
7. Expansion relief valve

1. Impact letdown valve
2. Tilt check valve
3. Tilt/trim separation valve
4. Trim check valve
5. Oil line fittings

Figure 12–27   **Integral power trim/tilt.** *(Courtesy of Outboard Marine Corporation)*

POWER TRIM/TILT

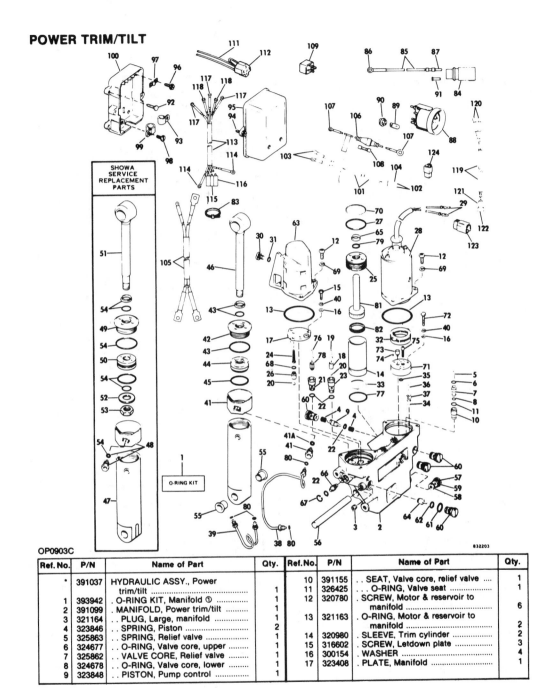

OP0903C

832203

Figure 12–28 Power trim/tilt (continued). (*Courtesy of Outboard Marine Corporation*)

Ref. No.	P/N	Name of Part	Qty.	Ref.No.	P/N	Name of Part	Qty.
*	391037	HYDRAULIC ASSY., Power trim/tilt	1	10	391155	. . SEAT, Valve core, relief valve	1
1	393942	. O-RING KIT, Manifold ①	1	11	326425	. . . O-RING, Valve seat	1
2	391099	. MANIFOLD, Power trim/tilt	1	12	320780	. SCREW, Motor & reservoir to manifold	6
3	321164	. . PLUG, Large, manifold	2	13	321163	. O-RING, Motor & reservoir to manifold	2
4	323846	. . SPRING, Piston	2	14	320980	. SLEEVE, Trim cylinder	2
5	325863	. . SPRING, Relief valve	1	15	316602	. SCREW, Letdown plate	3
6	324677	. . O-RING, Valve core, upper	1	16	300154	. WASHER	4
7	325862	. . VALVE CORE, Relief valve	1	17	323408	. PLATE, Manifold	1
8	324678	. . O-RING, Valve core, lower	1				
9	323848	. . PISTON, Pump control	1				

Ref. No.	P/N	Name of Part	Qty.	Ref. No.	P/N	Name of Part	Qty.
18	321347	. VALVE CORE, Relief and pilot valve	1	68	321542	. O-RING, Valve seat	1
19	321113	. SPRING, Relief and pilot valve	1	69	302290	. WASHER	6
20	160084	. BALL, Valve seat	2	70	321715	. RING BACKUP, End cap	2
21	323406	. VALVE SEAT, Relief valve	1	71	389836	. OIL PUMP ASSY.	1
22	202893	. O-RING	4	72	321219	. SCREW, Pump to manifold	1
23	321345	. VALVE SEAT, Impact sensor valve	1	73	322174	. DRIVE COUPLING	1
24	321165	. FILTER, Reservoir	1	74	322175	. BALL, Coupling	1
25	320788	. END CAP, Trim cylinder	2	75	321942	. SCREW, Socket head	2
26	321226	. VALVE SEAT, Filter	1	76	323407	. SPRING, Relief valve	1
27	319662	. O-RING, End cap	2	77	321283	. O-RING, Trim cylinder	2
28	393988	. TRIM/TILT MOTOR	1	78	389000	. VALVE CORE & SEAL ASSY.	1
29	511657	. . RING TERMINAL	2	79	328227	. QUAD-RING, Trim rod	2
•*	392202	. . END CAP & BRUSH REPAIR KIT	1	80	321117	. O-RING, Oil line	4
†*	390987	. . END CAP & BRUSH REPAIR KIT	1	81	322724	. PISTON AND ROD ASSY., Trim cylinder	2
30	321284	. PLUG, Reservoir	1	82	391622	. SEAL, Trim rod	2
31	307450	. O-RING, Fill plug	1	83	320107	CLAMP STRAP, Power trim cable	1
32	321350	. FILTER, Oil pump	1	*	582193	JUNCTION BOX & GAUGE ASSY.	1
33	321716	. RING, Backup sleeve	2	84	511465	. CONNECTOR	1
34	122757	. SEAL, Valve, pump & relief	1	85	582092	. LEAD ASSY.	1
35	321357	. O-RING	1	86	510151	. . TERMINAL	1
36	321282	. SPRING, Relief valve	1	87	511469	. . PIN	1
37	122759	. VALVE, Pump relief	1	88	387758	. INDICATOR GAUGE ASSY., Trim/tilt	1
38	390443	. OIL LINE ASSY., Upper	1	89	122552	. . BULB	1
39	387397	. OIL LINE ASSY., Lower	1	90	122553	. . SOCKET	1
40	306311	. LOCKWASHER	6	91	511496	. PLUG, Connector	2
*	387248	. CYLINDER ASSY., Tilt	1	92	310981	. SCREW, Mounting	5
□41	387584	. . ADPTR FITTING & NUT PKG.	1	93	317000	. CLAMP, Wiring harness	1
41A	173589	. . . O-RING, Fitting	1	*	582192	JUNCTION BOX & COVER	1
□42	173594	. . END HEAD ASSY.	1	94	328337	. SCREW, Cover	2
□43	174003	. . . O-RING PACKAGE, (Includes Item 45)	1	95	582191	. COVER & TRANSFER ASSY.	1
□44	387587	. . PISTON HEAD ASSY.	1	96	328337	. SCREW, Terminal board	6
□45	321446	. . O-RING	1	97	315004	. JUMPER, Terminal board	3
□46	387588	. . PISTON ROD ASSY.	1	98	321209	. SCREW, Clamp to junction box	1
*	387248	. . CYLINDER ASSY., Tilt	1	99	512066	. CLAMP	1
#47	390005	. CASE & BAND ASSY.	1	100	511870	. BASE, Junction box	1
#48	390004	. . BAND & NUT ASSY.	1	101	581960	. CABLE ASSY., Sending unit	1
#*	390002	. ROD ASSY. COMPLETE	1	102	309433	. . TERMINAL	2
#49	390006	. . CAP, O-ring & seal	1	103	510780	. . TERMINAL	2
#50	390000	. . PISTON ASSY.	1	104	511848	. . . CONNECTOR	1
#51	390001	. . . ROD & EYELET	1	105	582182	. CABLE ASSY., Battery	1
#52	324385	. . . WASHER, Rod end	1	106	582183	. LEAD & FUSE ASSY.	1
#53	324390	. . . HEX NUT, Rod end	1	107	510780	. . . TERMINAL	2
#54	390007	. . O-RING PACKAGE	1	108	511386	. . FUSE	1
55	321246	. BUSHING, Shock lower	2	109	582473	. RELAY & SEAL KIT	2
56	321202	. PIVOT PIN, Shock lower	1	111	582186	. CONNECTOR & LEAD	2
57	387517	. VALVE SEAT, Hsg & seal assy.	1	112	511866	. . CONNECTOR	1
58	304174	. . O-RING	1	113	582187	. CABLE ASSY., Trim switch & gauge	1
59	326649	. . O-RING	1	114	114208	. . RING TERMINAL	2
60	387376	. VALVE SEAT, Hsg & seal	4	115	309433	. . TERMINAL	3
61	326649	. . O-RING	1	116	309474	. . . CONNECTOR	1
62	304174	. . O-RING	1	117	510780	. . RING TERMINAL	3
63	391163	. RESERVOIR, Trim/tilt	1	118	511865	. . TERMINAL	2
64	321239	. PISTON, Letdown control	1	119	582188	. . CABLE ASSY., Trim/tilt motor	1
65	122494	. WIPER, Trim rod	2	120	511865	. . TERMINAL	2
66	387990	. VALVE AND SEAL ASSY., Manual release	1	121	325699	. . CONNECTOR	1
67	321362	. RETAINING RING, Manual release	1	122	511656	. . . TERMINAL	2
				123	325698	CONNECTOR, Trim/tilt motor	1
				124	511847	CONNECTOR, Sending unit	1

* Not Shown
# Showa Tilt Cylinder
□ Prestolite Tilt Cylinder
• Use Only on Motors With Part No. 20478 or 43728 Stamped on Bottom

† Use Only on Motors With Part No. PMP 4001 Stamped on Bottom
① Includes All O-Rings Required For Service of Complete Hydraulic Assy.

Figure 12–28   Power trim/tilt (continued).

REVIEW QUESTIONS

1. Large horsepower outboards use the single relay system (115 to 200 HP).
   a. true
   b. false

2. When the Up button is pushed , the relay_____.
   a. opens
   b. remains neutral
   c. closes

3. In the single relay system, the weight of the outboard and propeller thrust aid in trimming in the unit.
   a. true
   b. false

4. To place the outboard into the trailering position, which relay is activated?
   a. up
   b. down

5. What is the minimum MCA rating for the battery, when used with the power trim system? (88 thru 175 HP)
   a. 325 MCA
   b. 465 MCA
   c. 650 MCA

6. When troubleshooting the power trim system, the first step is to
   a. bleed the system.
   b. verify the customer's complaint.
   c. check the trim/tilt switch.

7. The limit switch limits operation
   a. within the trim range.
   b. for maximum lift in shallow water.
   c. above the trailering position.

8. If the manual valve is left open or leaking because of a broken O-ring, the unit trims out but not in.
   a. true
   b. false

9. When the manual valve is opened, the outboard may be tilted and the oil is allowed to move from the down side of the cylinder to the up side of the cylinder.
   a. true
   b. false

10. In the bounce system, if the lower unit strikes a submerged object with light steady pressure, the piston rod moves outward. This creates a vacuum between the shock valve and the floating piston. This condition momentarily exists until propeller thrust returns to the unit.
    a. true
    b. false

11.   The type of oil used in the older and late model systems is
      a.   transmission fluid
      b.   SAE 10W-30 automotive oil.
      c.   power steering fluid.
      d.   Both b and c are correct.

OMC Integral Power Trim and Tilt

12.   To trim the outboard, the trim cylinders push on the swivel bracket at the trim pads.
      a.   true
      b.   false
13.   The trim cylinders move the outboard through _____ degrees of movement and by mechanical and hydraulic advantage.
      a.   30
      b.   20
      c.   15
14.   The center tilt cylinder moves the outboard through _____ degrees of movement and into the trailering position.
      a.   40
      b.   50
      c.   60
15.   Torque of the pump motor shaft turns the pump, developing hydraulic pressure for the system.
      a.   true
      b.   false
16.   When the control button is held after the trim and tilt cylinders are retracted, hydraulic pressure will continue to rise, forcing open the _____ valve.
      a.   impact letdown
      b.   filter
      c.   pump relief
17.   When underway and running forward, the trim hydraulic oil must be held in a _____ condition.
      a.   static
      b.   balanced
      c.   free
18.   When operating in the shallow water mode, it is important to know
      a.   powerhead RPM.
      b.   position of the water inlet
      c.   that there is no possibility of ventilation at the propeller.
19.   Why are trail locks used while trailering the outboard?
      a.   to relieve the strain on the transom
      b.   to prevent damage to the transom brackets
      c.   to prevent possible damage in the hydraulic circuits

20. To troubleshoot the integral system for internal leakage,
    a.   dye is added to the reservoir.
    b.   a voltmeter is used to check voltage leakage.
    c.   two gauges are used.

21. Individual valves within the integral system may be checked by using _____.
    a.   adapters
    b.   dye
    c.   heavy oil

22. If the up relay clicks but the pump motor does not run, and there is no voltage at the relay pump terminal you should replace the_____.
    a.   pump motor
    b.   relay
    c.   up control switch

23. A mechanic's remote starter switch is used to jump from the relay battery terminal to the green wire on the pump motor.
    a.   true
    b.   false

24. On the single relay system, which relay is not used?
    a.   up
    b.   down

CHAPTER **13**

Propeller Performance

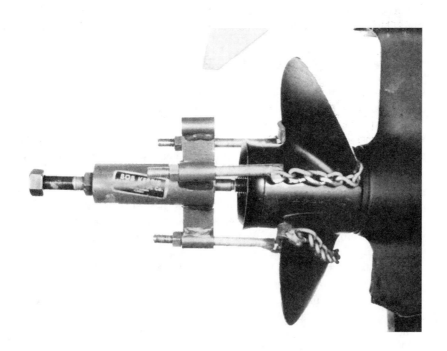

Objectives

After studying this chapter, you will know

- How the propeller works.
- Reasons for different hubs.
- What pitch and slip are.
- What blade rake is.
- Reasons for tilt angle adjustment.
- What cavitation is.
- What cupping is all about.
- Effects of propeller directional rotation.
- Trim/tilt and shock system.

13.1   Propeller Thrust

The purpose of the propeller is *thrust*, and nothing happens without it. How the outboard and boat perform are directly related to the propeller. It affects every phase of performance: handling, riding comfort, speed, acceleration, powerhead life, fuel economy, and safety. Care must be used in the selection of this vital part, so the correct propeller will be installed.

13.2   What the Propeller Does

The primary connection between the water and the outboard is the marine propeller. It moves the boat forward or backward; no propeller, no action (Figure 13–1).

The rotating propeller has blades designed to move the boat forward by creating a positive or pushing pressure on one side, and a negative or pulling pressure on the other side of the blades. Water is drawn into the propeller because of these pressures. Water moves to the front and accelerates out the back of the propeller.

You might wonder what type of a propeller to install. Should it be a three-bladed propeller of aluminum, bronze, or stainless steel? Of what design should it be ? Maybe a cleaver, a chopper, a Hi Rake, or a weedless propeller? (See Figures 13–2, 13–3, and 13–4.) The propeller (prop) must be selected for the job to be done with the boat. What are the primary uses for the boat—skiing, fishing, cruising, or is it a work boat? If the boat is going to be used for several activities, you may need a couple of propellers. The prop is something like a transmission in a truck. You select the gear (propeller) for the job to be done. When choosing the material used in the prop, consider the type of water in which you are going to operate. Will it be clear lake water, salt water, or shallow brackish water? The type of water the prop is run in can affect the longevity of the prop. Aluminum propellers usually come on the outboard. They are cheaper and do a good job. They have good strength and good resistance to corrosion and can be easily repaired. However, they are easily chipped or bent if they hit a sandbar or submerged tree stump or rock. The stainless steel prop is stronger, has thinner blades, and can resist corrosion and damage better, but it is more expensive. It is a choice for a replacement prop or for that second prop. Make sure you have the right tools in the boat to be able to change the prop if necessary.

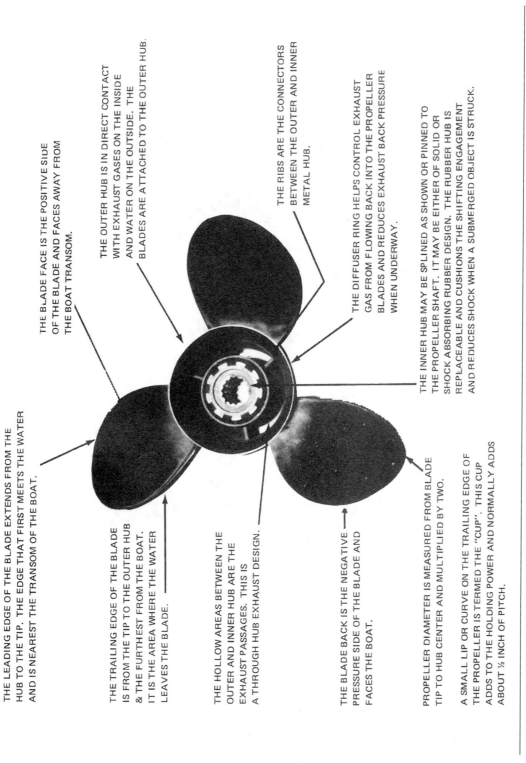

THE LEADING EDGE OF THE BLADE EXTENDS FROM THE HUB TO THE TIP. THE EDGE THAT FIRST MEETS THE WATER AND IS NEAREST THE TRANSOM OF THE BOAT.

THE BLADE FACE IS THE POSITIVE SIDE OF THE BLADE AND FACES AWAY FROM THE BOAT TRANSOM.

THE OUTER HUB IS IN DIRECT CONTACT WITH EXHAUST GASES ON THE INSIDE AND WATER ON THE OUTSIDE. THE BLADES ARE ATTACHED TO THE OUTER HUB.

THE RIBS ARE THE CONNECTORS BETWEEN THE OUTER AND INNER METAL HUB.

THE DIFFUSER RING HELPS CONTROL EXHAUST GAS FROM FLOWING BACK INTO THE PROPELLER BLADES AND REDUCES EXHAUST BACK PRESSURE WHEN UNDERWAY.

THE INNER HUB MAY BE SPLINED AS SHOWN OR PINNED TO THE PROPELLER SHAFT. IT MAY BE EITHER OF SOLID OR SHOCK ABSORBING RUBBER DESIGN. THE RUBBER HUB IS REPLACEABLE AND CUSHIONS THE SHIFTING ENGAGEMENT AND REDUCES SHOCK WHEN A SUBMERGED OBJECT IS STRUCK.

THE TRAILING EDGE OF THE BLADE IS FROM THE TIP TO THE OUTER HUB & THE FURTHEST FROM THE BOAT. IT IS THE AREA WHERE THE WATER LEAVES THE BLADE.

THE HOLLOW AREAS BETWEEN THE OUTER AND INNER HUB ARE THE EXHAUST PASSAGES. THIS IS A THROUGH HUB EXHAUST DESIGN.

THE BLADE BACK IS THE NEGATIVE PRESSURE SIDE OF THE BLADE AND FACES THE BOAT.

PROPELLER DIAMETER IS MEASURED FROM BLADE TIP TO HUB CENTER AND MULTIPLIED BY TWO.

A SMALL LIP OR CURVE ON THE TRAILING EDGE OF THE PROPELLER IS TERMED THE "CUP". THIS CUP ADDS TO THE HOLDING POWER AND NORMALLY ADDS ABOUT ½ INCH OF PITCH.

Figure 13–1 Basic propeller parts.

493

Figure 13–2   Three-blade
aluminum propeller.

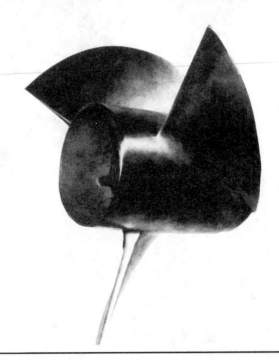

Figure 13–3   Cleaver propeller with solid-type hub.

13.3   How the Propeller Works

Consider how air moves through a fan. The electric fan pulls air in from behind it and
blows it out the front. The propeller of your boat does the same thing, except it moves

Figure 13–4   Chopper
propeller.

water instead of air (Figure 13–5). Water flows through the propeller as indicated by arrows as shown in Figure 13–6.

Water is pulled in from the front and through an imaginary tube a little larger than the propeller diameter. As the propeller spins, the prop water accelerates through this imaginary tube,

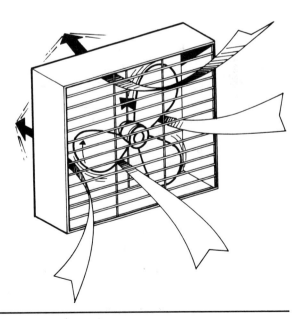

Figure 13–5   Air movement
through a fan.

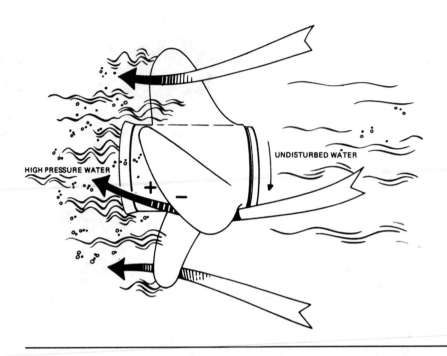

Figure 13–6   Water being pulled into and accelerated through propeller to create thrust.

creating a slightly smaller jet stream of higher pressure water behind the prop. Because of this prop water jet action, the imaginary tube diameter has become smaller than the prop. This jet action, which pulls in and pushes out the water very rapidly, results in a forward thrust. It is this pulling (-) and pushing (+) force that moves the boat forward (Figure 13–7).

As the prop rotates downward, water is pushed down and back naturally, just like with your hand when swimming. This results in positive pressure. At the same time, water is drawn in on the top (back) side of the blade as the blade turns. This results in the negative pressure. This action is taking place on all blades at the same time. So, negative pressure pulls the prop along, and at the same time, positive pressure pushes the prop (boat) along.

13.3.1   Solid Hub Versus Absorbing Rubber Hub Propellers

The solid type hub is generally used for racing (see Figures 13–4 and 13–8). Exhaust gases that are discharged from the powerhead are routed in such a manner that they do not disturb the action of the prop. There is no protection to the lower unit gear train for shifting or if a submerged object is hit. The splines are machined directly into the metal center bore of the hub. The shock-absorbing rubber hub type gives protection to the lower unit gear train when a submerged object is struck. It will torque, relieving some of the shock. This type also cushions the normal shock associated with shifting. The rubber hub may creep, but it

Figure 13–7 Negative pressure pulls the prop along, while positive pressure pushes it along.

is not considered a problem. A blade design is necessary that will perform properly with an exhaust gas enclosing outer ring. There are props with this rubber hub that are designed for a flow-through exhaust. Dropping the exhaust out through the prop aids in performance and makes for a quieter exhaust (Figure 13–8).

Figure 13–8 Mercury FLO TORQ® shock-absorbing hub propeller.

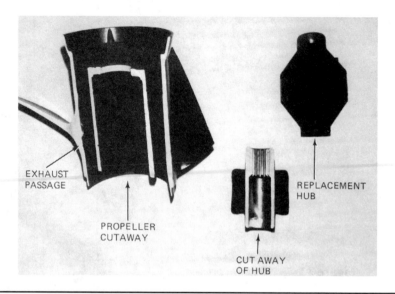

EXHAUST
PASSAGE

REPLACEMENT
HUB

PROPELLER
CUTAWAY

CUT AWAY
OF HUB

Figure 13–9   Propeller cutaway with FLO TORQ® shock-absorbing rubber hub cutaway.

13.4   Pitch and Slip

If you consider a wood screw going into a piece of wood, this gives you some idea of the lead (pitch) of the thread on the screw. In the same way, the propeller screws through water. Prop *pitch* is defined as the distance that a prop would move in one revolution if it were screwing through a soft solid (Figure 13–10).

There are two terms associated with pitch—progressive pitch and constant pitch (Figure 13–11).

Progressive pitch is a blade design that improves performance when rotation and forward speeds are high, and/or the water surface is broken by the blades. Progressive pitch starts at the lowest part of the leading edge and continuously increases to the trailing edge of the blade. Note that the blade has a curve when a straight edge is placed across the leading and trailing edge of the blade and is assigned an average pitch number. *Constant pitch* is a straight blade from the leading edge to the trailing edge, and is assigned a constant pitch number. It is straight, and therefore, pitch is constant.

Blade thickness varies from hub to tip. The design here is for strength and cantilever effect, with maximum strength near the hub and continuously reducing in strength and thickness, terminating at the blade tip. Blades should be as thin as possible, yet should retain their strength. This is one reason for popularity of stainless steel props. They are strong and thinner than aluminum propellers. Thin blades take less power, while thicker blades take more horsepower, and there is just so much horsepower available. (Figure 13–12).

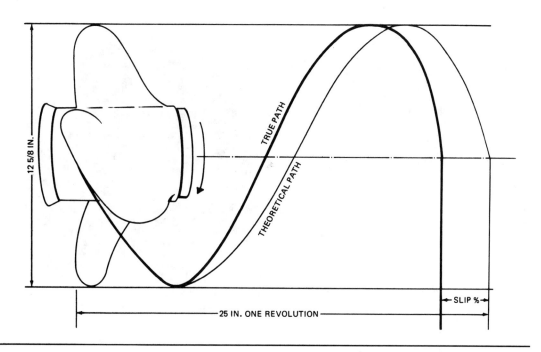

Figure 13–10   Pitch is the theoretical path of propeller travel.

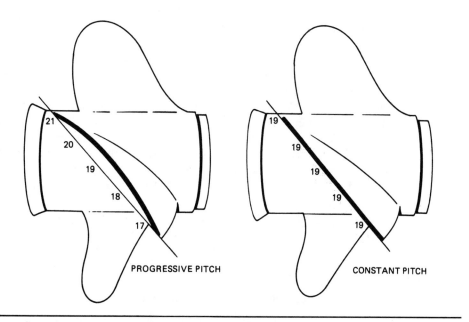

Figure 13–11   Progressive pitch improves performance.

Figure 13–12   Propeller pitch.

Numbers identify the prop, such as $12^5/_8$" diameter, 25" pitch. (***Propeller diameter*** is determined by measuring from the center of the propeller hub out to the tip of the blade and multiplying by 2.) If a single number is given, it is the pitch number. This is the distance in inches the propeller would screw through a soft solid in one revolution. We know that the prop does not turn through a soft solid but rather water, so ***slip*** is the percentage of designed pitch that the prop does not successfully travel in one revolution because of slippage in the water. Slip is necessary for a propeller to produce thrust and move the boat through the water. Too little or too much slip results in an inefficient prop. Slip is controlled by the proper selection of pitch, which allows for efficient engine operation within the operating RPM of the outboard. The most efficient amount of slip for the typical planing pleasure boat is between 10–15%, as shown in Figure 13–10. On a heavily loaded boat, possibly slower in planing, the slip increases to about 20–25%. There can be times of zero slip, for instance, when the boat is towed and the prop is wind-milling or on rapid deceleration. In general, slip can be considered the load on the prop. If slip rises to the critical point, the prop cavitates. The slip percentage formula is the following:

$$\frac{\text{Actual Boat Speed (MPH)} \times 1056}{\text{Prop Shaft RPM} \times \text{Prop Pitch (inches)}} \times 100$$

13.4.1   **What Pitch Is Correct?**

The correct pitch is dependent on several factors: boat hull design, size of outboard motor, and how you use the boat. There are also three propeller variables: diameter, pitch, and the number of blades. It takes a specific prop pitch for water skiing, possibly another pitch for cruising, and yet another for a work boat. The key that unlocks the correct pitch is the outboard motor revolutions per minute (RPM). Propeller selection is made to bring the outboard powerhead into the correct operating RPM, as shown in Figure 13–13.

In a truck transmission, the gear needed to pull the load is selected. The same thing applies with the propeller. Diameter, pitch, and the number of blades to pull the load must be selected. The dealer determines this combination for your motor and boat combination. The dealer's selection is not necessarily the correct propeller for your boat. The selection is theoretical, and has to be proven in the water while the boat is underway.

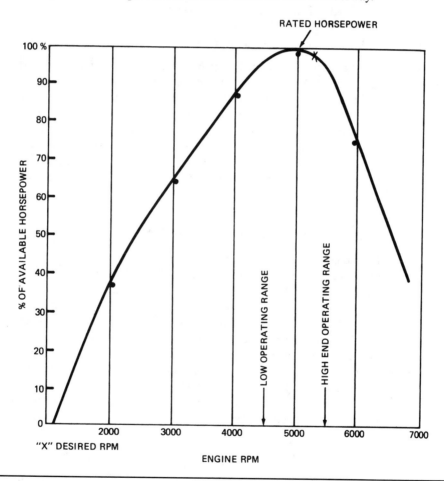

Figure 13–13   Example of available horsepower of one outboard.

A popular procedure for determining if the prop is correct is the RPM method. Before this method is used, make sure the powerhead is tuned, and that the boat hull (bottom) is in acceptable condition. Problems like hooks and rockers on the bottom will affect boat performance and overall MPH of the boat, but not necessarily powerhead RPM. To check for the correct prop you need an accurate tachometer. You may have a tach in your boat, but is it accurate? Check it out against a service tach at the dealership the next time the boat is in. With the tach installed and the average load you expect to have in the boat, you are ready to begin. Head for open water and ease it up to full throttle. With full throttle applied, the outboard powerhead should be in the operating range for your engine, possibly 4500 to 5500 RPM, with the heaviest load the boat will carry (see Figures 13–13 and 13–19). It should be running in the upper one-quarter (1/4) RPM limit for higher horsepower motors. Three general tachometer readings are possible:

1. RPM is in the recommended operating range (upper 1/4) = correct prop.
2. RPM is too high = increase pitch.
3. RPM is too low = reduce pitch.

Installation of a cupped propeller will usually reduce full throttle powerhead speed about 250 RPM below the same pitch prop without cupping. Propeller manufacturers generally design their props so that the next size in pitch will change motor RPM by 250 RPM. Remember that adjusting pitch is adjusting the load on the powerhead at wide-open throttle to bring the RPM into operating range. Operating below the recommended full throttle RPM range is laboring (lugging) the powerhead, and efficiency and power are lost. If operating beyond recommended operating RPM range, powerhead RPM is excessive, and efficiency and power are lost. Powerhead damage is possible in either situation.

13.4.2   Water Skiing? Change Pitch!

For the popular sport of water skiing, generally a lower pitch is best for popping the skier out of the water, gaining 10% more thrust for each pitch number dropped. You want to take advantage of available horsepower in the initial thrust, when the skier says "Hit it!" Because of the reduced pitch, the initial RPM will be higher for more thrust to pop the skier out of the water, and get the boat on plane faster. Once the skier is out of the water and the boat is on plane, the operator needs to keep an eye on the tachometer and not let the RPMs go continuously above the maximum operating range. Sustained operation above the maximum operating range can cause severe damage to the powerhead. When the skier is dropped or when just cruising around it is easy to exceed the maximum operating range, so watch that tachometer. If you are just doing some pleasure boating, then change the propeller back to the correct pitch for that type of boating.

 Note

Changing the propeller is like changing gears. It only takes a few minutes, if the propeller shaft and hub have had regular service.

13.5   Blade Rake

Rake is a measurement of tilt of the blade tip, toward or away from the lower unit. Blade rake is a bow raiser, increasing performance in the faster, lighter boats. A "Hi Rake" cup design propeller will allow a higher than normal outboard installation on the transom. This reduces drag by raising the propeller (lower unit) closer to the water surface. This design reduces the chance of ventilation in turns, and holds better than other designs. As the blade slants back from the vertical line, blade rake increases. The rake angle varies from 0–20 degrees for standard propellers (Figure 13–14).

13.6   Trim Angle Adjustment

Trim angle is an adjustment to be made once in the water and underway. The objective is to adjust the outboard lower unit in towards the transom or out (away) to maintain an angle of 90 degrees between the outboard exhaust housing (mid-section) and the surface of the water. This puts the prop shaft parallel to the surface. This gives the propeller stable water to push against and gives the boat a forward thrust. If the tilt angle is not correct, the boat tends to ***plow*** the bow down when the outboard is trimmed in toward the transom too far, or to run with the bow too high when the outboard is trimmed out too far (Figure 13–15). Trim angle depends on the hull, water condition, and load.

As the load of the boat changes with people or gear, the tilt angle should be trimmed in or out to have the anti-ventilation plate and prop shaft parallel to the surface of the water (mid-section 90 degrees to water surface). The best boat speed and operating economy are gained in this position. This adjustment is made by positioning the tilt pin in one of five

Figure 13–14   Propeller blade rake.

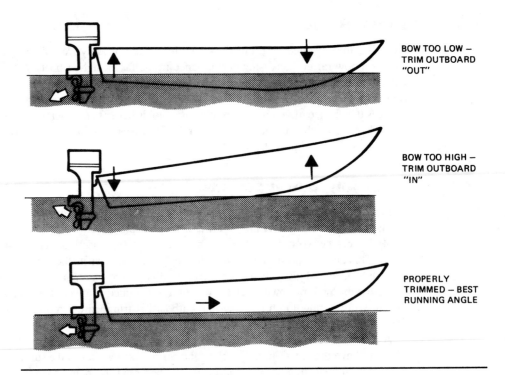

Figure 13–15   The effect of tilt angle on the boat.

holes on smaller horsepower motors, or by a power trim unit operated by an electric hydraulic pump and hydraulic cylinders. A boat that is properly trimmed out will have an ideal boat angle of 3–5 degrees in relationship to the water surface. This gives minimum drag, wetting enough of the bottom of the boat for good performance and stability (Figure 13–16).

Operating some boats at minimum trim in at planing speeds will cause unsafe steering conditions. Excessive trim out will also give stability problems on some high-speed hulls.

 Safety

Every boat should be tested for handling characteristics after each adjustment is made to the tilt angle. At normal tilt angle there should be little or no steering load. (See Figure 13–17.)

Steering torque is also affected by the tilt angle. With the outboard trimmed in, the prop shaft and therefore propeller blades are put on an increased angle to the surface of the

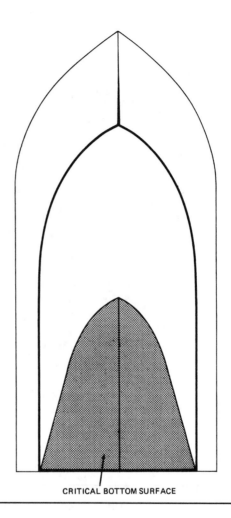

CRITICAL BOTTOM SURFACE

Figure 13–16 Wetted area when trimmed to 3 to 5 degrees.

water. The pitch of the downward moving propeller blade is increased, and there is greater load applied to the blade on the starboard side of the lower unit (Figure 13–18).

At the same time the load was increased on the downward moving propeller blade, it decreased pitch, unloading the upward moving port side blade. This created a left/right imbalance and pulled the boat to the starboard, which required correction by the operator. When the outboard is trimmed out, the exact opposite situation exists, as shown in Figure 13–19.

The port upward moving blade increased in pitch and load while the starboard blade decreased pitch and load, and the boat pulled to port. Therefore, a trimmed in or a trimmed out position increases steering torque. It is important to maintain the propeller shaft in an approximate horizontal position to the surface of the water. To help counteract hard steering left and easy steering right when trim is correct, many outboards are equipped

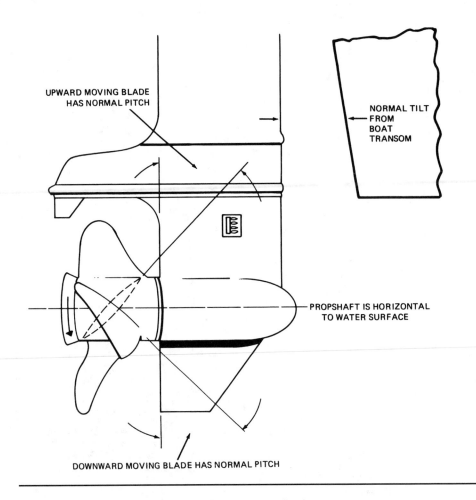

UPWARD MOVING BLADE
HAS NORMAL PITCH

NORMAL TILT
FROM
BOAT
TRANSOM

PROPSHAFT IS HORIZONTAL
TO WATER SURFACE

DOWNWARD MOVING BLADE HAS NORMAL PITCH

Figure 13–17   Tilt angle correct—little steering load.

with an adjustable trim tab, which can be set off-center to act as a rudder. This acts to balance the steering torque and give the necessary correction (Figure 13–20).

The trim tab may also be used to correct other minor hull conditions or designs. If the boat is hard to steer to the port (left), position the trailing edge of the trim tab to the port, as viewed from behind the boat. Reverse the procedure if it pulls to the starboard (right). The trim tab can act as a sacrificial anode to protect the lower unit from corrosion, when it is made of zinc and remains unpainted in the area of contact.

A shock-absorbing system must be coupled with this tilt system. Such a system allows the outboard to swing up on the pivot tube, if the lower unit strikes a submerged object. It can be as simple as a pivot tube, twin shock absorbers, or a hydraulic system that allows for the hydraulic lines and check valves to absorb the shock within the tilt cylinder, and the lower unit to swing up when hit (Figure 13–21).

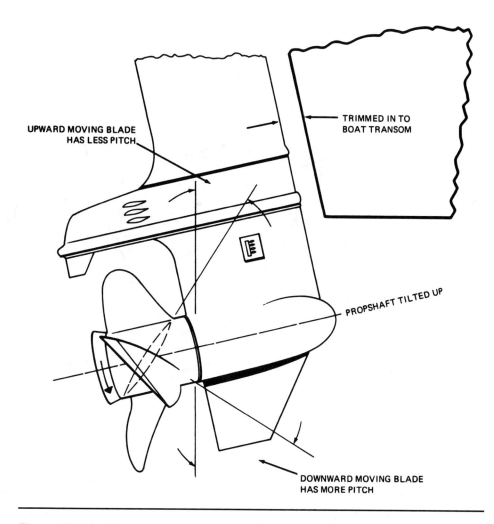

UPWARD MOVING BLADE
HAS LESS PITCH

TRIMMED IN TO
BOAT TRANSOM

PROPSHAFT TILTED UP

DOWNWARD MOVING BLADE
HAS MORE PITCH

Figure 13–18   Tilt angle in—left/right blade imbalance pulls lower unit to the right.

 **Warning**

If this shock system is not operational because the operator did not unlock it, or if the reverse lock out is engaged in forward gear, an accident is in the making. If the system is locked and the outboard lower unit strikes a stump or rock at high speed, the transom can be broken from the boat, with subsequent loss of boat and motor (Figure 13–22).

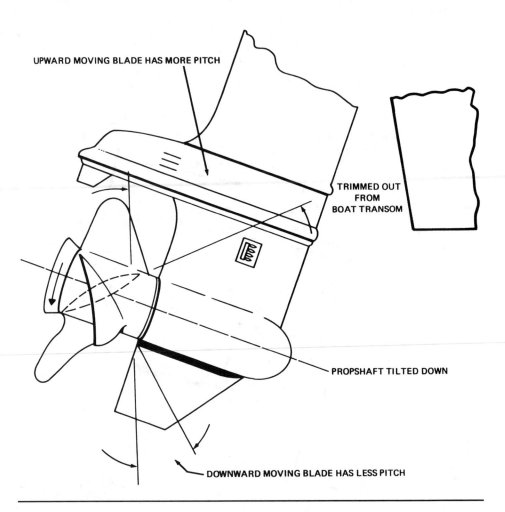

Figure 13–19   Tilt angle out—right/left imbalance pulls lower unit to the left.

13.6.1   Troubleshooting: Bow Too Low

When the outboard is trimmed (tucked) in too far, it forces the stern up and the bow down. The bow of the boat will tend to steer the boat, known as *bowsteering*. Fuel economy decreases and top speed decreases. The downward moving blade on the starboard side has more pitch and the upward moving blade on the port side has less pitch; therefore, the lower unit has a right/left imbalance and makes the boat want to steer right. Correct the condition by trimming out enough to place the prop shaft in the horizontal position (to water surface) so the thrust is directly to the rear. This movement will relieve steering torque.

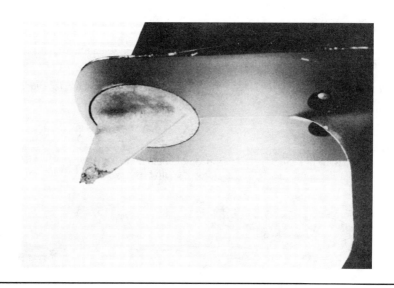

Figure 13–20   Trim tab used to correct steering condition.

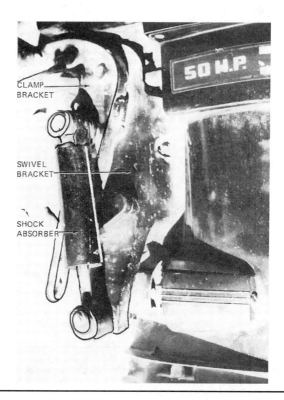

Figure 13–21   Shock system.

What's Important?

● Shock Absorption system must always be ready to Absorb some Blows to the lower parts of the motor.

● Motor must not Trim "In"   too far suddenly

What Can Happen?

● Without Shock protection,
    a Blow like this . . . . and Motor could break . . . or
                   Transom could break away . . . . . Motor may be lost overboard,
                                       . . . . . and Boat may . . . . . S
                                                            i
                                                            n
                                                            k

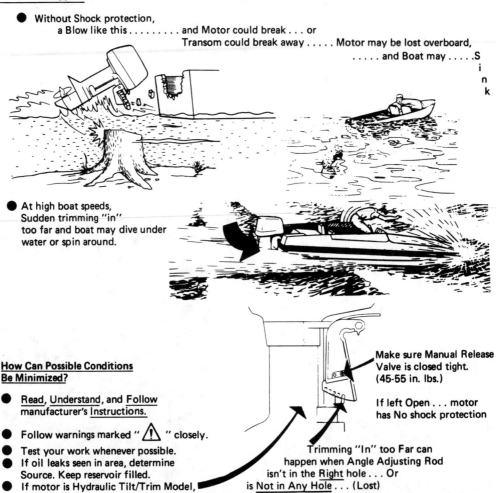

● At high boat speeds,
Sudden trimming "in"
too far and boat may dive under
water or spin around.

How Can Possible Conditions Be Minimized?

● <u>Read, Understand, and Follow</u> manufacturer's <u>Instructions.</u>

● Follow warnings marked " ⚠ " closely.

● Test your work whenever possible.

● If oil leaks seen in area, determine Source. Keep reservoir filled.

● If motor is Hydraulic Tilt/Trim Model,

Make sure Manual Release Valve is closed tight. (45-55 in. lbs.)

If left Open . . . motor has No shock protection

Trimming "In" too Far can happen when Angle Adjusting Rod isn't in the <u>Right</u> hole . . . Or is <u>Not in Any Hole</u> . . . (Lost)

Always return Rod to Hole Position determined earlier by Boat Operator . . . and make sure Angle Adjusting Rod Retainer is in locked position.

Figure 13–22   Outboard hydraulic/tilt/trim shock absorption system and persons safety. *(Courtesy of Outboard Marine Corporation)*

13.6.2   **Troubleshooting: Bow Too High**

When the outboard is trimmed out too far, it forces the stern down and the bow up. Getting on plane will be difficult and the propeller may lose its hold on the water. Depending on hull design, the boat may have the tendency to wobble (**walk**) left and right and porpoising may occur. The downward moving blade on the starboard side has less pitch and the upward moving blade on the port side has more pitch; therefore, the lower unit has a left/right imbalance and makes the boat want to steer left. Correct the condition by trimming in enough to place the prop shaft horizontal to the water surface, so the thrust will be directly to the rear.

13.7   **Cavitation**

What can be done to the water to make it boil at a lower temperature than 212° F? This is done by lowering the pressure applied to the water. The shape of the propeller blade moving through the water at high speed can cause a lowering of the pressure, which normally holds the water to the back and face of the blade. When this pressure reaches a sufficiently low level, boiling along the leading edge and face will occur. This occurs as the water attempts to make the sharp bend behind the leading edge of the blade, and may be aggravated by nicks or chips on the leading edge. As the water begins to boil, water vapor bubbles begin to form, moving downstream along the blade into a higher pressure area that won't sustain boiling. They condense back to water, and this condensing, collapsing action of the bubbles releases energy that chips away at the blades, causing a cavitation burn on the face of blade (Figure 13–23). A cavity in the prop means cavitation!

The moral of the story is to keep the leading edge of the propeller dressed, free of nicks, chips, and sharp leading edges. Also, use a prop that will allow the powerhead to run with the recommended operating RPM.

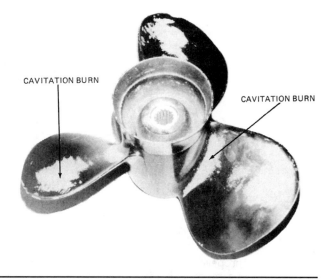

CAVITATION BURN

CAVITATION BURN

Figure 13–23   Propeller showing cavitation burns.

13.8   **Ventilation**

When discussing propeller problems you will hear the term ventilation. *Ventilation* is the result of air bubbles from surface air or exhaust gases being drawn into the blades. These pockets of exhaust/air cause a propeller to lose its grip or thrust. Engine RPM may accelerate wildly, but you do not gain or lose any speed. When an outboard that is mounted on a high transom and extreme trim-out is used or when making sharp turns, then ventilation may be noticed. Ventilation means air/exhaust is present at the propeller blades.

13.9   **Cupping of the Propeller Blades**

The cup is on the trailing edge of the blade, and is formed or cast with the edge curl outward (Figures 13–24 and 13–25).

Cupping the blade improves performance because the blades are able to hold better when operating in a cavitation or ventilating condition. With a better holding prop the outboard can be trimmed out further and mounted higher on the transom of the faster boats. If the blade pitch is increased, the load on the prop increases and, in the case of cupping, the cup has a load effect as if adding to the pitch, as well as to the rake. Usually a 150–300 RPM drop in wide-open throttle will be experienced, below the same pitch prop with no cupping. To have an effective cup, it should be completely concave and finished with a sharp corner. For total effectiveness of the cup, there should be no convex rounding. If you have a few chips or nicks, dress the trailing edge with a file, making sure you maintain the sharp corner and contour.

CUPPED BLADES IMPROVE PERFORMANCE.
THE TRAILING EDGE OF THE BLADE IS
CAST WITH THE EDGE CURLED INWARD.
IT REDUCES W.O.T. BY 150 TO 300 RPM.

Figure 13–24   Cupped propeller.

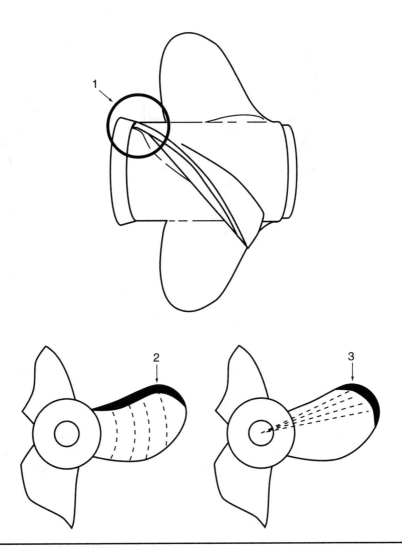

Figure 13–25 Propeller blade cup. (1) Blade trailing edge with an edge curl outwards. (2) Cupping in this area reduces RPM by adding pitch. (3) If cup is placed so that it intersects rake lines, it has the effect of increasing rake.

13.10 Propeller Directional Rotation

Most propellers have right-hand rotation on single outboard installations, causing a roll to the port. To offset this, the driver is placed on the starboard side, as shown in Figure 13–26.

When there is a dual installation, a left-hand counterclockwise rotating propeller can be installed. A more common dual installation probably is two right-hand rotating propellers. In this case, steering torque is offset by toeing in both outboards about 1/8 of an inch. By

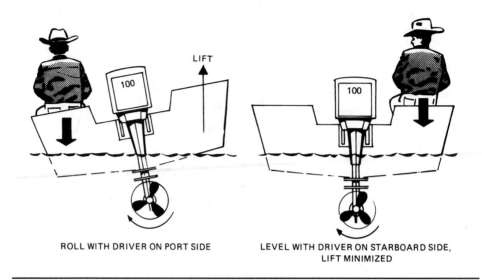

ROLL WITH DRIVER ON PORT SIDE   LEVEL WITH DRIVER ON STARBOARD SIDE,
                                              LIFT MINIMIZED

Figure 13–26   Boat roll, clockwise rotating propeller

doing this, steering torque effect is minimized by the opposite turning propellers or by toe-in, and the effect is self canceling.

To identify if a propeller has right-hand rotation, place the propeller with the diffuser ring down, and you will notice that the right-hand blade slants from the lower left to the upper right, pointing in the direction of rotation. The opposite is true for the left-hand rotating propeller. The left-hand blade slants from the lower right to the upper left. Thus, the opposite rotating propeller gives you a means of controlling some undesirable effects (Figure 13–27).

13.11   Damaged Propellers

Propellers that are damaged can sometimes be repaired. There are speciality shops that will repair your prop or offer you an exchange. Your local dealer can help you. If the pitch of the prop is incorrect for the type of boating you are doing, this same shop can change the pitch to meet your boating needs. Remember that minor nicks on the leading and trailing edge of the prop can be dressed with a file, so a good edge on the propeller blade is presented to the water (Figure 13–28).

At times you may suspect that the prop rubber hub is slipping. This can be checked by using a torque wrench and adapter made from an old prop shaft and nut. If the prop is still mounted, shift the lower unit into *neutral, remove the spark plug wires*, block the prop with a piece of wood against the anti-ventilation plate, and use the torque wrench on the prop nut and torque to hub specification. If the hub slips prematurely, exchange the prop.

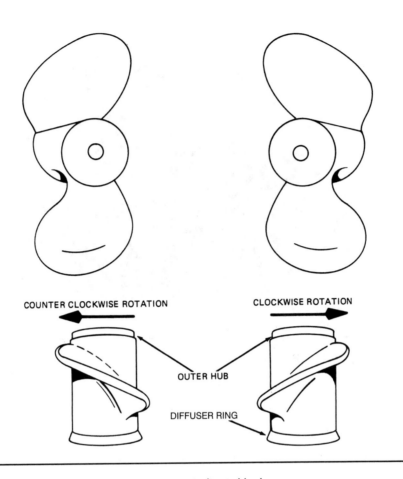

Figure 13–27 Pull is on lower unit as indicated by larger arrows.

 Note

Don't forget to reset the prop nut after torquing, so the prop can be changed easily. The rubber hub may creep with use and this is normal.

Blades that are severely damaged cause rotational and blade pressure imbalance. The propeller should be reworked in a propeller shop or exchanged for another one. Continued use of a damaged propeller will cause vibration and damage to the lower unit and/or the transom.

The propeller should be removed a couple of times a year at least to clean the propeller shaft splines and the propeller hub splines. With the splines clean, OMC Triple Guard grease or Merc Anti-Corrosion grease is applied to the splines to reduce corrosion. If this is not done, the splines will corrode and will seize the hub to the propeller shaft. The propeller will then have to be pulled (using a puller) or be burned off (see Figures 13–29 and 13–30).

Figure 13–28   Damaged propeller.

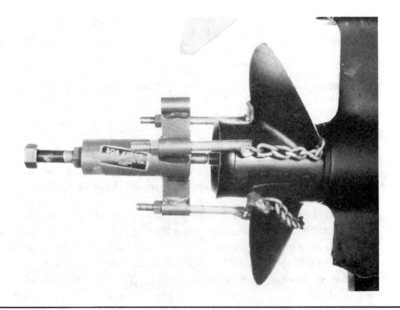

Figure 13–29   Pulling seized propeller from corroded propeller shaft splines.

Figure 13-30   Corroded propeller shaft/hub splines. No sealing compound or anti-corrosive grease was used.

Burning off means heating the inner propeller hub until the shock-absorbing rubber hub is hot enough to burn and release the propeller from the hub. The rubber from the hub has to be cut away and the splined hub drilled or chiseled from the propeller shaft; then the propeller can be rehubbed or exchanged. Of course, preventive maintenance on the splines using the proper grease is the way to go.

KNOW THESE PRINCIPLES OF OPERATION

- What the propeller does.
- How the propeller works.
- Function of the rubber propeller hub.
- Propeller pitch.
- How pitch affects engine RPM.
- Reasons for tilt angle adjustment.
- What causes cavitation.
- How cupping of the blades improves performance.
- How propeller rotation affects boat roll.
- Tilt/trim hydraulic shock absorption.

REVIEW QUESTIONS

1. What is the purpose of the marine propeller?
    a.   rotation
    b.   thrust
    c.   cavitation
2. The rotating propeller has blades designed to move the boat forward by creating a positive pressure on one side, and a negative pressure on the other side of the blade.
    a.   true
    b.   false
3. Forward thrust is the
    a.   pulling and pushing force that moves the boat.
    b.   negative and positive pressures working on the propeller blades.
    c.   rotating propeller creating a and b, so all are correct.
4. Propeller pitch is defined as the distance that a prop would move in one revolution if it were screwing through a soft solid.
    a.   true
    b.   false
5. Progressive pitch starts at the lowest part of the leading edge and continuously increases to the forward part of the hub.
    a.   true
    b.   false
6. Propeller blade thickness varies from hub to the tip to
    a.   increase pitch.
    b.   strengthen the through exhaust passage.
    c.   add strength and a cantilever effect, yet present a good edge to the water.
7. Propellers are identified by numbers, such as 12 5/8-25. The number 25 indicates the_____.
    a.   pitch
    b.   diameter
    c.   type of hub
8. The most efficient amount of slip for the typical planing pleasure boat is_____.
    a.   15–20%
    b.   5–10%
    c.   10–15%
9. Pitch, diameter, and the number of blades are three propeller variables.
    a.   true
    b.   false
10. To determine the correct pitch for a particular combination of outboard and boat, you should
    a.   measure transom height and pitch, and multiply by two.
    b.   use a tachometer and adjust the pitch until the RPM at wide-open throttle falls in the upper 1/4 of the operating range.

    c.   adjust tilt angle until the RPM is in the upper 1/2 of the operating range, and ventilation is occurring.

11.   When using the RPM method to determine propeller pitch for a given installation there are three possible readings; RPM is correct, RPM is too high, and RPM is too low. If RPM is too low you should_____.

    a.   reduce pitch

    b.   increase pitch

    c.   change the number of blades and use the same pitch

12.   For water skiing, you should change from your planing/cruising propeller to a _____ pitch propeller.

    a.   higher

    b.   lower

    c.   change types of hubs and increase pitch

13.   A Hi Rake cup design propeller will allow a higher than normal outboard installation on the transom.

    a.   true

    b.   false

14.   Tilt angle should be adjusted until the propeller shaft is tilted 15 degrees from the water surface.

    a.   true

    b.   false

15.   Operating some boats at minimum trim in at planing speeds will cause unsafe steering conditions.

    a.   true

    b.   false

16.   Steering torque is affected by the tilt angle.

    a.   true

    b.   false

17.   The trim tab may be used to correct steering conditions caused by minor hull problems.

    a.   true

    b.   false

18.   When the outboard is tucked in too far, this will cause the stern to _____ and the bow will _____.

    a.   rise/lower

    b.   lower/rise

    c.   walk/rise

19.   Cavitation is vapor bubbles forming near the leading edge of the propeller blade and flowing back along the blade and condensing back, releasing energy.

    a.   true

    b.   false

20. Cupping of the propeller blades improves performance because the blades are able to hold better when operating in a cavitation or ventilation condition.
    a. true
    b. false
21. Boat roll may be caused by
    a. a counterclockwise rotating propeller.
    b. increased tilt angle.
    c. decreased tilt angle.
22. Damaged propellers may cause
    a. vibrations in the boat.
    b. a change of powerhead RPM.
    c. Both a and b are correct.

Boat Performance Problems

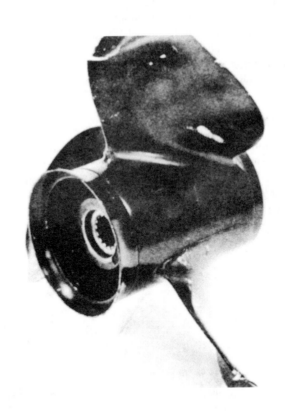

Objectives

After studying this chapter, you will know

- Propeller ventilation.
- Effects of a trim angle on how the boat planes.
- General problems that affect boat performance.
- How a damaged propeller affects performance.
- How marine fouling creates a drag.
- Reason for the boat capacity plate.
- Boat powering.

14.1   Propeller Ventilation

As described in Chapter 13, *Propeller Performance*, propeller ventilation can cause performance problems. Ventilation (sometimes incorrectly referred to as cavitation) occurs when air from the water's surface or exhaust gases from the exhaust outlet are drawn into the propeller blades. The normal water load is reduced and the propeller over-revs, losing much of its thrust. As the propeller momentarily over-revs, this brings on massive cavitation that can further unload the propeller and stop all forward thrust. This continues until the propeller is slowed down enough to allow the bubbles to surface. This action occurs most often in turns, particularly when trying to plane in a sharp turn or with an excessively trimmed out engine or drive unit (Figure 14–1).

14.2   Trim Angle

Another performance problem can be caused by trim angle. Trim angle of the outboard motor or stern drive unit has a significant effect on the way the boat planes, rides, and steers.

If trimmed in too far the top speed drops, fuel economy decreases, and the boat may oversteer in one direction or the other, called "bowsteering." Also steering torque will increase. Getting on plane should be easier with some V-bottom hulls. The ride in choppy water on plane at part-throttle should be smoother.

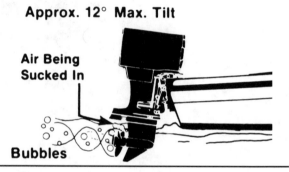

Figure 14–1   propeller ventilation. *(Courtesy of Boating Industry Association)*

If trimmed out too far the propeller may lose its hold on the water and steering torque will increase in the opposite direction to that of being trimmed in. Getting on plane may be difficult or labored. Porpoising of the boat may also occur. A fast V-bottom may start to "walk" from right to left, etc.

Although the engine is depicted as being in the best running angle with the anti-ventilation plate parallel to the boat bottom, this may not produce the best hull running attitude (Figure 14–2).

14.3   Weight Distribution Versus Boat Performance

Weight distribution is extremely important for proper boat performance. It affects a boat's running angle or attitude. For best top speed, all movable weight—fuel, battery, anchor, passengers—should be as far aft as possible for the bow to come up to a more efficient angle (3 to 5 degrees). The problem with this approach is that, as weight is moved aft, some boats will begin an unacceptable porpoise. Second, as weight is moved aft, getting on plane becomes more difficult. Finally, the ride in choppy water becomes more uncomfortable as the weight goes aft. With these factors in mind, each boater should seek out the weight locations to best suit his needs.

Weight and passenger-loading placed well forward increases the "wetted area" of the boat bottom and virtually destroys the good performance and handling characteristics of the boat. Operation in this configuration can produce an extremely wet ride from wind-blown spray, and could even be unsafe in certain weather conditions or where bowsteering may occur (Figure 14–3).

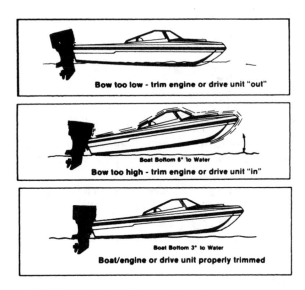

Figure 14–2   Trim angle. *(Courtesy of Boating Industry Association)*

Figure 14–3   Weight distribution versus boat performance.
(Courtesy of Boating Industry Association)

Weight distribution is not just confined to fore and aft locations, but also applies to lateral weight distribution. Uneven weight concentration to port or starboard of the longitudinal centerline can produce a severe listing attitude that can adversely affect the boat's performance, handling ability, and riding comfort in extreme rough water conditions. The safety of the boat and passengers may be in jeopardy.

14.4   General Problems Affecting Performance

The following problems always cause performance problems and should receive the boater's attention:

- Marine growth on prop or lower unit.
- Damaged propeller.
- Weeds on propeller or lower unit.
- Overload.
- Hot, muggy weather.
- Malfunctioning engine.
- Water in boat.

14.5   Damaged Propellers

As discussed in Chapter 13, *Propeller Performance*, damaged propellers always cause performance problems. Each time a product is serviced, the propeller should be inspected for damage.

Even slight propeller damage can mean a loss of one MPH. Greater damage can mean considerably more speed loss. Worse yet, the same amount of damage is usually not done to each blade and, therefore, sets up imbalance vibrations that can cause fatigue damage to other parts of the engine or drive (Figure 14–4).

14.6   Marine Fouling

Marine fouling can cause severe performance problems. Fouling is unwanted buildup (usually animal/vegetable-derived growth) occurring on the boat's bottom and drive unit.

Figure 14–4   Damaged propeller. *(Courtesy of Boating Industry Association)*

Fouling causes drag, which reduces boat performance. In freshwater, fouling results from dirt, vegetable matter, algae or slime, chemicals, minerals and other pollutants. In saltwater, barnacles, moss, and other marine growth can produce dramatic buildup of material quickly (Figure 14–5). It's important to keep the hull as clean as possible in all water conditions, to maximize boat performance.

Special hull treatments, such as anti-fouling paint, will reduce the rate of bottom fouling. However, the drive units (outboard or stern drive) are made primarily of aluminum, so be sure to select an anti-fouling paint with a copper-free organo-tin base. The BIS Tri Butyl Tin Adipate (TBTA) base paint will not set up a galvanic corrosion cell as it is completely compatible with aluminum and avoids any *electrolysis* problems connected with many other paints. Applied according to instructions, it is also very effective.

14.7   Powering the Boat

Certain requirements have been set by the American Boat and Yacht Council (ABYC) for manufacturers and boaters to abide by for safety and performance. These requirements in-

Figure 14–5   Marine fouling. *(Courtesy of Boating Industry Association)*

volve determination of boat maneuvering speed and maximum horsepower capacity for the outboard to be installed on the transom. Abide by these requirements.

14.7.1   Outboard Horsepower Capacity

The maximum horsepower outboard to be installed is stated on the boat's capacity plate. This should never be exceeded! The manufacturer determined and built the boat and transom (boats over 20 feet) to handle this maximum horsepower. For conventional boats under 20 feet, horsepower capacity is determined as follows:

1. Multiply the overall boat length in feet by the maximum transom width. If spray rails aft act as chimes or are part of the planing surface, they are to be included.
2. Next locate the factor and corresponding horsepower capacity in the table below. If the factor is over 52, the boat horsepower capacity is computed by use of the applicable formula in the table on the following page. If the horsepower capacity is not an even multiple of 5 it should be raised to the nearest multiple of 5. If the factor is 35, horsepower is 3.

Horsepower Capacity

Multiply overall length_____ × stern width _____= factor _____

							Remote wheel steering and 20" transom or equivalent	No remote wheel steering or transom less than 20" or equivalent	
								Flat bottom hard chine boats	Other boats
If this factor is	thru 35	36-39	40-42	43-45	46-52		over 52	over 52	over 52
H.P. capacity is	3	5	7½	10	15		(2 x factor) – 90	(1/2 factor) – 15	(.8 factor) – 25

H.P. capacity = _____ (raise to nearest 5 horsepower increment if factor is over 52)

Flat bottom hard chine boats—reduce horsepower capacity one increment for factors through 52.

14.7.2 Boat Maneuvering Speed

Another safety consideration is the conventional boat's maneuvering speed. It is determined by using one of two methods: the Quick Turn and Test Course methods. These tests are used on boats using a remote steering wheel.

The boat maneuvering speed is determined for boats 20 feet and longer in length and able to exceed 40 MPH, using the rated outboard horsepower indicated on the certification plate. Boats under 20 feet with remote steering, capable of exceeding 35 MPH or more, are also tested using the maximum rated horsepower listed on the boat certification plate. When a test is made, certain requirements must be met: trim angle, amount of gasoline in tank, recommended propeller, and the outboard installed in the lowest position on the transom. Either of the following tests may be used:

Quick Turn Test

At a given throttle setting of 40 MPH or higher, the operator steers the boat straight ahead. With calm water conditions the operator turns the boat steering wheel 180° in one-half second and holds there without changing trim or throttle setting. The boat successfully passes the test if a 90° turn is accomplished and the operator remains confident in maintaining control. When the test is successful, turn entry (throttle) speed is incrementally increased, running additional tests until the operator does not feel comfortable completing the test. *Maneuvering speed* is set at the highest speed the boat successfully completes the test.

Test Course Method

The tilt angle is set so ventilation is limited and, therefore, no loss of directional control occurs. The throttle speed is set at 35 MPH and the test course is run a minimum of three consecutive times in each direction, passing outside the designated course markers without hitting them.

If the 35 MPH run is successful, repeated tests are made at increased speeds of 5 MPH increments until the test cannot be completed without making contact with any of the markers. The highest speed run without making marker contact is the maximum maneuvering speed. While negotiating the course there shall be no oscillating motion in the roll or yaw axes exhibited.

If a maneuvering speed less than the top speed on the boat is determined, a warning shall be posted on the boat in a location that is highly visible to the operator, such as the following:

 Warning

Maneuverability above 45 MPH is limited. Sudden turns may cause loss of control. *Read your owner's manual!*

Speedometers are installed when maneuvering speed is less than the top speed of the boat. Speedometers are also found in boats that pass maneuvering specifications and are part of the normal equipment.

Below is the basic layout of the Test Course. Buoy markers are placed in the water at prescribed footage intervals and with the boat properly loaded. The operator attempts to run the course at a prescribed speed.

Test Course—35 to 50 MPH

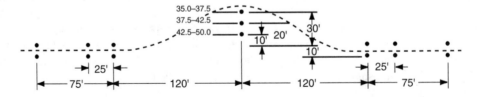

The information in this chapter comes from the American Boat and Yacht Council (ABYC) and the National Marine Manufacturers Association (NMMA) handbook. A complete handbook of information on marine boating standards is available (for a fee) by contacting these organizations. The following table lists additional performance problems and possible causes.

Boat Problem		Specific Possible Causes	
1.	Poor speed—light load.	a.	Prop pitch too low or far too high.
		b.	Load too far forward.
		c.	Lower unit too low in water.
		d.	Trim too far in.
2.	Poor speed—heavy load.	a.	Under-powered.
		b.	Incorrect propeller selection.
		c.	Trim too far in or out.

(continued on next page)

Boat Problem	**Specific Possible Causes**
3. Slow to plane—heavy load.	a. Trim too far out.
	b. Propeller pitch too high.
	c. Too much load in stern.
4. Hard ride in rough water.	a. Too much load in stern.
	b. Trim too far out.
	c. Poor speed management.
	d. Flatter bottom boat design.
5. Runs wet in rough water.	a. Load too far forward.
	b. Trim too far in.
	c. Poor speed management.
6. Lists on straight when heavily loaded.	a. Load not evenly distributed.
	b. Trim too far in.
	c. Hull has more hook on one side.
7. Lists or rolls on straight when lightly loaded.	a. Loose steering.
	b. Trim too far in.
	c. Incorrect transom height.
	d. Load too far forward.
	e. Hull has a hook.
8. Nose heavy—catches on waves and in turns.	a. Trim too far in.
	b. Load too far forward.
	c. Hull has a hook.
9. Porpoises on straight run.	a. Trim too far out.
	b. Too much load in stern.
	c. Hull has a rocker.
10. Porpoises on turns only.	a. Trim too far out.
	b. Boat has less hook or more rocker toward chine.
	c. Over-powered for hull design.
11. Banks too much in turns.	a. Load too far forward.
	b. Trim too far in.
	c. Hull has a hook.
12. Excessive ventilation (or cavitation).	a. Incorrect propeller selection.
	b. Outboard mounted too high on transom.
	c. Trim too far out.
	d. Keel extends too far after thru-hull fittings disturb water flow.
13. Excessive fuel consumption.	a. Cruising in high headwinds.
	b. Overloaded boat or improper load distribution.
	c. Misadjusted engine.
	d. Leaking fuel line fittings.
14. Chine walking.	a. Engine too low.
	b. Trim too far out.

(continued on next page)

Boat Problem	**Specific Possible Causes**
	c. Load too far adrift.
	d. Loose steering
15. Excessive steering torque.	a. Trim too far in, or out.
	b. Misadjusted trim tab.
	c. Engine mounted too high.
	d. Bent skeg.

KNOW THESE PRINCIPLES OF OPERATION

- How water-surface air and exhaust gases are drawn into the propeller blades.
- How trim angle affects the way a boat planes.
- What affects the boat running angle or attitude.
- Problems that adversely affect boat performance.
- What can be done to avoid marine fouling.
- Boat powering requirements.
- What the boat capacity plate tells you.

REVIEW QUESTIONS

1. Ventilation occurs when air from the water's surface or exhaust gases from the exhaust outlet are drawn into the propeller blades.
    a.   true
    b.   false
2. When does ventilation generally occur?
    a.   in turns
    b.   with an excessively trimmed out outboard
    c.   Both a and b are correct.
3. If the trim angle is trimmed in too far,
    a.   RPM increases.
    b.   top speed drops and fuel economy decreases.
    c.   it will not cause the boat to oversteer.
4. If the trim angle is trimmed out too far, steering torque will increase in the opposite direction to that of when it's trimmed in.
    a.   true
    b.   false
5. Marine fouling adds up to_____.
    a.   a smoother ride
    b.   drag
    c.   increased powerhead RPM
6. When selecting a paint for the bottom of the boat, it should contain a copper base.
    a.   true
    b.   false

7. Maximum horsepower capacity for a boat is listed on the capacity plate.
   a. true
   b. false

8. Certification standards for boats are given
   a. in the service manual.
   b. on the capacity plate.
   c. in the NMMA handbook.
   d. Both b and c are correct.

9. The manufacturer of the boat is listed on the capacity plate of newer boats.
   a. true
   b. false

10. A damaged propeller will create imbalance vibrations and a loss of miles per hour.
    a. true
    b. false

11. The maximum capacity for persons or weight in a boat is listed
    a. on the bow of the boat.
    b. on the size of boat tag attached to the outboard.
    c. on the boat's capacity plate mounted on the transom.

12. When installing equipment in a boat, you should follow
    a. National Marine Manufacturers Association guidelines.
    b. US Coast Guard requirements.
    c. Both a and b are correct.

13. What causes excessive fuel consumption?
    a. overloaded boat or improper load distribution
    b. misadjusted engine
    c. leaking fuel line fittings
    d. All of the above are correct.

14. Weight and passenger loading placed well forward increases the "wetted area" of the boat bottom and virtually destroys good performance.
    a. true
    b. false

15. What two tests are used to determine boat maneuvering speed? _____ and
    _____
    a. Quick Turn
    b. Test Course method
    c. High Speed Turn method
    d. Either a or b.

Rules of the Road

The information in this appendix includes safety information, navigational information, weather and distress signals, information on boat lighting, and Coast Guard information on obtaining assistance and reporting accidents. By following these rules, you keep yourself, your passengers, and fellow boaters safe. Remember safe boating is no accident!

Personal Flotation Devices

All boats, powered or nonpowered, must carry at least one Coast Guard approved personal flotation device *(PFD)* for every person aboard (see Figures A–1 and A–2). Failure to have a sufficient number of approved devices aboard consititutes a violation of state and federal law.

Such devices must be kept in serviceable condition. If straps are broken, hardware is missing, or the approval number cannot be read, the device is considered to be unserviceable. Having devices that are in an unserviceable condition also constitutes a violation.

Inflatable PFDs

Some newer Inflatable Personal Flotation Devices (PFDs) are now approved by the Coast Guard. This is indicated by UL testing and Coast Guard approval numbers that are affixed to the inflatable PFDs. PFDs come with a booklet to explain to the customer the use and care, as well as who the intended user should be. These PFDs can be carried aboard recreational vessels as acceptable life jackets. There are numerous inflatable PFDs on the market but not all are Coast Guard approved.

What You Need To Know

- Manual approved PFDs are Type III, regardless of their buoyancy factor.
- User must perform required maintenance on all inflatables PFDs.
- Where water impact is expected, such as water skiing or riding personal water craft, inflatables should not be used. Instead use a Type III four-belt water ski vest.

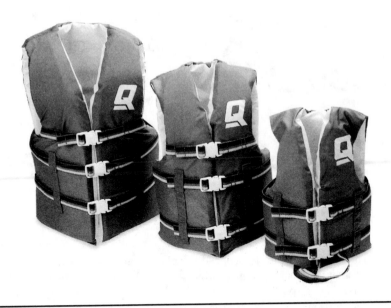

Figure A–1   Quicksilver three-belt boating vests with belts, nylon zipper, and lace side adjustments. USCG approved. *(Courtesy of Mercury Marine)*

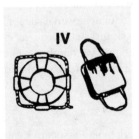

Figure A–2   This Type IV throwable flotation device has 16.5 pounds of flotation and is designed to be thrown.

- Non-swimmers or weak swimmers should not use inflatable PFDs unless the PFDs are already inflated.
- Persons weighing less than 80 pounds should not use inflatable PFDs.
- Many brands of inflatables do not float without inflation.
- After each inflation, inflatable PFDs must be rearmed unless inflated orally.
- Status of each inflatable PFD and cylinder must be checked before each trip.
- Children under 16 should not use an inflatable PFD.
- Inflatable PFDs should not be worn under restrictive clothing, because accidental or intended inflation could restrict breathing and cause injury.

Tips On Using Your Personal Flotation Device (PFD)

- Buy your own PFDs
- Look at the label. Is it Coast Guard approved? What is the weight and size limitation?
- Try it on and check the fit. Once the straps and buckles are secured, it should not slip over your head or come above your ears.
- Throw away a PFD if you find mildew, rot or hear air leaking.
- Do not alter a PFD.
- Annually check your PFDs for fit and flotation.
- Wear a PFD to set an example for children.
- Never use water toys in place of a life jacket.

REMEMBER, A PFD FLOATS AND YOU DON'T!

Fire Extinguishers

All motorboats, unless exempt, must carry a serviceable Coast Guard approved fire extinguisher (see Figure A–3).

Open-construction, outboard motorboats less than 26 feet long, without permanently installed fuel tanks, are not required to carry fire extinguishers if the boats do not carry passengers for hire. *Fire extinguishers are recommended, however.*

Each fire extinguisher is classified by letter and Roman numeral according to the type of fire it can extinguish and the size of the extinguisher.

Unless exempt, boats under 26 feet long must carry one B-I fire extinguisher. Boats 26 to less than 40 feet must carry two B-Is or one B-II. Boats 40 to 60 feet must carry three B-Is or one B-I and two B-IIs.

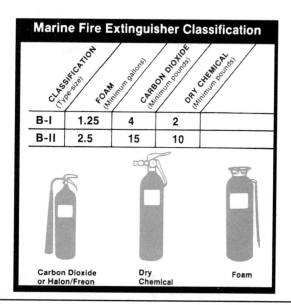

Marine Fire Extinguisher Classification

CLASSIFICATION (Type-size)	FOAM (Minimum gallons)	CARBON DIOXIDE (Minimum pounds)	DRY CHEMICAL (Minimum pounds)
B-I	1.25	4	2
B-II	2.5	15	10

Carbon Dioxide or Halon/Freon   Dry Chemical   Foam

Figure A–3 Marine fire extinguisher classification.

Federal Channel Marking System

The diagram in Figure A–4 shows the course a boat will take following the lateral system of buoyage.

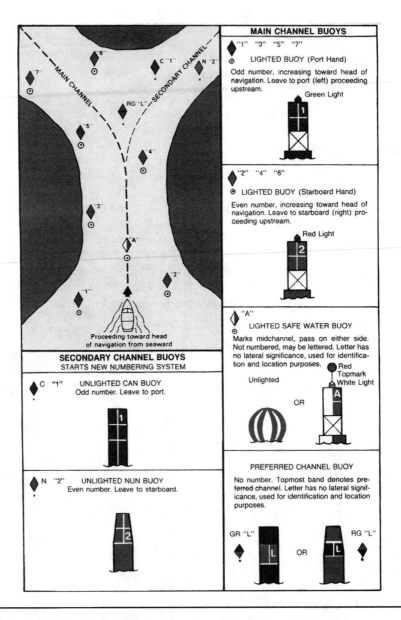

Figure A–4   Main channel and secondary channel buoys.

Weather Displays

At selected locations in and near boating areas, storm advisories are displayed by flag hoists or lights (see Figure A–5). Display points are usually Coast Guard stations, yacht marinas, or municipal piers. A boater should become familiar with the display stations in the area and the meanings of the signals.

Recognized Distress Signals

The diagrams in Figure A–6 show some of the signals that are recognized as indicating distress and need of assistance. On coastal waters, boaters must carry Coast Guard approved visual distress signaling devices. Table A-1 shows the required number and acceptable use for visual signaling devices. Table A-2 lists additional essential and desirable items for safe boating.

Running Lights: Inland and International

All vessels must show required running lights between sunset and sunrise and during periods of restricted visibility (see Figures A–7 and A–8). The following illustration shows the inland and international light requirements for boats less than 20 meters (65 ft. 7 in.). In most cases, the lights prescribed for a particular vessel are the same under both inland and international rules. Any exceptions are noted. A sailboat operating under power or under power and sail must display the proper lights for a powerboat.

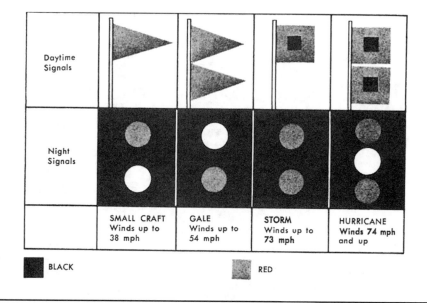

Figure A–5 Flag and light signals indicating weather conditions

Figure A–6 Distress signals for boaters.

Table A-1 Visual distress requirements for devices and their acceptable use (see Table C on next page).

TABLE C—VISUAL DISTRESS REQUIREMENTS

Boaters may select a group or any combination as long as it meets the specific requirement for their boat.

Number on Device	Device Description	Accepted use for	Number Required to be carried
160.021	Hand red flare, distress signals	Day and night [1]	3
160.022	Floating orange smoke distress signals	Day only	3
160.024	Pistol-projected parachute red flare distress signals	Day and night [2]	3
160.036	Hand-held rocket-propelled parachute red flare distress signals	Day and night	3
160.037	Hand-held orange smoke distress signals	Day only	3
160.057	Floating orange smoke distress signals	Day only	3
160.066	Distress signal for boats, red aerial pyrotechnic flare	Day and night [3]	3
160.072	Distress signal for boats, orange flag	Day only	1
161.013	Electric distress light for boats	Night only	1

[1] These signals must have a date of manufacture of October 1, 1980 or later to be acceptable.
[2] These signals require use in combination with a suitable launching device approved under 46 CFR 160.028.
[3] These devices may be either self-contained or pistol launched, and either meteor or parachute assisted type. Some of these signals may require use in combination with a suitable launching device approved under 46 CFR 160.028.

Power-Driven Vessels

Note: A sail vessel under machinery power and sails is considered a power-driven vessel. In inland and international waters, a power-driven vessel must exhibit navigational lights as shown in Boxes 1 or 2 of Figure A–7. Vessels less than 12 meters (39 ft. 5 in.) in international and inland waters may, in lieu of the lights depicted in Boxes 1 or 2 use those lights shown in Box 3 of Figure A–7. In international waters, a power-driven vessel of less than seven meters (23 ft.) in length and whose maximum speed cannot exceed seven knots may, in lieu of the lights prescribed in Boxes 1 and 2, exhibit an all-round white light. Such vessel must, if practicable, also exhibit side lights.

Sailing Vessels and Vessels Under Oar

In inland and international waters, sailing vessels under sail alone must exhibit navigation lights shown in Boxes 4, 5, or 6 of Figure A–8. The tri-colored lantern shown in Box 5 and the all-round green and red lights shown in Box 6 should never be used together. Side lights and stern lights shown in Box 4 should never be used with the tri-colored lantern.

Recommended Additional Equipment

Items E = essential D = desirable	Less than 16 ft.			16 ft. to under 26 ft.			26 ft. to under 40 ft.			40 ft. to 65 ft.		
	Open waters	Semi-protected	Protected	Open waters	Semi-protected	Protected	Open waters	Semi-protected	Protected	Open waters	Semi-protected	Protected
Anchor, cable (line, chain, etc.)	E	E	E	E	E	E	E	E	E	E	E	E
Bailing device (pump, etc.)	E	E	E	E	E	E	E	E	E	E	E	E
Boat hook	–	–	–	D	D	D	E	E	E	E	E	E
Bucket (fire fighting/bailing)	E	E	E	E	E	E	E	E	E	E	E	E
Compass	E	E	D	E	E	D	E	E	E	E	E	E
Distress signals *	E	E	E	E	E	E	E	E	E	E	E	E
Emergency drinking water	E	D	–	E	D	–	E	D	–	E	D	–
Fenders	D	D	D	D	D	D	D	D	D	D	D	D
First-aid kit and manual (10- to 20-unit)	E	E	E	E	E	E	E	E	E	E	E	E
Flashlight	E	E	E	E	E	E	E	E	E	E	E	E
Heaving line	–	–	–	–	–	–	D	D	D	D	D	D
Light list	D	D	–	E	E	D	E	E	E	E	E	E
Local chart(s)	E	D	–	E	E	E	E	E	E	E	E	E
Mirror (for signaling)	D	D	–	D	D	–	D	D	–	D	D	–
Mooring lines	E	E	E	E	E	E	E	E	E	E	E	E
Motor oil and grease (extra supply)	–	–	–	D	D	D	D	D	D	D	D	D
Oars, spare	E	E	E	E	E	E	–	–	–	–	–	–
Radio direction finder	–	–	–	D	–	–	D	–	–	D	–	–
Radio, telephone	D	–	–	D	D	–	D	D	–	D	D	–
Ring buoy(s) (additional)	D	D	D	D	D	D	D	D	D	D	D	D
Shear pins (if used)	E	E	D	E	E	D	–	–	–	–	–	–
Depth sounding device, (lead line, etc.)	D	D	–	D	D	D	E	E	E	E	E	E
Spare batteries	D	D	D	D	D	D	D	D	D	D	D	D
Spare parts	E	D	–	E	E	D	E	E	D	E	E	D
Tables, current	–	–	–	–	–	–	–	D	D	–	E	E
Tables, tide	–	–	D	–	–	D	–	D	D	–	E	E
Tools	E	D	–	E	E	D	E	E	D	E	E	D

* Distress signal devices are required on coastal waters on certain sized boats or during certain times.

Table A–2   Additional equipment recommendations for boaters.

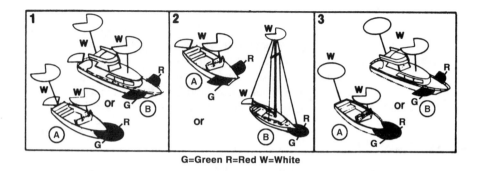

G=Green R=Red W=White

Figure A–7   Requirements for running lights for power-driven vehicles.

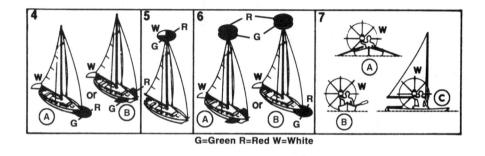

G=Green R=Red W=White

Figure A–8   Requirements for running lights for sailing vessels or vessels under oar.

RANGE AND DEGREE OF VISIBILITY OF LIGHTS

	Visible Range		
Locations	Vessel less than 12 meters	Vessel 12 meters or more but less than 20 meters	Degrees
Masthead Light **W**	2 Miles	3 Miles	225°
All-round Light **W**	2 Miles	2 Miles	360°
Side Lights or	1 Mile	2 Miles	112.5°
W Stern Light	2 Miles	2 Miles	135°

W=White

Table A–3   Requirements for visibility and location of running lights.

A sailing vessel of less than seven meters (23 ft.) in length must: (a) if practicable, exhibit those lights prescribed in Boxes 4, 5, or 6 of Figure A–8, or (b) have ready at hand an electric torch or lighted lantern showing a white light which must be exhibited in sufficient time to prevent collision (see Box 7c).

A vessel under oars may: (a) display those lights prescribed for sailing vessels or (b) have ready at hand an electric or lighted lantern showing a white light which must be exhibited in sufficient time to prevent collision (see Box 7 of Figure A–8).

Anchor Lights

An anchor light is an all-round white light exhibited where it can best be seen and is visible for two miles.

Power-driven vessels and sailing vessels at anchor must display anchor lights. Exceptions are: (a) vessels less than seven meters (23 ft.) in length are not required to display anchor lights unless anchored in or near a narrow channel, fairway, or anchorage, or where other vessels normally navigate and (b) vessels less than 20 meters in inland waters when at anchor in a special anchorage area designated by the Secretary of Transportation are not required to exhibit an anchor light.

Additional Coast Guard Tips

For further tips, see Figure A–9.

To Get Help Use:

- Channel 16 VHF/FM
- 2182 kHz HF/SSB
- Visual distress signaling devices (examples: flares, signal mirror)

How to Avoid Trouble:

For your safety and the safety of your passengers:

- Take a nationally recognized boating course.
- Get a free Courtesy Marine Exam from the Coast Guard Auxiliary
- Insure your motor is properly tuned.
- Fill your fuel tanks.
- Check your motor compartment for fumes.
- File a float plan.
- Instruct your passengers on basic safety procedures.
- Always have everyone wear a life jacket (PFD).
- Check your safety equipment.
  - Radio
  - Life jackets (PFDs)
  - Fire extinguishers

(continued on page 544)

GETTING HELP ON THE WATER

On the water, minor problems can rapidly develop into a situation beyond your control. For this reason, let someone know even when you are experiencing relatively minor difficulties, **before** your situation turns into an emergency.

The Coast Guard serves as Search and Rescue (SAR) coordinator for all maritime emergencies and is the appropriate point of contact whenever you are concerned for your safety. If you are in distress (distress is defined as a situation where you or your boat are threatened by grave or imminent danger requiring assistance), the Coast Guard will take immediate steps to help you. Normally, Coast Guard or Coast Guard Auxiliary rescue boats and/or aircraft will be sent, but assistance from any available source will be arranged to expedite your rescue.

How To Signal For Help

If you are in distress use "MAYDAY, MAYDAY, MAYDAY" on the radio. If your situation is not a distress, simply call "Coast Guard." Channel 16 VHF/FM and 2182khz HF/SSB are dedicated distress and calling frequencies we monitor at all times. Citizen's Band (CB) is not dependable and is not monitored at most Coast Guard stations. If you do not have a radio, attempt to signal a fellow boater who can assist or call the Coast Guard for you. In a distress situation, use flares or any other distress signalling device to catch the attention of another boater.

What To Tell The Coast Guard

While arranging help, we will ask for the following:
- Your location or position.
- Exact nature of the problem (special problems).
- Number of people on board.
- Your boat name, registration and description.
- Safety equipment on board.

When It's Not A Distress

The Coast Guard's primary search and rescue role is to assist boaters in distress. If you are not in distress and alternate sources of assistance are available, we will normally coordinate the effort to assist you. If you have a friend, marina, or commercial firm that you want contacted, we will attempt to do so. You may also contact them directly on Channel 16 VHF/FM or through the marine operator.

If this effort is unsuccessful, we will make a Marine Assistance Request Broadcast (MARB) on your behalf. This announces that you need help, gives your location, and invites others to come to your aid.

If you do not accept services offered in response to the first MARB, we will:
- Provide information on other commercial firms, if available, so you may contact them directly, or
- If you request, make a second MARB to see if any other help is available.

Who Will Answer Your Call When You're Not In Distress

- A Commercial firm may offer help. In order not to interfere with commercial enterprise (you will have to pay for these services), we normally do not provide direct on-scene assistance if a commercial firm is available to help you safely in a reasonable time. If you agree to the assistance of a commercial firm and then refuse this service when it arrives, you still may be legally obligated to pay a fee.
- If the Coast Guard or Coast Guard Auxiliary arrives to assist you and you require a tow, they normally will tow you to the nearest location where you can either arrange for repairs or a tow back to your home port.
- In addition to Coast Guard, Coast Guard Auxiliary and commercial firms, others that may be available to assist you include a fellow boater, local fire or police department, or another public

Figure A-9   Procedures for getting help on the water (continued).

agency. Keep in mind that a Good Samaritan, although well-meaning, may not have the equipment or skills needed to help you safely and effectively.

When To Call Back

Keep in contact with the Coast Guard at regular intervals. Call us when help arrives. If someone offers help but cannot get to you within a reasonable time, usually not to exceed one hour, contact the Coast Guard to arrange other assistance. We also need to know if conditions change sufficiently to cause alarm— for example:

- A medical emergency develops.
- A storm approaches.
- You begin to take on water.
- Your last reported position changes.

Tips On Non-Distress Assistance

Unless you are familiar with the person(s) offering you help, clearly understand the type and quality of the assistance offered before accepting help or entering into a contract. Consider the following before accepting any assistance:

- Large physical stresses can occur in towing and salvage operations, risking damage to one or both boats, and personal injury.
- Does the provider have the proper equipment to handle your problem safely?
- Does the provider have the proper insurance to protect you and your vessel if he/she should cause damage or injury?
- Can the crew handle the situation safely, given the conditions and the nature of the problem?
- If a fee is being asked, does the operator have a Coast Guard license? All operators must have a license if they charge for towing services.

Figure A–9   Procedures for getting help on the water continued.

- Anchor and line
- Basic tools and spare parts
- Flares and sound signaling device
- Compass and chart
- First aid kit
- Paddle and bailer
- Lights and flashlight

Think safety while boating and practice good seamanship.

Accident Reporting Procedures

Follow the procedures listed in Figure A–10 when reporting an accident.

As the operator of a vessel you are *required by law* to file a formal, written report of an accident with local authorities.

1. *When to Report* —

- There is damage by or to the vessel or its equipment*
- There is injury or loss of life
- There is disappearance of any person on board a vessel (under circumstances indicating death or injury)

*Damage is determined by federal regulation to be reportable when it **exceeds $500** or there is **complete loss** of the vessel. **NOTE:** Many states have set a limit less than $500 — contact the local boating authority to determine the amount.

2. *What to Report* —

BOATING ACCIDENTS INCLUDE:
- Grounding
- Capsizing
- Falls Overboard
- Collision
- Sinking
- Struck by Boat/Prop
- Swamping
- Flooding
- Fire
- Explosion
- Disappearance (Other than Theft)

3. *Time Limits* —

WITHIN:
- **48 hours** if there is loss of life
- **48 hours** if there is injury requiring medical treatment beyond first aid
- **48 hours** of the disappearance of a person from a vessel
- **10 days** if there is only damage to the vessel and/or property

4. *How to Report* —

BOATING ACCIDENT REPORT FORM:
The Coast Guard Boating Accident Report Form may be used in all cases of accident reporting regardless in which state the report is to be filed. The state form may also be used. You can use or reproduce this form (located on back of this pamphlet) as a guide to filling out the official form.
Copies of Boating Accident Report Forms, information on state dollar damage reporting thresholds, and other valuable boating safety information is available through your local boating law agency or the Coast Guard Hotline **1-800-368-5647.**

5. *Who Must Report* —

The form is usually filled out by the operator of the boat unless the operator is physically unable to complete the form. In that case, then the owner(s) of the vessel must submit the form.

6. *Why Report?* —

A report must be filed because the information you supply is used to develop safety regulations and manufacturing standards for the benefit of the boating public. The information is also used in boating safety education programs and other boating safety initiatives. Without good data, a boating safety hazard might be completely overlooked and other boaters could be hurt or killed.

Figure A–10 Your responsibilities for reporting accidents. *(Courtesy of United States Coast Guard)*

Boat Certification

Figure A–11 reviews the reasons for certifying outboard motor boats. This is another important consideration for safe boating!

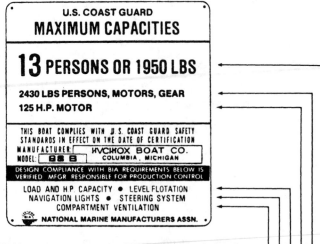

WHY CERTIFIED BOATS?

The marine industry has long been concerned with safety. The Boating Industry Association's voluntary safety standards program was begun more than two decades ago. Certification was born in 1968 from a desire to put more "teeth" into this program. To earn Certification, the manufacturer must prove to a team of inspectors that every boat in his line meets the applicable industry standards and federal regulations. Until all models in the line pass inspection, none can.

ARE THEY BETTER?

Probably. The components that go into each system covered by Certification must meet rigid ABYC specifications. This means more durable components and more careful installation.

ARE THEY MORE EXPENSIVE?

Again, probably. But not very much. An approved fuel hose, for example may cost the boat builder only a few cents more to buy and no more to install. But shop around. Those few extra bucks may be the best bargain you've ever gotten.

WHAT ARE THE STANDARDS?

The NMMA Certification Handbook gives the certification standards builders use covering all items of standard factory equipment. When you install any of this equipment, you should follow them too.

WHY THESE POINTS?

Elementary. Each of the Certification inspection points was chosen after analysis of years of boating accident reports. Horsepower capacity combats overpowering; weight capacity guards against overloading; the fuel system and ventilation points minimize the danger of fire or explosion; steering and lighting requirements help prevent collisions; flotation helps prevent sinking.

WHO SAYS SO?

Certification is backed by the 900-member National Marine Manufacturers Association, largest marine manufacturer group in the world. Inspections are performed by a nationally-recognized, independent testing organization. These inspectors visit the plant before the model year begins. Only after a thorough inspection and correction of any deficiencies can th manufacturer attach the certification plate. The inspectors return unannounced to the plant throughout the model year to insure continued compliance with certification requirements. If a competitor spots backsliding, he can challenge formally.

WHO MAKES 'EM?

Firms that represent nearly 75 per cent of the nation's production are sized up in NMMA's Boat Certification Program. They have produced millions of certified boats.

Figure A–11   Boat certification.

APPENDIX B

Trailer Lighting

TRAILER LIGHTING

STANDARDS BASIS

SAE J1239
49 CFR 571.108

TRAILER LIGHTING

Lighting installation or assembly instructions shall be provided for each trailer model. All lamp housings, lenses and reflex reflectors shall bear design conformance markings as shown in Table IV, unless, separate certification and test reports are provided.

Where practicable, combination fixturers may incorporate 2 or more of these items together except that clearance lamps may not be combined optically with tail or identification lamps.

Lenses of combination clearance and side marker lamps shall be marked PC and must be mounted to meet the photometric requirements for both services. Other combination fixtures must bear the lens markings for all functions combined, except that license plate lamps need not be marked L when combined in tail lamp housings.

With the exception of clearance, identification and license plate lamps, all lamp centers must be a least 15 inches above the road.

Lamps and reflectors shall be located so that their visibility shall not be obstructed by any part of the trailer throughout the photometric and visibility angles specified in the applicable SAE Standard (See Table IV). Except for license plate lamps, these standards provide photometric requirements for all lights and reflectors 10° up and down from the horizontal axis of the fixture. Reflectors, stop, tail and turn signal lamps must meet these requirements 20° to the right and left of the lamps vertical axis. Clearance (See Note, Table IV), side marker and identification lamps must meet these requirements 45° to the right and left of the vertical axis. In addition stop, tail and turn signal lamps must show at least 2 sq. in. of illuminated lens 45° to the right and left of the lamp's vertical axis.

The requirements for side markers used on vehicles less than 30 ft. in overall length may be met for inboard test points at a distance of 15 ft. from the vehicle on a vertical plane that is perpendicular to the longitudinal axis of the vehicle and located midway between the front and rear side marker lamps.

License plate lamps shall be located so that at no point on the license plate will the incident light make an angle of less than 8 degrees to the plane of the plate.

WIRE-SIZE TYPE INSULATION

All wire and insulation shall conform to standards of the Society of Automotive Engineers (See Table IV).

As a precaution against rupture by vibration all conductors shall be of the stranded type and no single conductor smaller than No. 16 AWG shall be used. Multiconductor cable circuits shall be no smaller than No. 18 AWG.

Wire sizes should be in accordance with those specified in Table III. These values are calculated with the intention of providing safe wiring with not over a 3 percent line loss.

INSTALLATION

Where lighting is not factory installed, adequate installation instructions shall be provided.

Wiring and related devices should be installed in a workmanlike manner, mechanically and electrically secure. Devices, lamps and so forth, requiring periodic service shall be readily servicable and accessible. The trailer connector ground wire shall be attached to the trailer frame.

The edges of holes punched or drilled through metal members, through which cable passes, shall be deburred or bushed with suitable grommets. Shielding over cables may be substituted for grommets.

Exposed wiring shall be secured in a workman like manner to stop sidewise movement and prevent rubbing or chafing.

Clips for retaining cables and harness shall be rigidly attached to body or frame member and cable or harness.

Wiring should be located to afford protection from road splash, stones, or abrasion. Wiring exposed to such conditions should be further protected by the use of – or a combination of – additonal tape application, plastic sleeving, nonmetallic or other suitable shielding or covering.

NOTE: Beginning January 1992, over 80 inch wide trailers will require stop lamps and turn signal lamps not less than 12 square inches in area on each side of the trailer. Specific DOT 108 Regulation specifications are not available at time of this printing.

TRAILER LIGHTING

CERTIFIED
NMMA
®

STANDARDS BASIS

SAE J1239
49 CFR 571.108

TABLE IV

REQUIRED MOTOR VEHICLE
LIGHTING EQUIPMENT
PER MOTOR VEHICLE SAFETY STANDARD NO. 108
TRAILERS OF LESS THAN 80 INCHES OVERALL WIDTH

Item	Req.	Lamp Marking	Applicable SAE Std Ref. in FMVSS108	Location
Tail Lamps 3, 9	2 red	T	J585e Sep. 1977	**On the rear,** one on each side of the vertical centerline, at the same height and as far apart as practicable. Not less than 15 inches nor more than 72 inches above road surface for tail and stop lamps and 83 inches for turn signal lamps.
Stop Lamps 3, 9	2 red	S	J586c Aug. 1970	
Turn Signal Lamps 9	2 red or 2 amber	I	J588e Sep. 1970	
Reflex Reflectors 3, 4, 8, 9, 11	4 red & 2 amber	A	J594f Jan. 1977	**On the rear,** one red on each side of the vertical centerline, at the same height and as far apart as practicable. **On each side,** one red as far to the rear as practicable and one amber as far to the front as practicable. Not less than 15 inches nor more than 60 inches above road surface.
Side Marker Lamps 4, 8, 9, 11	2 red & 2 amber	P2 or PC	J592e July 1972	**One each side,** one red as far to the rear as practicable and one amber as far to the front as practicable. Not less than 15 inches above road surface.
License Plate Light	1 white	L	J587 Oct. 1981	**At the license plate,** to illuminate the plate from the top or sides.
Intermediate Side Marker Lamps 1, 9	2 amber	P2 or PC	J592e July 1972	**On each side,** one amber lamp located at or near the midpoint between the front and rear side marker lamps. Not less than 15 inches above road surface.
Intermediate Side Reflex Reflectors 1, 9	2 amber	A	J594f Jan. 1977	**On each side,** one amber located at or near the midpoint between the front and rear side reflex reflectors. Not less than 15 inches nor more than 60 inches above road surface.

Trailers of 80 or more inches overall width shall be equipped as above and include the following:

Item	Req.	Lamp Marking	Applicable SAE Std Ref. in FMVSS108	Location
Identification Lamps	3 red	P2	J592e July 1972	**On the rear,** three lamps as close as practicable to the top of the vehicle at the same height, as close as practicable to the vertical centerline, with lamp centers spaced not less than 6 inches or more than 12 inches apart.
Clearance Lamps 2, 7, 9, 10, 11	2 red & 2 amber	P2 or PC	J592e July 1972	**On the front and rear,** two amber lamps on front, two red lamps on rear, to indicate the "overall width" of the vehicle, one on each side of the centerline, at the same height and as near the top thereof as practicable.
Optional Clearance Lamps 2, 10 Boat Trailers Only	2 red & 2 amber	P2 or PC	J592e July 1972	**One on each side,** near the midpoint at the extreme width showing amber to front and red to rear.

See 5, S4.1.3   See 6, S4.3.1

TRAILER LIGHTING

CERTIFIED
NMMA ®

STANDARDS BASIS

SAE J1239

49 CFR 571.108

REQUIRED MOTOR VEHICLE
LIGHTING EQUIPMENT
PER MOTOR VEHICLE SAFETY STANDARD NO. 108

DEFINITION – SPECIAL CONDITIONS

Item

1. S 5.1.1.3 Intermediate Side marker devices are not required on vehicles less than 30 feet in overall length.

2. S 5.1.1.9 Boat trailers need not be equipped with both front and rear clearance lamps provided an amber (to front) and red (to rear) clearance lamp is located at or near the midpoint on each side of the trailer so as to indicate its extreme width.

3. S 5.1.L.14 A trailer that is less than 30 inches in overall width may be equipped with only tail lamp, stop lamp, and rear reflex reflector which shall be located at or near its vertical centerline.

4. S 5.1.3 A trailer that is less than 6 feet in overall length including the tongue need not be equipped with front side marker laps and front side reflex reflectors.

5. S 5.1.3 No additional lamp, reflective device or other motor vehicle equipment shall be installed that impairs the effectiveness of lighting equipment required by this standard.

6. S 5.3.1 Except as provided in succeeding paragraphs of S 5.3.1 each lamp, reflective device and item of associated equipment shall be securely mounted on a rigid part of the vehicle other than glazing that is not designed to be removed except for repair.

Item

7. S 5.3.1.1. Clearance lamps may be mounted at any location other than on the front and rear if necessary to indicate the overall width of vehicle or for protection from damage during normal operation of the vehicle and at such location they need not be visible at 45 degrees inboard.

8. S 5.3.1.3 On a trailer, the amber front side reflex reflectors and amber front side marker lamps may be located as far forward as practicable exclusive of the trailer tongue.

9. S 5.4.1 Two or more lamps, reflective devices or items of associated equipment may be combined if the requirements for each lamp, reflective device and Item of associated equipment are met, except that no clearance lamp may be combined optically with any tail lamp or identification lamp.

10. Definition The term "overall width" refers to the nominal design dimension of the widest part of the vehicle exclusive of signal lamps, marker lamps, outside rear view mirrors, flexible fender extensions and mud flaps. Source: 49CFR571.3

11. Definition The trailer tongue is considered to begin where the frame side rails begin to angle inward toward the trailer centerline. Source: NMMA.

TRAILER LIGHTING

STANDARDS BASIS

SAE J1239

49 CFR 571.108

TRAILERS LESS THAN 80 INCHES IN WIDTH

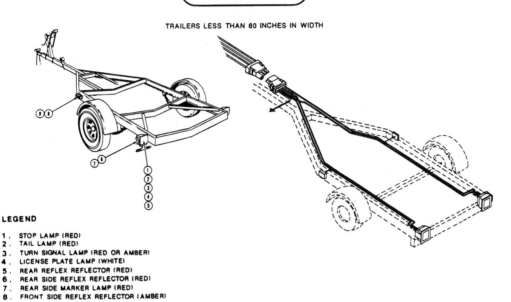

LEGEND

1 . STOP LAMP (RED)
2 . TAIL LAMP (RED)
3 . TURN SIGNAL LAMP (RED OR AMBER)
4 . LICENSE PLATE LAMP (WHITE)
5 . REAR REFLEX REFLECTOR (RED)
6 . REAR SIDE REFLEX REFLECTOR (RED)
7 . REAR SIDE MARKER LAMP (RED)
8 . FRONT SIDE REFLEX REFLECTOR (AMBER)
9 . FRONT SIDE MARKER LAMP (AMBER)

TRAILERS MORE THAN 80 INCHES IN WIDTH

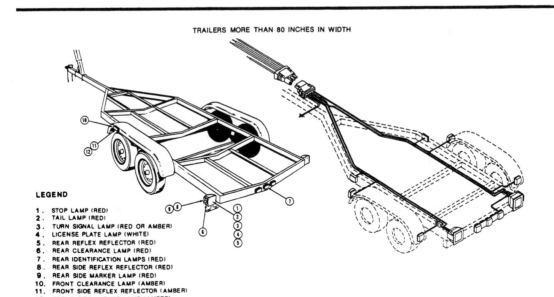

LEGEND

1 . STOP LAMP (RED)
2 . TAIL LAMP (RED)
3 . TURN SIGNAL LAMP (RED OR AMBER)
4 . LICENSE PLATE LAMP (WHITE)
5 . REAR REFLEX REFLECTOR (RED)
6 . REAR CLEARANCE LAMP (RED)
7 . REAR IDENTIFICATION LAMPS (RED)
8 . REAR SIDE REFLEX REFLECTOR (RED)
9 . REAR SIDE MARKER LAMP (RED)
10 . FRONT CLEARANCE LAMP (AMBER)
11 . FRONT SIDE REFLEX REFLECTOR (AMBER)
12 . FRONT SIDE MARKER LAMP (AMBER)

TRAILER LIGHTING

CERTIFIED
NMMA
®

STANDARDS BASIS

SAE J1239

49 CFR 571.108

TABLE V

CONDUCTORS

(SAE Types)

TYPE	INSULATION
GPT	General Purpose, Thermoplastic Insulated
HDT	Heavy Duty Thermoplastic Insulated
GPB	General Purpose, Rubber or Thermoplastic Insulated, Braided
HDB	Heavy Duty, Rubber or Thermoplastic Insulated, Braided
HDB-X	Extra Heavy Duty, Rubber or Thermoplastic Insulated Double Braid
STS	Standard Duty, Thermoset Elastomer, Insulated
HTS	Heavy Duty, Thermoset Elastomer, Insulated
SXL	Special Purpose, Cross-linked Polyethylene, Insulated

TABLE VI

STRANDED CONDUCTORS FOR 12 VOLT CIRCUITS

3% VOLTAGE DROP

Wire Size	20	18	16	14	12	10	8
Stranding	7x28	16x30	19x29	19x27	19x25	19x23	19x21
Circular Mil Area	1072	1537	2336	3702	5833	9343	14810
Circuit Current in AMPS							

Circuit Current in AMPS	MAXIMUM LENGTH OF CONDUCTOR IN FEET FROM POWER SOURCE TO LOAD						
1.	36.4	52.3	78.0				
2.	18.2	26.1	39.0	63.0	99.0		
3.	12.2	17.4	26.0	42.0	66.0		
4.	9.1	13.1	19.5	31.5	49.5	78.8	
5.	7.3	10.4	15.6	25.2	39.6	63.0	
6.	6.1	8.7	13.0	21.0	33.0	52.5	83.8
7.	5.2	7.4	11.1	18.0	28.2	45.0	72.0
8.		6.5	9.8	15.8	24.8	39.4	63.0
9.		5.8	8.6	14.0	22.0	35.0	56.0
10.		5.2	7.8	12.6	19.8	31.5	50.4
15.			5.2	8.4	13.2	21.0	33.6
20.				6.3	9.9	15.8	25.1

CONNECTORS & CIRCUIT IDENTIFICATION

Trailer manufacturers shall provide plugs on their wiring harnesses for connection to tow vehicles according to Figure 1. For ease of tracing, repair and hookup to the tow vehicle, the wire color code indicated shall be used, except that the colors indicated in Figure 1 for circuits other than those required for running lights are considered advisory only.

Stranding shown above to be considered minimum. More strands of a smaller size wire are permitted provided the circular mil area of the conductor is maintained.

These current carrying capacities are for 1 thru 3 conductor cables or bundles; for 4 thru 6 conductor cable or bundles use 80 percent of the maximum length; for 7 thru 9 conductor cable or bundles, use 70 percent.

TRAILER LIGHTING

STANDARDS BASIS

SAE J1239

49 CFR 571.108

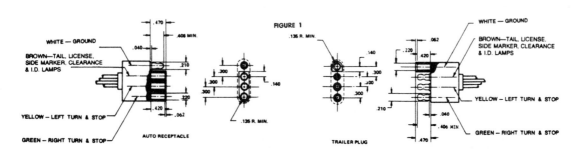

FIGURE 1

WHITE — GROUND

BROWN—TAIL, LICENSE, SIDE MARKER, CLEARANCE & I.D. LAMPS

YELLOW — LEFT TURN & STOP

GREEN — RIGHT TURN & STOP

AUTO RECEPTACLE

TRAILER PLUG

WHITE — GROUND

BROWN—TAIL, LICENSE, SIDE MARKER, CLEARANCE & I.D. LAMPS

YELLOW — LEFT TURN & STOP

GREEN — RIGHT TURN & STOP

SPECIAL EXPLANATION FOR INSTALLING BOAT TRAILER SIDE MARKER AND SIDE CLEARANCE LAMPS

The photometric requirements for clearance and side marker lamps are identical, so whether the lamp is a clearance or a side marker lamp depends only on whether it is mounted perpendicular or parallel to the trailer's longitudinal axis. (Front to Rear Centerline).

Combination clearance/side marker lamps, with a lens marking of PC, add the angular requirements together for both functions making them provide the required light through a 180 degree arc. In order to do this the combination clearance/side marker lamp must be mounted on a 45 degree angle from the trailer's longitudinal axis. The exception (see item 7. S 5.3.1.1.1) has the effect of providing some flexibility in the mounting angle. For example, if the trailer tongue blocks seeing an amber combination lamp the required 45 degrees across the trailer centerline, and there is no better practical location for the lamp, that is all right, but nothing must block seeing the light from dead ahead. If the blocking tongue allows this light to be seen only 15 degrees across the front, then the mounting angle for this PC lamp may be any angle between 15 and 45 degrees from the trailer's longitudinal axis.

If the lens of a PC lamp incorporates a reflector, the reflector only functions as a side marker reflector if the lamp is mounted parallel to the trailer's longitudinal axis; therefore, such installed PC lamp is being used only as a side marker lamp. Whenever a combination clearance/side marker function is used, a separate reflector is always required.

GLOSSARY

A

ABYC: American Boat and Yacht Council.

ADI: Alternator Driven Ignition system.

aft: Toward the rear.

air bleed: A small opening designed to meter air added to fuel flow in the carburetor.

ampere: A measure of the rate of electrical current flow.

anti-syphon valve: Prevents flow of fuel until a pressure drop exists.

ATDC: After Top Dead Center.

atmospheric pressure: Fourteen and seven-tenths pounds per square inch at sea level.

B

bearing: A part that transmits a load to a journal and in so doing absorbs most of the friction wear of the moving parts. Bearing types are ball bearing, needle bearing, plain bearing, and insert (sleeve) bearing.

BIA: Boating Industry Association (now ABYC and NMMA).

blow out: A phenomenon that limits top speed below what would otherwise be possible with the available horsepower. Blow out occurs when the very low pressure cavitation bubbles eventually reach back to the aft end of the lower unit torpedo in sufficient quantity to suddenly pull in or connect up with the engine exhaust.

booster port: A third port in the cylinder wall, working with the power port in the piston.

boundary lubrication: A rupture of the oil film, allowing metal-to-metal contact to occur.

bow: Front of the boat.

bowsterring: Oversteering in one direction. May be caused by being trimmed in too far.

BTDC: Before Top Dead Center.

BTU: British Thermal Unit; a quantity of heat.

C

camshaft: A shaft with a series of precisely spaced lobes used to operate engine valves.

capacitance: Property of capacitor (condenser) which allows it to accept and retain an electrical charge.

capacitor (condenser): Unit used in breaker point ignition to store surges of electricity to prevent arcing across the points.

cavitation: The rapid formation and collapse of water vapor bubbles that generally occurs in a low-pressure area on the leading edge of the prop. May be caused by chips and nicks on the leading edge of the prop.

CDI: Capacitor Discharge Ignition.

CEMF: Counter Electromotive Force.

circuit: A path for electrical current to flow within.

compression: Reduction in the volume of gas (fuel mix) by squeezing it into the combustion chamber. Increasing the pressure reduces the volume and increases the temperature and density of the gas.

compression ring: Number one and two piston rings used to seal the piston to the cylinder wall preventing combustion leakage.

condenser: See Capacitor.

conductor: A wire and a path in which electricity can flow.

convection: Heat transfer by means of air passing over a hot surface.

crab angle: Trimming the unit out will cause lower pressure on the underside of the torpedo, around the skeg, and/or the effect of a surfacing prop pulling the aft end of the lower unit torpedo to the starboard with a right-hand rotation prop. This causes lower pressure on the port side because of the crab angle in which the gear case is forced to run through the water. The typical combination of a surfacing right-hand rotation prop, exhaust gases, and trimming out for best speed creates an extra low-pressure pocket on the lower left side of the torpedo.

cross-flow scavenging: Intake and exhaust ports located directly across from each other causing a cross-flow of gases within the cylinder.

crosshatch pattern: A pattern of 20–40 degrees installed on the cylinder wall to aid in ring seating and oil retention.

cupping: When the propeller's trailing edge of the blade is formed or cast with an inward curl.

current: Flow of free electrons.

D

detonation: Occurs when the anti-knock value of the fuel does not meet powerhead requirements. A portion of fuel begins to burn spontaneously from increased pressure and heat. This causes two flame fronts, resulting in an explosion hammering the piston.

DFI: Direct Fuel Injection

dielectric: A non-conductor of direct electric current. Dielectric silicone in grease form used on electrical connectors.

diffuser ring: On a propeller it aids in reducing exhaust back pressure and in preventing exhaust gas from feeding back into the propeller blades.

diode: An electrical device that will permit current to flow in one direction.

dry well: An area just forward of the transom that allows for outboard installation and tilting.

E

ECU: Electronic Control Unit

EFI: Electronic Fuel Injection.

electrolysis: Chemical change, especially decomposition, produced in an electrolyte by an electric current.

electrolyte: An electrically conducive solution, such as salt water or brackish water, with a high mineral content. Sulfuric acid and water solution in a lead acid battery.

electromagnetic induction: A voltage can be induced by physically passing a permanent magnet (magnetic field) past a conductor. (Flywheel magnet passing ignition coil inducing a voltage in primary and secondary windings.)

EMF: Electromotive Force (voltage).

F

FFI: Ficht Fuel Injection (OMC): The injector hammer-pulses gasoline into the combustion chamber.

field strength: A measurement of the effective size of a magnetic field.

flux: The force of magnetism (magnetic field).

flywheel: A metal wheel attached to the crankshaft that rotates with it; helps smooth out power surges from power strokes; may have magnets set in place for production of ignition and alternator voltage; and may include a ring gear for engagement with starter drive gear.

four-stroke cycle: Intake, compression, power and exhaust accomplished in two revolutions of the crankshaft.

fractured connecting rod: A rod that does not have a smooth finished parting line, but instead is broken (fractured) at the parting line producing an irregular break.

fuel mix: A ratio of gasoline and oil, 50 parts of gasoline to one part of oil; for example, 50/1.

fuel stabilizer: An additive added to gasoline to prevent varnish, gum deposits and fuel degrading when outboard is in storage.

fuse: Protective part installed in an electrical circuit. It will burn out when amperage of fuse specification is exceeded.

fusible link: Wire type electrical device designed to burn out in the event of electrical overload in a circuit.

G

galvanic action (corrosion): An electrical process where atoms of one metal are carried in a solution and deposited on the surface of a dissimilar metal.

ground: An electrical path or connection to the powerhead.

H

helix: A splined spiral; machined on starter armature shaft.

hook: A concave condition in the bottom of the boat in the fore-and-aft direction.

horsepower (HP): A unit of measurement of engine power. One horsepower is the equivalent of raising 33,000 pounds one foot in one minute.

hydrostatic lock: Water is trapped (static) in a cylinder on top of the piston.

I

induction: The process of voltage being induced in a conductor as it moves through a magnetic field.

J

jet: A metered hole in a carburetor circuit designed to control air or fuel flow.

journal: A machined bearing surface on the crankshaft.

L

labyrinth seal: A seal made up of annular rings (grooves) machined into the cylinder assembly or reed valve block, controlling crankcase pressures between cylinders (using puddled fuel).

lapping: A process using a lapping compound to remove high spots between tow surfaces.

load: An electrical device that draws current.

loop scavenging: A method using three ports to direct three air-fuel streams which enter the cylinder from opposing directions and converge at the top of the cylinder, then back down, forcing out the spent exhaust gases.

M

magnetic field: The flux (field) around a magnet or electromagnet.

magnetism: A property produced by certain materials by which these materials can exert mechanical force on neighboring masses of magnetic materials.

micron: A particle having diameter dimensions; used for between .01 and .0001 when measuring outside diameter.

N

NMMA: National Marine Manufacturers Association.

negative: Terminal or side of a circuit from which negatively charged electrons flow. Battery negative terminal.

O

octane number: Numerical system of the octane rating of gasoline.

octane rating: An indication of the ability of the gasoline to resist detonation. The anti-knock rating of gasoline.

OEM: Original Equipment Manufacturer.

ohm: A unit of electrical resistance.

ohmmeter: Test instrument used to measure resistance in an electrical circuit and components.

OHV: Over Head Valve.

oil ring: A three-piece piston ring designed to control four-stroke engine oil from going into the combustion chamber.

open: The electrical circuit is broken so that a gap exists and there is no current flow.

operating RPM: The desired RPM at which the outboard should run at wide-open throttle.

P

parallel circuit: An electrical circuit formed when two or more electrical devices have their terminals connected, positive to positive and negative to negative, so that each may operate independently of the other circuits using the same power source.

parting line: The line of separation between the rod cap and the connecting rod.

PFD: Personal Flotation Device: life preserver, buoyant vest, buoyant cushion, ring life buoy.

piston boss: The area of the piston into which the piston pin is pressed/slipped.

pitch: The distance that a propeller would move in one revolution when screwed through a soft solid (like a wood screw into wood).

plain bearing: A bearing surface machined directly into the aluminum cylinder assembly.

plow: Incorrect tilt angle is forcing bow down.

porpoising: Up and down movement of the bow.

port: Left of the boat as you face forward.

power porting: Uses a modified piston head, a hole in the piston skirt, and a transfer port in the cylinder wall, to give an additional source for fuel delivery.

preignition: The ignition of the fuel-air charge prior to the timed spark. Any hot spot within the combustion chamber can cause preignition.

pressure differential: The difference between atmospheric and negative pressure (vacuum).

propeller diameter: Measure from the center of the propeller hub out to the tip of the blade and multiply by 2.

propped out: The outboard is "propped out" more toward the middle of the recommended RPM limit.

PSI: A measurement in pounds per square inch.

puddled fuel: Two-cycle fuel mix which has fallen out of the air stream and puddled in the bottom of a crankcase.

R

rake: The propeller blade slants back from vertical. Generally standard props have a rake varying between zero to 15 degrees.

rectifier: A device that changes alternating current to direct current. In alternator circuitry, a diode.

reed valve: A flat spring that flutters with crankcase pressures, emitting fuel and air mixture into the crankcase (two-cycle).

resistance: The opposition to current flow.

resistor: A device used in an electric circuit to produce work or lower voltage in a circuit.

ring gap: The gap between ring ends when the ring is in the cylinder.

ring side clearance: The clearance measured between the piston ring and the piston ring land.

rocker: A bulge or convex condition in the bottom of the boat in the fore-and-aft direction.

rocker arm: Four-stroke engines (OHVs); a device that rocks on the rocker arm shaft as the camshaft lobe moves it to open the intake or exhaust valve.

S

SAE: Society of Automotive Engineers.

scavenging of gases: The bringing in of fuel and expelling of exhaust gases from the cylinder.

seal rings: Rings installed around the crankshaft to seal one crankcase cavity from an adjacent crankcase using puddled fuel.

series circuit: An electrical circuit in which the devices are wired end to end, positive terminal to negative terminal. The same current flows through all devices in a given circuit.

short: An electrical path that goes back to the battery without going through the electrical load.

silicon controlled rectifier (SCR): A semiconductor that does not allow current flow in either direction. However, when the SCR is triggered by placing a positive voltage on its gate, it allows current flow from anode to cathode.

slip: The percentage of the designated pitch that the prop does not successfully travel in one revolution. The difference between the prop's actual path of the blade and the theoretical path of the blade.

solenoid: A relay that connects a solenoid to a current source when it closes; specifically the starter motor solenoid relay.

spalling: The loss of the bearing surface; it resembles flaking or chipping.

specific gravity: A unit of measurement for determining the sulfuric acid content of electrolyte.

starboard: Right side of the boat as you face forward.

steel bearing liner: A steel liner installed into the large end of an aluminum rod on which needle bearings will roll.

stern: Rear of the boat.

T

TC-W3: Two-cycle water-cooled (oil).

TDC: Top dead center; in reference to piston location.

tensile: Of or involving a force that produces stretching.

test wheel: A specially designed propeller (wheel) to load the outboard when tested in a test tank.

tilt range: Using mechanical and hydraulic advantage, the trim cylinder(s) move the outboard through 50 degrees of movement after the trim range is completed.

timing/synchronization: The coordination of ignition timing to opening of the carburetor throttle plate.

torque: A twisting motion.

transistor: Electrical device that acts as an electrical switch.

transom: A part of the hull across the stern of the boat.

trim angle: How far out from the transom the lower unit is tilted. This has an effect on the planing angle of the boat and steering torque.

trim of the outboard: A movement or partial movement through approximately 15 degrees aft from horizontal.

trim planes: A pair of flat, movable surfaces that extend aft from the boat bottom, one on each side of center.

trim range: Using mechanical and hydraulic advantage, the trim cylinder(s) move the outboard through the first 15 degrees of movement. (See Figure 12-12.)

two-stroke: In reference to intake, compression, power, and exhaust; two piston strokes completed in one revolution of the crankshaft.

U

under propping: Propping the outboard at or a little above the maximum recommended RPM limit.

V

vacuum: A pressure less than atmospheric.

valve clearance: Clearance between the rocker arm and the valve stem tip, when the valve is closed.

valve guide: A cylindrical part or hole in the cylinder head in which a valve is installed and in which it moves in and out.

valve overlap: Condition during which intake valve is still not fully closed and exhaust valve has begun to open.

valve seat: Contact surface area for the valve face to seal against.

valve spring: A coiled device installed with a retainer onto the valve stem to close/seat the valve against the valve seat.

valve timing: The timing of opening and closing of the valves in relation to piston position.

ventilation: Occurs when air from the water's surface or exhaust gases from the exhaust outlet are drawn into the propeller blades.

volt: A measurement of electrical pressure (voltage).

voltage: Difference in electrical pressure between two points in a circuit.

voltmeter: Test instrument designed to measure voltage.

voltage drop: A test made to determine voltage loss caused by resistance in an electrical circuit. (Corrosion of loose connections.)

W

walk: Trimmed out too far, and the boat bottom walks left to right.

WOT: Wide-open throttle.

Y

yaw: To deviate from intended course.

INDEX